NONLINEAR DYNAMICS OF A WHEELED VEHICLE

Advances in Mechanics and Mathematics

VOLUME 10

NONLINEAR DYNAMICS OF A WHEELED VEHICLE

By

RYSZARD ANDRZEJEWSKI
The Technical University of Łódź, Poland

JAN AWREJCEWICZ
The Technical University of Łódź, Poland

Library of Congress Cataloging-in-Publication Data

A C.I.P. record for this book is available from the Library of Congress.

ISBN 0-387-24358-5 e-ISBN 0-387-24359-3 Printed on acid-free paper.

Printed in the United States of America.

9 8 7 6 5 4 3 2 1 SPIN 11161615

springeronline.com

Contents

Preface

On average, 60% of the world's people and cargo is transported by vehicle that move on rubber tires over roadways of various construction, composition, and quality. The number of such vehicles, including automobiles and all manner of trucks, increases continually with a growing positive impact on accessibility and a growing negative impact on interactions among humans and their relationship to the surrounding environment. This multiplicity of vehicles, through their physical impact and their emissions, is responsible for, among other negative results: waste of energy, pollution through emission of harmful compounds, degradation of road surfaces, crowding of roads leading to waste of time and increase of social stress, and decrease in safety and comfort.

In particular, the safety of vehicular traffic depends on a man-vehicle-road system that includes both active and passive security controls.

In spite of the drawbacks mentioned above, the governments of almost every country in the world not only expect but facilitate improvements in vehicular transport performance in order to increase such parameters as load capacity and driving velocity, while decreasing such parameters as costs to passengers, energy resources investments, fuel consumption, etc.

Some of the problems have clear, if not always easily attainable, solutions. For example, fuel waste can be reduced through engine capacity increase, driving optimization (for engine-transimission gear-driving axle systems), improvement of car aerodynamics, and decrease of vehicle specific weights. Also, there is no doubt that further motorization development depends on high-tech electronic systems applied to drive transmission and load suspension systems, security and diagnostics, information, and communication.

The following, revolutionary, solutions are expected to result soon from current research and development efforts:

(i) electronically controlled servo-breaking systems;

(ii) electronically controlled mechanics aiding steering and gear systems;

(iii) electronically controlled suspension systems (with stiffness and damping elements);

(iv) electronically controlled drive units (engine-clutch-gearbox-driving axle - differential mechanisms);

(v) devices for control of distances between vehicle.

Note that future research will be focused not only on improvement of efficiency and stability of separate process, as for instance braking or steering, but also on optimization processes coordinating all forces and influences on vehicle dynamics.

In spite of numerous past investigation and vast investments of research time and money, *vehicle motion stability* remains one of the most important unresolved problems of road vehicle transportation. In other words, vehicle stability is the ultimate goal of the study of vehicle resistance to various motion perturbations. Our monograph was conceived as a doorway between the world of the classical and current approaches to attaining the stability in wheeled vehicle dynamics and that of the requirements and expectations of future motorization developments.

A major part of this monograph is devoted to the analysis of various components of the dynamical behaviour of a wheeled car and its control, e.g. longitudinal, transversal and vertical dynamics. The studied processes are modelled by lumped mechanical objects, i.e. they are governed by non-linear ordinary differential equations. The last two chapters address modelling of both mono-cylinder combustion engines and girling duo-servo brake, being the fundamental parts of any wheeled vehicles.

In this monograph we propose new theoretical, numerical and experimental approaches to the analysis of the dynamics and control of various types of either complete wheeled cars, or their elements (engine and brake).

In Chapter 1 (Introduction) we describe the state of the art of motorization, putting an emphasis on, among others, the application of *integrated units, a dispatcher-vehicle communication system, motion stability control and improvement* (i.e. passive systems including circumferential force controllers like ABS or ASR, and automatic control). This chapter includes a historical reference background, the beginnings of the wheeled vehicle, and frequently occurring stability problems.

In Chapter 2, stability concepts are briefly reviewed, with an emphasis on technical stability aspects. First, definitions of various stability concepts, i.e. technical, Lyapunov, Poincaré, Poisson, Bogusz and Szpunar ones, are introduced, illustrated and discussed. Then, the general stability estimation is addressed.

Special attention is paid to suitability of technical stability problems to wheeled cars investigation developed by two Polish scientists, L. Bogusz and

K. Szpunar, not very widely known in the Western countries. Many practical examples illustrate theoretical considerations. Recent advances and directions in general stability estimation, with a special stress put on engineering applications, are rigorously discussed in the invited section prepared by B. Radziszewski.

In Chapter 3 our attention is focused on the stability of a wheeled vehicle. In particular, stability of various mathematical models of a road vehicle and its sub-assemblies is analysed.

Note that for any wheeled vehicle (for instance, a car) the requirement for being in operation consist of the following aspects: (i) efficiency (proper performance of given tasks); (ii) reliability and controllability (possibility to achieve defined states); (iii) stability (mutual closeness of solutions (motions)); (iv) boundedness and restrictions of solutions (motions).

Stability characterizes the ability of the system to preserve a required functional property and is closely related to the sensitivity of solutions of differential equations to various changes of initial conditions, structural parameters and disturbances.

The most celebrated stability, Lyapunov stability, states that any solutions (motions) that are close to each other at a certain time instant t_0 become arbitrarily close during the whole observation time. Stability in the sense of Poincaré (or orbital stability) means, roughly speaking, that each solution that is close to an investigated one remains arbitrarily close later on (observe that contrary to the Lyapunov stability, there is no requirement for the solutions to be close to each other at the same time instant). Finally, a system is stable in Lagrange's sense if all its solutions are bounded.

First of all, it should be emphasized that in engineering practice, and particularly in respect of *car stability* concepts, very often the mentioned stability does not have any relation to stability in the meaning of control theory. Presently, two clear paths can be distinguished while analysing wheeled cars:

- the Lyapunov concepts are widely used in *simulating investigations* of stability with respect to mathematical models of cars, in particular where an investigated motion exists for a relatively long time (a power transmission system or a straight line vehicle movement);

- the so-called *'technical stability'* concepts are very frequently used in *experimental investigations*, in which different kinds of motion observed in relatively short time intervals are analysed. As an example one may refer to the standard stability studies of a braking vehicle moving on a stationary circled trajectory. In many laboratories the increment of rotational car velocity measured around a vertical axle one second after the distrubance (i.e. braking) has begun, is taken as a criterion . The mentioned time interval of one second possesses an important practical meaning, since it corresponds strictly to the beginning of the potential driver reaction to any dangerous parameter changes of the vehicle motion.

Stability concepts originating from the Lyapunov stability are certainly important for models governed by differential equations, but not necessarily for various other technical objects. In particular, in many engineering systems the motion duration is bounded and sometimes very short, and often a priori constraints are attached which describe a safe system (vehicle) behaviour (for instance, a width of a public road). Therefore, especially in the investigation of car movement, the so-called technical stability definition is much more appropriate. This point of view is applied to formulate stability conditions of a wheeled car and its subsystems for various types of motion.

A wheeled car stability is addressed in section 3.1, pneumatic type properties are illustrated and discussed in section 3.2. The road surface – car wheel – car body dynamics is considered in section 3.3, and a brief introduction to a moving car stability is given in section 3.4.

Chapter 4 of this monograph is devoted to two cases of longitudinal wheeled vehicle dynamics. First, the longitudinal dynamics of a biaxial car which depends significantly on circumferential horizontal forces at the points of contact of a tire with a road surface, is studied. The mathematical model is derived, and then analysed. The originally constructed and designed ABS device is presented, and then braking on either homogeneous or non-homogeneous road surfaces of either front or rear wheels are studied.

In section 4.2 a longitudinal tank vehicles dynamics with a fluid is modelled and analysed. Two cases are considered, i.e. when the fluid part does not move with respect to the tank walls and when the fluid part is modelled as a mass-spring system.

The theoretical considerations are supplemented by an example of the longitudinal dynamics of a semi-trailer with tanks. Taking into account semi-trailer and tank models, an ABS model, as well as braking mechanism models for the tractor and the semi-trailer road wheels, a computational model is obtained. The tank is filled with liquid up to a certain height. The tired, wheel-road surface friction pair is approximated with Wagner's model.

Various cases are analysed and many useful conclusions are given.

In Chapter 5, dynamics of biaxial and multi-element transversal vehicles are analysed. In section 5.1 the so-called *effective* methods are applied to analyse stability of a biaxial wheeled vehicle. The name refers to methods that do not require any knowledge of solutions to the mathematical models, but they yield stability estimation of a *system*, an *object* or a *process*. The stabilities in both the Lyapunov and the Bogusz sense are applied, and in addition the *steerability characteristics* as well as the so-called *first* and *second technical stability factors* are used, thus eliminating a drawback of many other existing approaches.

It is worth noticing that the technical stability concepts in the Bogusz sense are suitable for an analysis of wheeled vehicles, and include adequate system parameters while defining the so-called *admissible solutions zone*. The technical stability concepts include: (i) occurrence of input data motion monitoring scatter; (ii) continuous perturbations; (iii) steering depending on turn angles of steered road wheels.

In section 5.2 the stability of wheeled articulated vehicles is studied. Articulated vehicles, where two vehicles, a pulling one and the one being pulled, are coupled by a ball-and-socket joint, are universally applied as a truck tractor-trailer, an *articulated bus*, or a passenger car with a sidecar.

Unfortunately, the stability of *linked* vehicles is not satisfactorily solved yet. In particular, the so-called *jack knifing* seems to be a very dangerous phenomenon, hence a simplified one-truck vehicle model valid for small transversal angle displacements is examined.

Applying a physical model and appropriate mechanical relations one may derive the equations of motion. The mathematical model of a truck tractor-trailer consists of three second order non-homogeneous differential equations.

Again, the stability in the Lyapunov (matched with the Routh–Hurwitz criterion) and the Bogusz sense is applied. Many useful conclusions, examples, stability diagrams for given norms, tolerances and deviation zones occurring in real conditions are reported.

It is certain that a universal human body model cannot be satisfactorily designed. In addition, the behaviour of the brain and nervous system cannot be predicted. Psychological and emotional changes depending on the time of day can lead to serious complexity modelling of the brain. A driver's reaction exerted on a brake pedal or on a steering gear may depend strongly on the driver's current psychological state, alcohol or drugs taken, or on his general physical condition. The circumstances mentioned above indicate the practical impossibility to obtain a proper *driver model*. Constructed models are usually limited to interaction modelling for given conditions and include mainly a driver's kinematic reaction on a steering gear.

In what follows the so-called *anticipative* driver model with delay is considered. The following main model hypothesis is taken into account. A driver always observes a point on a road. The *observable point* lies on a required car trajectory a distance of L_0 from the driver. L_0 depends on the car velocity and the trajectory curvature. The steering driver's quantities are the visual ones: a sight angle of the observable point and a rotational velocity. Hence, the stability of the car driver system is studied.

Shimmy, i.e. the phenomenon of self-excited wheel car vibrations around the relieving axle is analysed in section 5.3, including road wheel properties and applying the Nyquist stability criterion. A few examples analysing stability of road wheel-vehicles for the shimmy model are examined.

In Chapter 6 the dynamics of a suspension model of two bodies vibrating system with a passive and an active constraint is considered. Two constraint systems are studied, i.e. *passive* and *active* ones. Various criteria for optimization of the active suspension are used. In addition, the stability of an active suspension system with various control functions is analysed. The influence of active suspension parameters on the Lyapunov and the technical stability is provided. Algorithms for an analysis of suspension vibrations are proposed. An example of a seven degree-of-freedom vehicle model is examined. Various aspects of non-linear regular and chaotic dynamics, stability and bifurcation are studied using the model of a standard European passanger car of the lower middle class. Many useful and practical conclusions are given.

In Chapter 7 the dynamics of a transversal tilt is illustrated by means of an analysis of two fundamental cases. In the first case a mechanical system consisting of a vehicle body supported by massless stiffness and damping lumped elements and a transversal pendulum with friction are studied. The pendulum models a suspended load. Small system vibrations and their stability in the senses of Lyapunov and Bogusz are analysed.

In the second case, a model composed of a vehicle body and a road wheels suspension is investigated. An influence of the height of the vehicle body mass centre on both vehicle transversal and transversal tilt dynamics is analysed.

In section 7.1 the dynamics of a vehicle body with the pendulum type load is analysed. The Lagrange formalism is applied to obtain the governing differential equations, and the stable and unstable zones are derived. The periodic, quasi-periodic, and chaotic dynamics are illustrated and discussed. Particular attention is paid to the so-called *semi-critical dynamics*.

A curvilinear vehicle trajectory taking into account a vehicle body displacement, roll and the motion around an instantaneous roll centre is considered in section 7.2. Stability concepts in the senses of Lyapunov and Bogusz are applied, zones of admissible solutions are shown, stability factors are derived and an example of a real vehicle is given.

A longitudinal vehicle tilt occurs either when a vehicle runs on a road surface with transverse roughness like "waves" or "humps" or during the force excitation in starting or braking processes. This phenomenon is particularly important when inside the vehicle there is a body that can shift with respect to the vehicle body. In Chapter 8 the longitudinal vehicle tilt dynamics is studied for two vehicle systems: (i) vehicle body – shifting body; (ii) vehicle with a trailer – a shifting body is inside the trailer.

First, in section 8.1 the dynamics of the vehicle body-movable body system is analysed. An example including three different constraints of the pair: movable mass-car body is studied. Then, in section 8.2, the dynamics of a longitudinal tilt of a road vehicle with a semi-trailer (trailer) is examined. The differential equations governing the dynamics of a pulling car with a semi-trailer and a movable mass are derived and analysed. A jump type displacement of the movable body and a real pulling vehicle model are taken and numerically analysed, and three different cases of small (medium and large) stiffness of the rear road wheels suspension and a standard distance of the joint linking two vehicles to the rear wheels axle of the pulling car are considered.

In Chapter 9 we provide modelling and numerical simulation of a road wheel rotational dynamics. In section 9.1 the dynamics of the road wheel rotational motion during the braking process is studied. Various stability investigations for different wheel type characteristics are given.

The driven road wheel rotational motion is studied in section 9.2. First, an experimental rig is described, where (among others) a delay (occurring in real systems) between the generated controlling moment and the controlling signal is accounted. Next, the driven road wheel rotational motion (a biaxial vehicle) is studied. An illustrative example of the introduced methodology application is given.

As known, even a single harmonically or parametrically excited pendulum may exhibit a rich spectrum of non-linear phenomena including various local and global bifurcations, attractors and repellers, stable and unstable manifolds, scenarios leading to chaos and out of chaos, symmetry breaking and crisis bifurcations, steady-state and transitional chaos, oscillatory-rotational attractors, etc.

On the other hand, many real processes can be modelled via coupled pendulums. It is clear that a coupled system of pendulums may exhibit more complex non-linear dynamics, and hence it attracts the attention of mathematicians, physicists and engineers, who show a particular interest in the examination or control of various systems modeled by coupled pendulums. In addition, it may be expected that many unsolved theoretical problems of non-linear dynamics can be explained using a model of rigid multi-body coupled pendulums.

It turns out that although many technological and design-oriented details have been neglected, the inverted triple pendulum can be used to model a real piston – connecting rod – crankshaft system of a mono-cylinder combustion engine. This is the subject of Chapter 10.

It is clear that in a case of a motor-cycle as well as a car with axles, their "life" and possibilities depend on an engine. In this chapter the inverted triple pendulum is proposed to model a real piston-connecting rod-crankshaft system of a mono-cylinder combustion engine. After a short introduction, a model of a rigid multi-body mechanical system with unilateral frictionless constraints is proposed. Then, a generalized impact law is described. Sliding states along some obstacles are discussed and a computational model is introduced. The dynamics of a triple physical pendulum with barriers is illustrated and discussed. The piston-connecting rod-crankshaft system is then analysed, and the numerical examples that follow show good agreement with a real mono-cylinder combustion engine behaviour.

The basic piston positions include: (i) four piston positions in the cylinder barrel: two skew positions (with the contact between one corner of the piston and one side of the cylinder and between the opposite corner and the second side of the cylinder barrel), and two positions of the piston adjoining one of the two sides of the cylinder surface; (ii) four displacements (turns) of the piston with one of the four corners being in contact with the cylinder.

In addition, in each of the four piston positions three equilibrium states of dynamic forces are distinguished, whereas in each of the four piston displacements two such states are distinguished. Therefore, the piston can be in one of the twenty equilibrium states of the dynamic forces. The piston movement from one side of the cylinder to the opposite side has been assumed to consist of two piston turns and one skew piston position. A direct piston movement with loss of contact with the cylinder has not been analysed.

Assuming a constant rotational speed of the crankshaft, a schedule of the forces acting on the piston and the connecting rod has been made, including tangent forces of interaction between the cylinder and the piston surfaces, forces of interaction between the piston and the cylinder via the piston rings, and with the assumption of a more real friction model in the bearings. In that way a system of six equations of equilibrium of the dynamic forces has been obtained for the piston with the connecting rod system.

The obtained equations can be solved for one of the possible piston states for each crankshaft position. The obtained values of both normal and friction forces verify the admissibility of the assumed piston state. If the piston state is not admissible, the next piston state is assumed and the calculations are repeated until an admissible piston state is found. In that way, by varying the crankshaft position with a small angle step, for each crankshaft position one admissible piston state can be found.

It is clear that without an engine a car (or a motor-cycle) cannot run. On the other hand, it can not stop without a brake. Therefore, Chapter 11 is devoted to modelling and analysis of one examplary brake, namely, a duo-servo brake.

The 2-DOF self-excited system with friction is analysed using numerical methods. A special numerical scheme based on the Hénon approach and exhibiting good suitability for investigations of non-smooth dynamical systems is applied. Many interesting dynamical non-linear behaviors are reported and analysed, including stick-slip periodic, quasi-periodic and chaotic motions. In the analysis, all standard techniques are applied, i.e. time histories, phase planes, Poincaré maps, bifurcation diagrams and the Lyapunov exponents. The calculation of Lyapunov's exponents from an interpolated time series offers sufficient accuracy and correct values of its spectrum. The expected estimation accuracy of the Lyapunov exponents for each type of motion yields different relations between the number of the trajectory points solved by the Hénon method and its equivalent Lagrange interpolation.

In addition, the numerical analysis is supported by the investigation of a real laboratory object modeling the feedback reinforcement of the friction force (model of T_- branch) and the friction force without the feedback (model of T_+ branch). The numerical solution obtained using the T_- branch model has not proved a transition, which can be observed in our experimental measurement. The sticking velocity is almost the same, but in the sliding phase some distinguishable differences are observed.

It is suggested that the T_- friction force model should be used after an analysis of the friction effects occurring in the systems where the normal force acting between cooperating surfaces fluctuates. The application of the branch friction force model leads to rapid entries on the stick phase and rather smooth backslides from it.

Since friction during a braking process plays a key role, a brief review of friction modelling is given in the Introduction. In section 11.2 a modelled system with friction is presented including the governing differential equations. A numerical analysis is carried out in section 11.3. Section 11.4 is focused on experimental investigations of the introduced physical model. The last section 11.5 is devoted to conclusions.

To conclude this chapter, a new idea for the friction pair modelling using both laboratory equipment and numerical simulations is proposed enabling observation and control of the friction force. The experimental data are compared with those obtained via numerical simulations showing good agreement.

This monograph is addressed to practising engineers, as well as to graduate students and researchers in Applied Mechanics, Mechanical, Industrial and Aerospace Engineering, Applied Mathematics and Applied Physics.

Both authors greatly appreciate the help of Mr W. Dziubiński with a final manuscript preparation. Some chapters devoted to non-smooth dynamics have been financially supported by the Polish National Research Committee grant No. 5 T07A 019 23.

LODZ
DECEMBER 2003

RYSZARD ANDRZEJEWSKI
JAN AWREJCEWICZ

Chapter 1

INTRODUCTION

In this chapter some history of wheel vehicle transport is reviewed. Also, progress or current and future solutions in the field of communications and projections for its future development are explained.

1. Motorization [61, 64, 75, 92, 101, 110]

In our Preface we summarized the concepts on which our book is based and commented on expected future solution to vehicle dynamics problems. We begin this chapter whit consideration of informatics and communication. Recent development in informatics have provided us with an application on the concept of *integrated units*; so-called Integrated Safety Systems (ISS). An example of an integrated system is given in Figure 1.1.

It is expected that owing to the development of motorization, safety will improve but energy consuming will grow. In road vehicles, internal nets of information control linked to the central control net will appear.

In Figure 1.2 a scheme of the *communication system dispatcher-vehicle* is shown. It allows for control of vehicle groups, sending information to drivers, efficient control of vehicles, movements, vehicle velocity control, and prevention of accidents by delivering information on weather and by-pass roads.

Beginning from the 1980 control systems improving *vehicle dynamics* have been successfully applied. In particular, they improve the following vehicle properties:

(i) braking process efficiency (for instance Antiblocking System ABS, Brake Assist System BAS, Sensotronic Brake Control SBC, Brake-by-Wire BbW);

(ii) efficiency and security of the driving process (for instance ASR Antriebschlupfregelung, TCS - Traction Control System, TRC - Traction Control, ETC - Electronic Traction Control, TCL - Traction Control, ASC+T - Au-

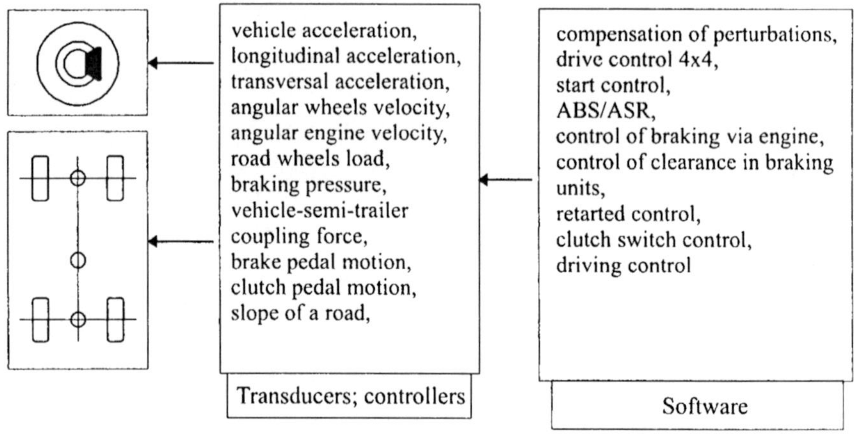

Figure 1.1. Integrated system of driving and braking vehicle control.

tomatische Stabilität - Control plus Traction, EDS Elektronische Differential-Spere);

(iii) efficiency and stability of a change (or not) of a motion direction (the various 4WS units, as for instance HICAS, Active 4WS, FDR Fahrdynamik Regelung and ESP Electronic Stability Program);

(iv) motion fluency (for instance the so-called active suspensions like ARS Active Roll Stabilisation, ECAS, EAS, CCS, Hydroactiva).

Note that future research will focus not only on improvement of efficiency and stability of the separate processes, as for instance braking or steering, but also on an optimization processes co-ordinating the actions of all units influencing vehicle dynamics.

One of the most important properties of a modern road vehicle is motion stability. In other words, *vehicle stability* characterizes a vehicle resistance to various motion perturbations. For example, a stable vehicle (with respect to its transversal dynamics) is able to maintain a straight line vehicle run while braking on a non-homogeneous road surface, or to help a driver avoid error and improve his movement on a road turn, or be resistant to the action of a transversal wind.

The applied system *improving motion stability* includes controllers of the circumferential F_{xi} and vertical F_{zi} forces on the separated road wheels, as well as of the turn road wheels angles δ_i when all of them are steered (4WS) via mutually coupled or individual units. The stability improving systems can be divided into two categories (see Figure 1.3).

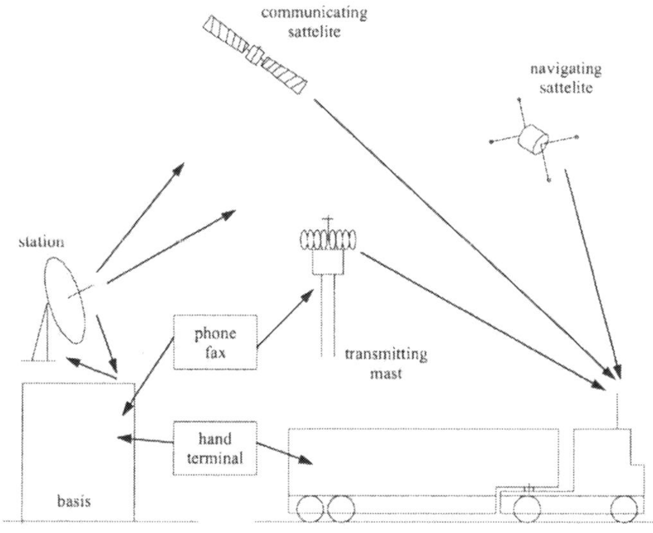

Figure 1.2. A dispatcher-vehicle communication system.

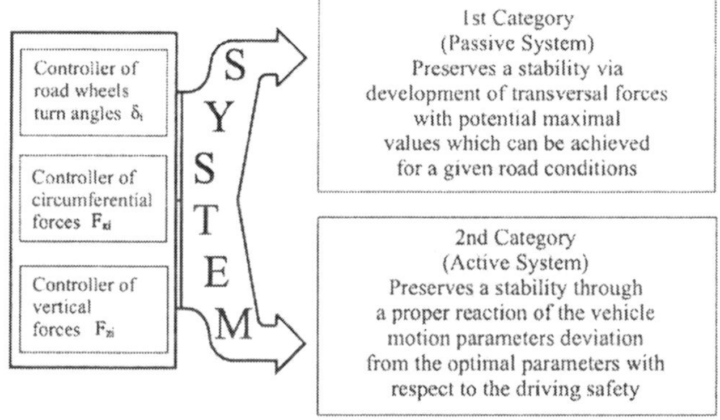

Figure 1.3. Two categories improving stability of a vehicle motion.

The first category (passive systems) includes the systems with circumferential forces controllers, as in ABS or ASR. A control of the braking (ABS) and driver (ASR) systems takes place in the optimal zones of the (circumferential slip of a road wheel) in both longitudinal and transversal vehicle motion. For given road conditions such a control relies on a use of the potential friction force F_r to preserve a possibly high value of the longitudinal circumferential friction force F_{xi}. As it has been show in Figure 1.4, such a control allows for development of large, circumferential friction forces of either braking or driving process, simultaneously preserving a development of large transversal friction values. Note that the transversal forces acting on the road wheels decide the vehicle motion stability. For example their action allows for keeping a straight vehicle motion direction while braking on a non-homogeneous road surface. The 1st category systems do not allow for a large slip of a given road wheel, and hence they improve the interaction between a wheel tire and a surface road. However, this category of systems do not possess the devices of automatic control, but a driver makes a decision on a run direction change.

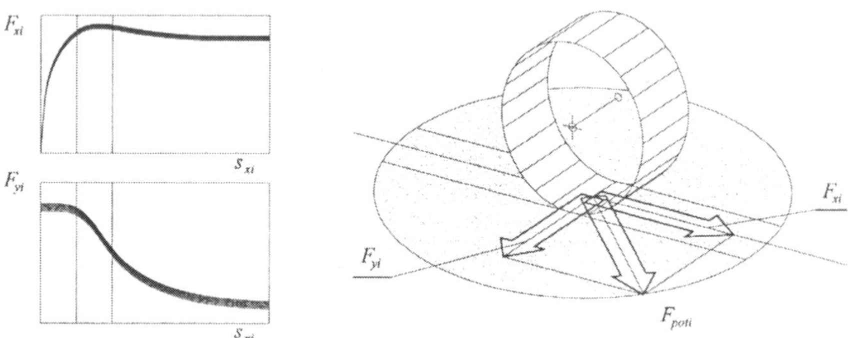

Figure 1.4. The friction characteristic of the road – road surface system with the marked optimal values on the F_{xi} (s_{xi}) and F_{yi} (s_{yi}) drawings (F_{poti} – frictional potential force of the i-th road wheel; F_{xi} – frictional circumferential force longitudinal, braking or driving; F_{yi} – lateral frictional force; s_{xi} – longitudinal slip).

The 2nd category systems have properties of automatic control, since they react on the motion deviations. They may introduce a proper motion trajectory correction, if it differs from that expected by a driver. In order to explain the 2nd category systems, the kinematical quantities describing a vehicle motion and the forces acting on a road wheel are shown in Figure 1.5. A driver acts on a vehicle via a steering wheel and steering engine, clutch or brake pedals. Owing to this and the appropriate control units actions, the defined turn angles as well as the circumferential forces on the road wheels appear. The latter implies a vehicle motion with defined linear and angular velocities $(v, \dot{\Psi})$, and in direction of the appropriate slip angle α with respect to the longitudinal vehicle axle and with the proper accelerations (a_x, a_y, a_x).

The measuring devices of the 2nd category systems trace steering processes associated with the steering wheel and pedals movements. On the other hand also other motion parameters are traced by additional sensors. In the system memory the proper (required) parameters are located. If the forced parameters diverge from the required ones, then the system starts to react by a generation of the appropriate values of δ_i, F_{zi} and F_{xi} (see Figure 1.5). In addition, the 2nd category system can work under the scheme reported in Figure 1.6.

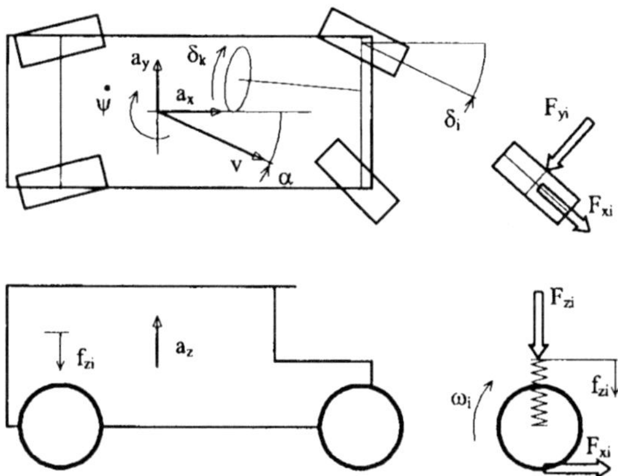

Figure 1.5. The parameters characterizing a vehicle dynamics (δ_k – angle of steering wheel rotation; v – vehicle velocity; α – slip angle of vector ν; a_x, a_y, a_z – components of vehicle acceleration; δ_i – angle of a road wheel turn; $\dot{\Psi}$ – angular vehicle velocity around a vertical axle; f_{zi} – vertical suspension displacement; F_{zi} – vertical force yielded by a road wheel).

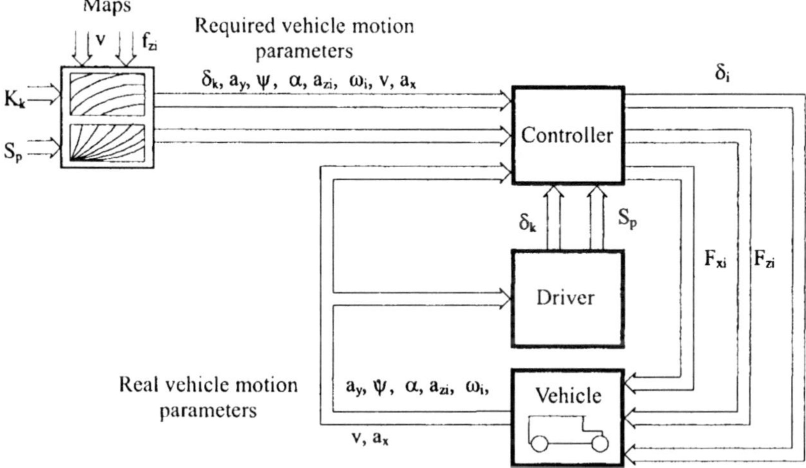

Figure 1.6. Regulation and control of a vehicle motion via a driver action (S_p – drivers action on the brake, clutch and accelerator pedals; K_k – drivers action on the steering wheel).

The sensors applied in the described systems of control (Figure 1.6) should deliver information on the longitudinal, transversal and vertical (a_x, a_y, a_z) vehicle acceleration, should allow for exact measure for both liner v and angular $\dot{\Psi}$ velocities, should measure the steering wheel rotational angles (δ_k), the steering angles (δ_i) and the angular velocities (ω_i) of the steered road wheels.

In order to define the transversal vehicle dynamics (a transversal slip) the special sensors to measure the slip angle α should be applied (see Figure 1.5).

The *internal sensors* are used for linear acceleration measurement, the *impulse inductive sensors* are used to measure the road wheels, angular velocities, and the *angular-code sensors* are applied to measure the steering wheel rotation angle or road wheel turn angles. The sensors measuring the slip angles belong to more complicated devices. The exemplary solution is shown in Figure 1.7.

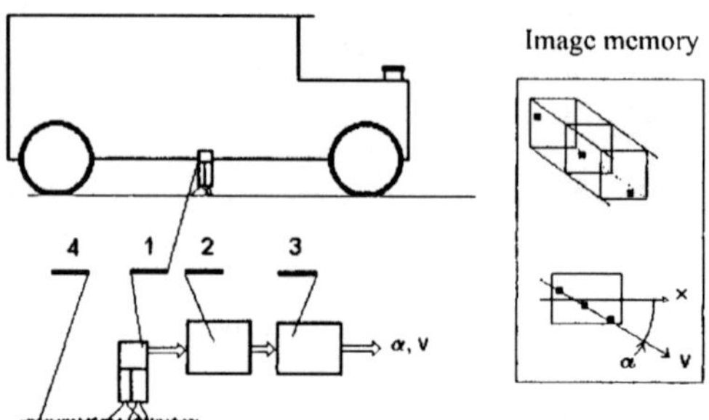

Figure 1.7. A construction of the sensor measuring a slip angle (1 – camera; 2 – digital image memory; 3 – block of image processing; 4 – road roughness; x – longitudinal direction; v – car velocity; α – slip angle of the velocity with respect to x direction).

The road surface image (micro-non-homogeneities) is traced via the camera 1. Owing to the memory image 2 and image processing 3 blocks, the successive positions of a non-homogeneity spot are compared yielding the values of the drift angle α and the driving velocity. For example, as the angular velocity $\dot{\Psi}$ sensors, the tuning fork shaped vibration type gyro or Fiber Optical Gyro can be used.

The inertial sensor applies the Coriolis effects generated during vehical rotation around a vertical axle. The resonator shown in Figure 1.8 is excited electrically and vibrates with a constant frequency. In the case when the sensor housing rotates with velocity $\dot{\Psi}$, the Coriolis force acts on the resonator plates with a value proportional to this velocity. Hence, the measured electrical signal is proportional to $\dot{\Psi}$.

A principle of a *angular velocity laser sensor* action is presented in Figure 1.9.

Two waves of the laser radius reach the light detector 3. One of the mentioned waves is in agreement with the rotation direction Ψ, and the other one has opposite direction. They have shifted phase in detector 3 owing to the real (attached to a vehicle) angular velocity $\dot{\Psi}$. The phase shift value is proportional to $\dot{\Psi}$, and it can be found from the relation

$$\Psi = \frac{\Delta f \cdot \lambda \cdot c}{4\pi \cdot L \cdot R},$$

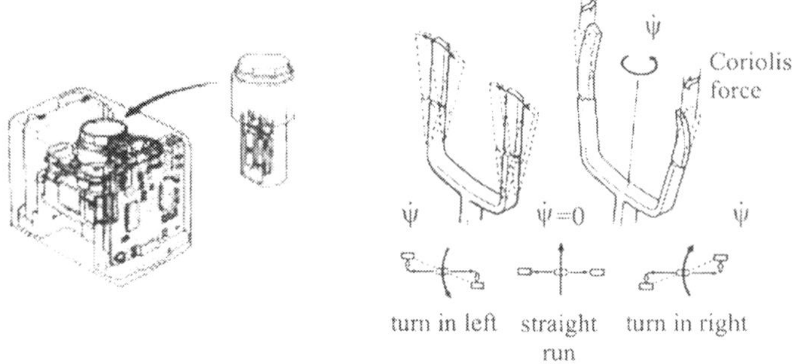

Figure 1.8. General view and a principle of operation of an angular velocity $\dot{\Psi}$ sensor (yaw velocity sensor).

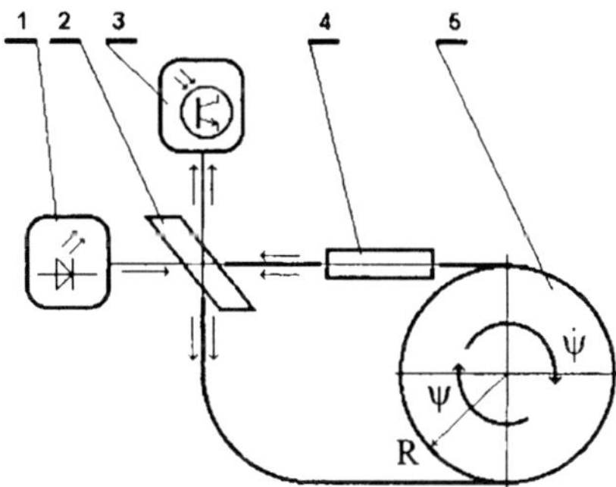

Figure 1.9. A laser sensor of an angular vehicle velocity (1 - laser; 2 - prism; 3 - light detector; 4 - phase modulator; 5 - light pipe winded on a reel with R radius).

where: Δf - phase shift, λ - light wave length, c - light velocity, L - light pipe length, R - radius of the light pipe reel.

2. Historical Reference

The first signs of man's existence come from the Paleolithic era (3 million years ago). Man (Homo habilis, Homo erectus) used tools made of stone, animal and perhaps human bones. Homo sapiens appeared probably in the interglacial era, about 350000-200000 years ago. Homo sapiens neandertaliensis (150000 years ago) is known today for the custom of burying the dead tribesmen, and also for being able to make clothes of animal skin, produce simple tools and light fire.

For man, who in those days was as wild as the animals he treated as a source of food and skins for clothes, the Earth was merely a few-hundred-hectare piece of what we know now as the globe. The existence of cosmic space with uncountable stars and planets was far beyond man's comprehension. It took a long time before man evolved to become Homo faber who constructed a boat, a wheel carriage, a balloon, an aeroplane, a rocket, a hovercraft and a television set among many other devices. Yet, the car – a land vehicle powered by a driving engine and equipped with 4 (2, 3, 6, 8...) road wheels – has remained man's favourite invention. The car became so important in man's life that soon a new species developed - *the motor people.*

Figure 1.10. Ford T from the 20th century.

At the beginning of 1900 engineers constructed a 1200-kilo automobile that was equipped with a 20-KM engine and a transmission with two gears and a wooden bodywork painted black. People could drive it on bad, dusty or muddy roads, and it soon became obvious that the more efficient vehicle would inevitably supersede horse carriages, which had served as the basic means of transportation for 6000 years. Today, motor people may enjoy driving a 1600-kilo car with a 8000 cm^3, 736 kW engine, a 7-speed transmission, an all-wheel-drive system, with maximum speed of 406 kph and bodywork in any chosen colour.

Cars are commonly used everywhere today. Average people living in economically highly developed countries use cars to travel distances of 1-1000 kilometres. If the distance is smaller than one or two kilometres an average traveller walks. The distances of 200-12000 kilometres are travelled by plane. In the cities, tram or underground train systems are used to cover distances of 1-30 kilometres. In passenger traffic, various means of transportation are used to various extents, depending on the travel's purpose. Cars are used for about 90% of business trips, whereas public transportation (buses) is used by young people for daily commuting to schools.

Figure 1.11. Bugatti Veyron De EB 16.4, 736 kW (1001 HP), 406 kph.

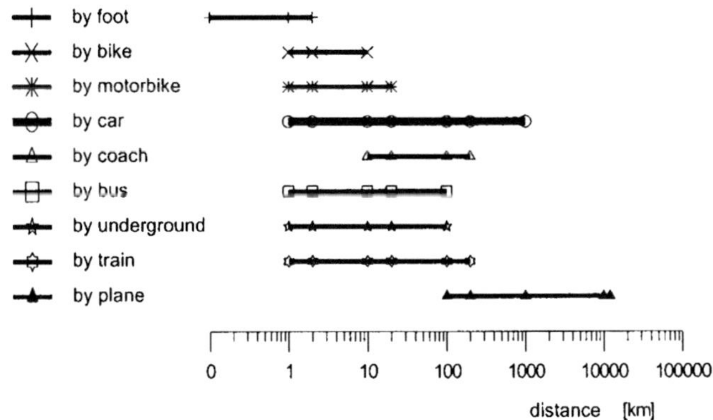

Figure 1.12. Ways and means of transportation used for covering distances of different lengths.

Today we seem to regard the car as a mature, incomparable and almost perfect construction. However it is necessary to admit that there is a huge number of enthusiasts of another land vehicle which reveals its superiority over cars in many aspects. It is a bicycle: a technical wonder that uses the phenomenon of the gyro effect and only two wheels to go. It is very economical: at the speed of 15 kph the ratio of energy consumption per unit does not exceed 15 cal/kg/km. It has fantastic load capacity – it can carry loads up to ten times heavier than its own weight. Although bicycles certainly have some drawbacks, their users in many countries truly appreciate them.

The Beginnings of the Wheeled Vehicle

It is said that man invented the first machine about 10000 years BC. Obviously it was a weapon – a bow. In the period between 10000 and 3000 BC man developed agriculture and livestock breeding. Most likely at that time a road wheel was invented and a wheeled vehicle was constructed. That ancient vehicle must have been primitive, uncomfortable, dangerous and undurable. Perhaps it was even frightening. Nevertheless it made a start for the development of modern wheeled vehicles.

Figure 1.13. A rival to the car? You cannot ride it on motorways. (Antoni Falat, The Cyclist, 1975).

The historical facts concerning wheeled vehicles date back to 3000 BC. The discoveries were made in the lands that used to be inhabited by Sumerians. About 3000 years BC Sumerians used wheeled transport carriages with wheels made of wood. They also built roads and bridges. Surprisingly enough, the spacing of ruts in Sumerian roads was the same as the wheel track of today's cars: 125 cm.

Sumerians lived approximately 5500–4300 years ago. They lived in Mesopotamia (a Greek name), between the rivers Tigris and Euphrates. At that time it was one of the most populated and civilised regions in the world, with the oldest known state structure. Sumerians, who according to the oldest surviving writings called themselves "the blackheaded", arrived in Mesopotamia from the East. Their history is filled with wonderful achievements in art and craft and their country flourished due to the impressive level of its political and social development and people's abundant religious life as well. Unfortunately, the state of Sumerians lasted only about one thousand years. In 1979 BC, repeatedly raided by enemy Semitic tribes from the West, Sumerians lost their last ruler. The magnificent city of Nineveh was ultimately turned into ruins and burnt in 612 BC and disappeared for many years covered by desert sand. Other Sumerian cities suffered the same misfortune. Their enemies destroyed them and the desert buried their ruins obliterating the picture of the ancient glamour of a leading civilisation from the memory of its contemporaries. The fact that the desert hid the remains of Sumerian civilisation had a significant meaning for historians. The desert preserved thousands of clay slates covered with cuneiform writings, ruins of walls and foundations of buildings, cemeteries, many precious artefacts, and many other objects that served as sources of knowledge about Sumerians and their times. Among them was a seven-centimetre statuette made of copper that resembled a wheeled carriage – it was the oldest (dating back to the year 2500 BC) artistic evidence of the practical use of wheeled vehicles in those days. As time passed by, people entirely forgot about Sumerians and their country. Only the Bible mentions the splendour of a legendary country between the Tigris and the Euphrates, with the cities of Nineveh and Nemrod. The Bible also calls it the cradle of Abraham's people and describes it as the location of the mythical Tower of Babel.

It was not until the 19th century AD that some peculiar hills between the Tigris and the Euphrates were explored. Marvellous ruins of ancient cities were brought to daylight again. The clay slates were found and . . . Sumerians were discovered. Jules Opport, a French linguist, discovered, analysed and deciphered the cuneiform writings carved in the clay slates. He published his discovery in 1869. It took many years of hard work of a great number of outstanding scientists and experts to translate the ancient texts. One of them – "School Life" – is particularly beautiful, although there are many versions of it. It was written about 2000 years BC by an unknown poet. The author describes quite an unfortunate day of a schoolboy who was flogged as punishment for being late to school. Alas, it is only the beginning of his misfortune. However, the boy and his father's actions eventually lead to a happy end. Here is a part of the poem:

"School Life" - written 4000 years ago - [32]

"I am thirsty - give me something to drink,
I am hungry give me some bread,
wash my feet, make my bed, I wish to rest;
wake me up early in the morning,
I cannot be late or may teacher gives me a flogging."
When I woke up early in the morning
I saw my mother
and I said these words to her: "Give me my breakfast, I want to go to school..."
My mother gave me two pies and I went to school.
At school they asked me: "Why are you late?"
I felt scared, my heart beat faster.
I stood before my teacher, he sent me to my seat.
My school patron read my slate,
he grew angry (?), he flogged me...

The one who is responsible for the courtyard said:
"Write!", ... a quiet place.
I took my slate...
I filled my slate with writing...
The one who was responsible for... [said] "Why did you talk without my permission?", he flogged me.
The one who was responsible for... [said] "Why did you raise your head so high without my permission?", he flogged me.
The one who was responsible for drawing [said] "Why did you stand up without my permission?", he flogged me.
The one who was responsible for the gate [said] "Why did you go out without my permission?", he flogged me.
The one who was responsible for... [said] "Why did you take it without my permission?", he flogged me.
The one who was responsible for Sumerian [said] "You said...", he flogged me.
My teacher [said] "Your hand is not good", he flogged me.
What son told, his father heard.
The teacher was brought from school.
When he entered the house, a seat of honour was offered to him.
The schoolboy took..., and sat humbly in front of him.
Everything he learnt in the art of writing
he presented to his father.
With contented heart, his father spoke gladly to the teacher:
"My young son opens his hands and you make them grasp knowledge;
show him entire beauty of the art of writing.
You explain to him the results of calculations on mathematical tables,
you reveal the mysteries of learning."
"Pour fragrant olive onto his neck and stomach,
may the good olive fill his... , as water fills a vase,
I'll dress him with a new robe, I'll offer him a gift, I'll put a ring on his finger."
"My young boy, for you have not ignored my word, have not denied it,
may you reach the height of the art of writing, master it completely.
For you have given me something you didn't have to give,
you have offered me a gift more precious than my payment, you have showed me your respect,
may Nidebah, the queen of guardian gods, be your guardian goddess..."

Having read that poem written so many years ago, we must admit that it has taken thousands of years of strenuous work of countless numbers of people before we reached the present level of technological development. For ages our children have been going to schools to learn "the art of writing", to acquire "the results of calculations on mathematical tables" and "reveal the mysteries of learning". Knowledge and creativity enabled mankind to build a wheeled vehicle a few thousand years ago and come up with an idea that a strong tameable animal – a horse, for instance – might pull it. An image of that ancient vehicle has been found. It was preserved till our days only because man needed something more than creating reality. Man created visions, images.

The image of a Sumerian chariot from 2500 BC presented in Fig. 1.14 is a 7-cm statue made of copper – today known as "the copper quadriga from Tell Agrab" exhibited in a museum in Baghdad.

Figure 1.14. The copper quadriga from Tell Agrab. A two-wheeled carriage drawn by four oxen (or onagers). Height: 7 cm. Iraqi Museum, Baghdad.

Another most beautiful work of art – a priceless historical artefact – is the banner from Ur (about 2500 BC), which displays images of Sumerian wheeled vehicles of another type (Fig. 1.15). They are four-wheeled carriages used for civilian and military transportation purposes.

Figure 1.15. One of the findings from the royal cemetery in Ur (about 2500 years BC), the so-called mosaic banner. The scene presents chariots in combat. Speeding horses trample a warrior. British Museum, London.

Wheeled carriages were used also in many other countries in different regions of the world. Between 1700 and 1300 BC, light uniaxial combat chariots drawn by two horses or wild donkeys were used in Egypt, Palestine or China. Since

1500 BC, wheeled carriages were used in Greece and Georgia, and later in Italy and England as well.

Figure 1.16. A banquet with musicians from Ur – Sumer ca. 4600 BP.

Long before our era roads with hardened surface had been laid down to connect cities. One of the most famous and still existing roads is Via Appia that was built in about 312 BC to link Rome with Capua. The road was built so solidly that it resisted the weight of thousands tanks riding it during World War II. Roman roads were about one metre thick since they were made of a few layers of stones and gravel fixed with lime mortar.

The first wheeled vehicles were very unstable

The wheels of the first types of carriages were rigidly connected to the axle. Both of the wheels span together with the axle at the same angular velocity. Such a system made it difficult to take a turning. Moreover, at high speed the vehicle would even easily fall over.

Disengaging the wheels from the axle soon eliminated that obvious drawback.

Year 1559 marked another major step in the history of wheeled vehicles. Historical resources claim it was the time when a bold experiment was made aiming at finding an alternate power, other than a horse, an ox or a man, to move a vehicle. It was Holland's Simon Stevnus's idea to use a sail for that purpose. Throughout the next centuries there were also other more or less successful experiments in using the power of wind to put a vehicle in

Figure 1.17. An Egyptian combat chariot.

Figure 1.18. M.Hacquet's sail-vehicle from 1834.

motion, for example M.Hacquet's sail-vehicle with a twelve-metre-high sail mast, constructed in 1834. So far it has been impossible to realise the idea of a land vehicle propelled by wind power like a sailboat and thus provide a cheap means of land transportation. A land sail vehicle without a stabilising centreboard can be very unstable and susceptible to sudden gusts of wind that might easily knock it over.

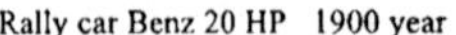

Rally car Benz 20 HP 1900 year

Rally car Benz 35 HP 1903 year

Figure 1.19. Two cars made by Daimler Benz: the short model - produced until the fatal car race accident, and the modernised longer model.

By the end the 19th century a car was constructed; a wheeled vehicle propelled by its own source of power.

The first cars had problems with stability.

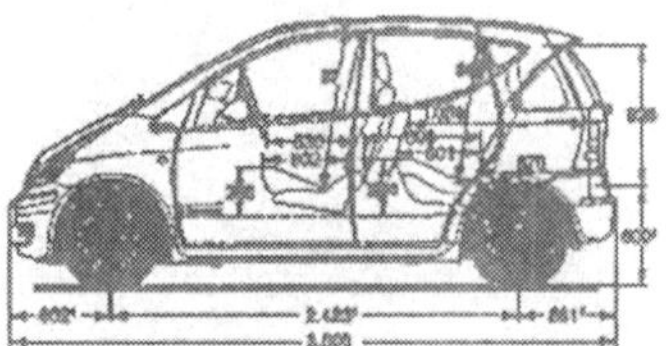

Figure 1.20. Little Mercedes A Class. Axle base $2423\,mm$. Famous debut in 1997 year. Today as standard: airbag, ABS, brake assist, ESP, electronic power steering.

The 1899 Daimler's axle base was as long as the axle base of contemporary chaises. In 1900 there was an accident during a car race, in which the wrecked car's driver died. Experts believed (and they were correct, as it later turned out) that "bad stability of movement" due to too short length of the car's axle base was the major cause of the accident. In the 1901 model the axle base was 1.6 times longer. The new Daimler was safer and its shape resembled modern cars.

Chapter 2

THE PRINCIPLES OF THE THEORY OF STABILITY

In this chapter, stability concepts are briefly reviewed with an emphasis on their technical aspects. First, definitions of various stability concepts, such as those proposed by Lyapunov, Poincaré, Poisson, Bogusz and Szpunar, are introduced, illustrated and discussed. Finally, a general method for stability estimation is addressed.

1. The Definitions of Technical Systems Stability[1]

The term "stability" is used to describe a system's capability of maintaining balance despite the affects of disturbances, including them due to initial conditions. In reality, there may be slight or considerable disturbances. When a system (in particular a non-linear system) is stable at slight disturbances, it may be unstable at bigger disturbances. The notion of stability is not a notion related to the physics of a phenomenon, but it is defined depending on what qualities an ideal model should possess.

The beginnings of the theory of stability

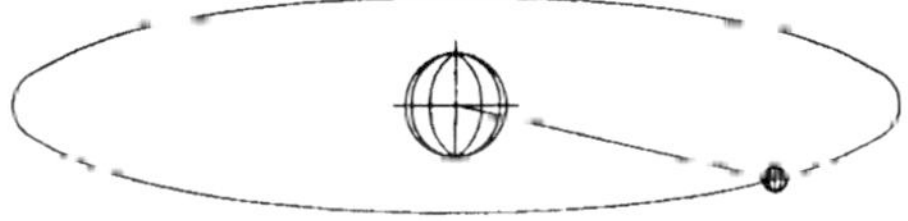

The theory of differential equation was developed in the 18th century, a time when understanding of natural and technical sciences was considered indispensable. Astronomers studying the mechanics of celestial bodies concentrated on the most accurate description of their movement, especially those of the Solar

[1]based on, among others, [47] - Demidowicz B. P. and [37] - Bogusz Wl.

system. Isaac Newton (1643–1727) conceived a peculiar conception according to which it was impossible for the orbits of the Solar system's planets to maintain their existing positions in relation to one another without "a divine intervention" at a proper time. Newton believed that reciprocal attraction of celestial bodies should cause chaos. A strong wish to refute that view led his contemporaries to the discovery of the laws of planetary movement. The basis for the description of that phenomenon was the theory of differential equations. Long before, though, Isaac Newton had discovered a brilliant principle and noted it in the form: *"It is very useful to solve differential equations"*.

Leonard Euler (1707–1783) used differential equations to describe the movement of a planet around the Sun. Here is the equation:

$$\frac{d^2x}{dt^2} - 2n \cdot \frac{dy}{dt} - n^2 \cdot \left(1 + x^2\right) = \frac{-n^2 \cdot (1 + x)}{\left[(1 + x)^2 + y^2\right]^{3/2}},$$

$$\frac{d^2y}{dt^2} + 2n \cdot \frac{dx}{dt} - n^2 \cdot y = \frac{-n^2 \cdot y}{\left[(1 + x)^2 + y^2\right]^{3/2}},$$

where: x, y are the planet's co-ordinates; $n = 1/(a\sqrt{a})$; a is the average distance from the planet to the Sun.

Facing the problem of the practical use of the derived equations of planetary movement, we need to solve two significant problems. One is to develop a method for finding solutions of the equations. The other is to define the initial conditions and the values of constant parameters (such as the mass of the Sun) as accurately as possible. The level of science in the 18th century was not high enough to solve the second problem easily; accurate distances between the planets and their masses were unknown. Moreover, it turned out that even the slightest changes of those values affected the results of some differential equations to such as extent, that the calculated trajectories of planets significantly differed from one another. Usefulness of such equations, that were called "unstable", for describing planets' trajectories seemed extremely doubtful. Therefore, more effort had to be put into developing a method of examining the sensitivity of differential equations to initial conditions and to the values of constant parameters. As a result a new field of science was created: *"stability of differential equations"* that was important not only for astronomy but for every science that involved bodies in motion.

2. Stability in Lyapunov's[2] Sense [47]

Consider a system of differential equations (2.1), and a function f_i that meets the conditions of existence and uniqueness of the solutions:

$$\dot{x}_i = f_i\left(t, x_1, x_2, \ldots, x_n\right), \;\; x_i(t_0) = a_i, \;\; i = 1, 2, \ldots, n, \qquad (2.1)$$

where: $a_1, a_2, \ldots, a_n$ are given numbers, time $t_0 \in \langle 0, +\infty)$.

Definition 1 — Stability in Lyapunov's sense

Solution $x_i(t)$ of system (2.1) is stable in Lyapunov's sense, when for any given number $\varepsilon > 0$, there is a number $\delta > 0$, such that for every solution $x_i^*(t)$ at initial conditions $x_i^*(t_0) = b_i$ that meets the condition

$$\|\mathbf{a} - \mathbf{b}\| = \sum_{i=1}^{n} \left[(a_i - b_i)^2\right] < \delta^2,$$

$$\text{for } t \geq t_0, \quad \text{it holds that } \|\mathbf{x}(t) - \mathbf{x}^*(t)\| = \sum_{i=1}^{n} \left\{[x_i(t) - x_i^*(t)]^2\right\} < \varepsilon^2.$$

The field defined by the number 2δ and the initial conditions of solution $x_i(t_0)$ we shall call the *initial conditions field* and notate as ω. The field defined by the number 2ε and the trajectory of solutions $x_i(t)$ we shall call the *field of acceptable solutions* and notate as Ω. In other words; solution $x_1(t), x_2(t), ..., x_n(t)$ of system (2.1) may be called stable in Lyapunov's sense, if solution $x_1^*(t), x_2^*(t), ..., x_n^*(t)$, sufficiently close to $(\pm\delta)$ for $t = t_0$, is also close to $(\pm\varepsilon)$ within time $t_0 < t < \infty$. Figures 2.1 and 2.2 present geometric interpretations of a stable solution in Lyapunov's sense.

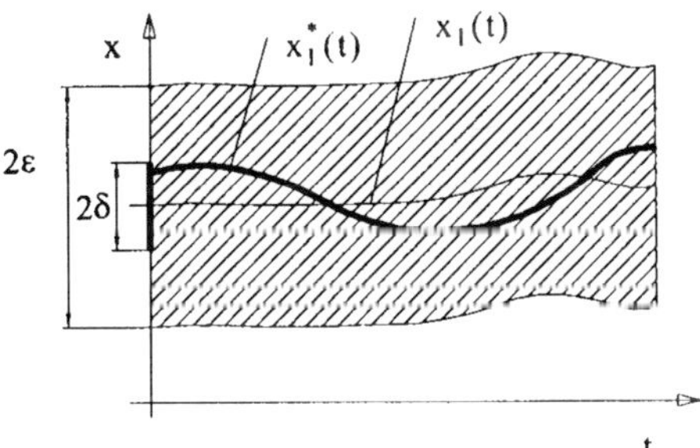

Figure 2.1. Stable solution in Lyapunov's sense on plane xt.

[2] Aleksander Lapunow - a Russian mathematician 1857-1918

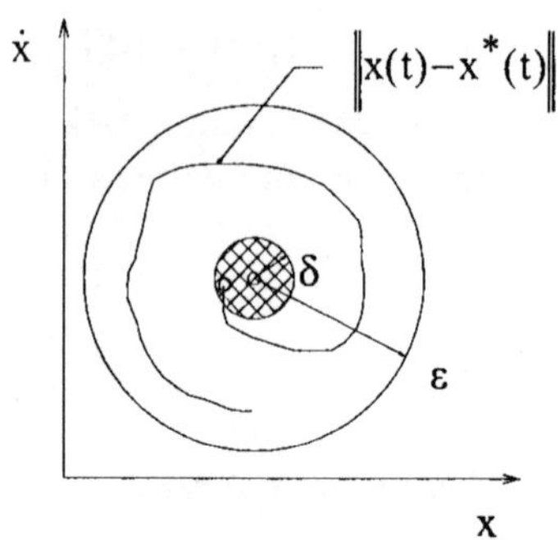

Figure 2.2. Stable solution in Lyapunov's sense on a phase plane.

Definition 2 — Asymptotic stability in Lyapunov's sense

Solution $x_i(t)$ of system (2.1) is considered to be asymptotically stable in Lyapunov's sense, if it is stable in Lyapunov's sense and if there is a number $\delta > 0$, such that for every solution $x_i^*(t)$ and $x_i^*(t_0) = b_i$ that meets the condition

$$\|\mathbf{a} - \mathbf{b}\| = \sum_{i=1}^{n} \left[(a_i - b_i)^2\right] < \delta^2,$$

it holds that

$$\|\mathbf{x}(t) - \mathbf{x}^*(t)\| = \lim_{t \to \infty} \left\{ \sum_{i=1}^{n} \left\{ [x_i(t) - x_i^*(t)]^2 \right\} \right\} = 0 \text{ for every } t \geq t_0.$$

Figure 2.3 presents a geometric interpretation of an asymptotically stable solution in Lyapunov's sense.

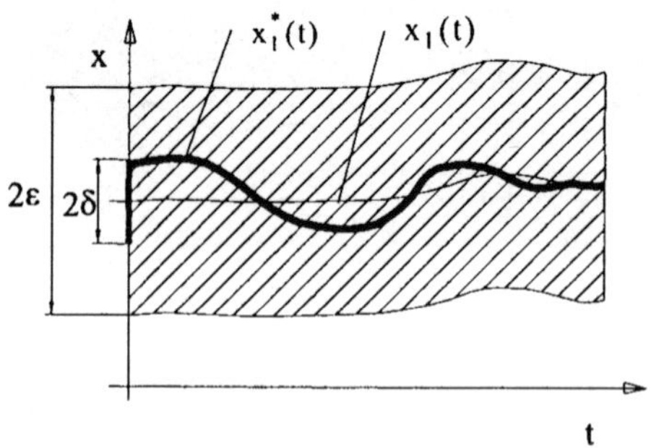

Figure 2.3. Asymptotically stable solution in Lyapunov's sense on plane xt.

EXAMPLE 2.1 *There is a given system:* $\dot{x} + x = 0$; $x(0) = 1$. *Its solution is:* $x(t) = e^{-t}$. *Let us take any other solution (for comparison):* $x^*(t) = a \cdot e^{-t}$, $x^*(0) = a$. *Then there is a number* $\varepsilon > 0$ *that fulfils the inequality*

$$[x(t) - x^*(t)]^2 = e^{-2t} \cdot (1 - a)^2 \leqslant \varepsilon^2.$$

There is also a number δ *that fulfils the dependence*

$$[x(0) - x^*(0)]^2 - \delta^2 \;\; \textit{for}\; x(0) = 1, \;\; x^*(0) = a, \;\; \textit{thus} \;\; [1 - a]^2 < \delta^2 = \varepsilon^2.$$

Thus we have proved, that solution $x(0) = 1$ *is stable in Lyapunov's sense. Solution* $x(0) = 1$ *is asymptotically stable as well, since:* $\lim_{t\to\infty} \left(e^{-t} - a \cdot e^{-t}\right)^2 = 0$.

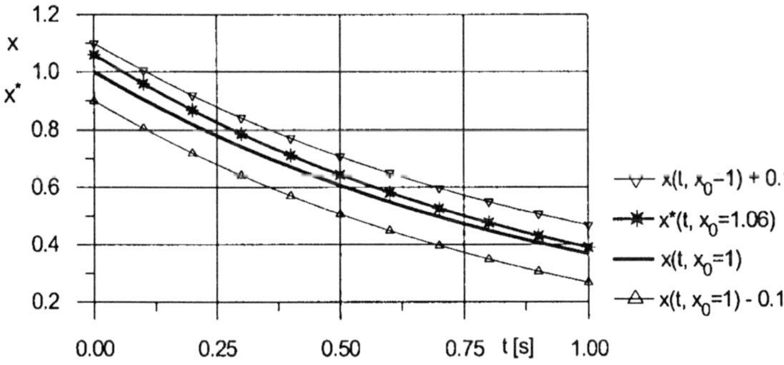

Figure 2.4. Solutions of system $\dot{x} + x = 0$ in relation to time.

Figure 2.4 presents graphic interpretations of the solutions of system $\dot{x} + x = 0$, $x(0) = 1$ *and* $x(0) = a$ *in relation to time.*

EXAMPLE 2.2 *There is a given system:* $\dot{x} - x = 0$, $x(0) = 1$. *Its solution is:* $x(t) = e^t$. *For any given solution* $x^*(t) = a \cdot e^t$, $x^*(0) = a$, *there is a number* $\delta > 0$ *that fulfils*

$$[x(0) - x^*(0)]^2 < \delta^2, \;\; \textit{for} \;\; x(0) = 1, \; x^*(0) = a: \;\; [1 - a]^2 < \delta^2.$$

We cannot find any number e that fulfils the inequality: $\|\mathbf{x}(t) - \mathbf{x}^*(t)\| \leq \varepsilon^2$, *because the norm of solutions* $x(t)$ *and* $x^*(t)$ *for* $t \to \infty$ *approaches infinity:*

$$[x(t) - x^*(t)]^2 - e^{2t} \cdot (1 - a)^2 \to \infty.$$

Solution $x(0) = 1$ *of the system is unstable in Lyapunov's sense.*

Definition 3 — Stability in Lyapunov's sense at persistent disturbances

Let us consider the system of equations:

$$\dot{x}_i = f_i(t, x_1, x_2, \ldots, x_n) + R_i(t, x_1, x_2, \ldots, x_n), \; i = 1, 2, \ldots, n, \; x_i(t_0) = a_1, \tag{2.2}$$

where: $a_1, a_2, \ldots, a_n$ are given numbers, time $t_0 \in \langle 0, +\infty)$, and functions $R_i(t, x_1, \ldots, x_n)$ are persistent disturbances that for $t \geq 0$ fulfil the inequality:

$$\|\mathbf{R}(t, x_1, \ldots, x_n)\| < r,$$

where r is a sufficiently small number.

Solution $x_1(t), x_2(t), \ldots, x_n(t)$ of system (2.2) is considered to be stable in Lyapunov's sense, when for any given number $\varepsilon > 0$, there are two numbers $\delta > 0$ and $r > 0$, such that for every solution $x_1^*(t), x_2^*(t), \ldots, x_n^*(t)$ of system (2.2) with the initial conditions $x_i^*(t_0) = b_i$ that meet the condition

$$\|\mathbf{a} - \mathbf{b}\| = \sum_{i=1}^{n} \left[(a_i - b_i)^2 \right] < \delta^2,$$

and for the disturbance functions which meet the condition

$$\|\mathbf{R}(t, x_1, \ldots, x_n)\| < r^2,$$

it holds that

$$\|\mathbf{x}(t) - \mathbf{x}^*(t)\| = \sum_{i=1}^{n} \left\{ [x_i(t) - x_i^*(t)]^2 \right\} < \varepsilon^2 \text{ for every } t \geq t_0.$$

The field defined by the number δ and initial conditions we shall call the initial conditions field and notate it with ω. The field defined by the number ε and the trajectory of solution $x_1(t), x_2(t), \ldots$ we shall call the field of acceptable solutions and notate as Ω.

A few remarks on stability in Lyapunov's sense

The definitions of stability in Lyapunov's sense quoted above referred – as it was clearly indicated – to individual solutions. Systems are stable when stability of all possible solutions is examined. In the case of a non-linear system with a constant matrix of coefficients, all solutions are either stable or unstable.

Linear equations systems usually take the form of: $\dot{\mathbf{x}} = \mathbf{A}(t) \cdot \mathbf{x}(t)$, where $\mathbf{x}(t_0) = \mathbf{x}_0$. While analysing the system's stability, we consider stability of relocations $\mathbf{x}$ and velocity $\dot{\mathbf{x}}$, as well. If matrix $\mathbf{A}(t) = \mathbf{A} = const$, then stability of relocations denotes stability of velocity. In case of a variable matrix such a relation may not occur. In case of a non-linear system, some solutions may be stable, while other may be unstable.

EXAMPLE 2.3 *There is a differential equation:* $\dot{x} + x - x^2 = 0$, $x(t_0) = x_0$. *The general solution of this equation takes the form of:* $x = x_0 \cdot \left[x_0 + (1 - x_0) \cdot e^{t-t_0}\right]^{-1}$. *Figure 2.5 presents the solutions for various initial values* x_0. *The graph clearly proves that for our system some solutions are stable and some are not. The solutions with the initial conditions* $x_0 < 1$ *are stable, since for* $t \to \infty$ *they approach* 0, *and the solutions with the initial conditions* $x_0 > 1$ *are unstable since for* $t \to \infty$ *they approach* ∞, *while the base solution* $x(t, x_0 = 0) = 0$.

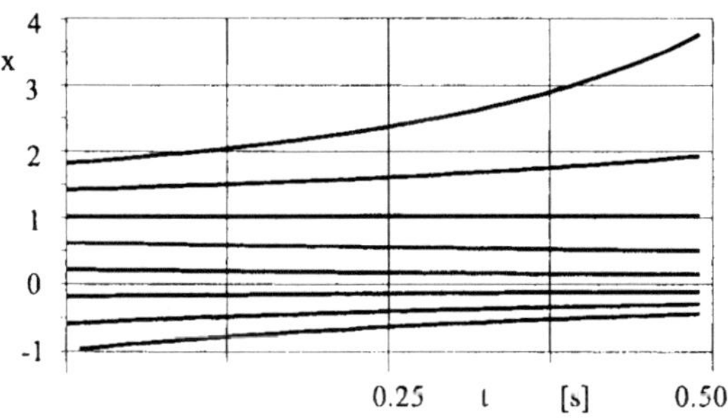

Figure 2.5. Graphic interpretation of a few solutions of system $\dot{x} + x - x^2 = 0$ among which some are stable and some are not.

For linear systems it is essential to invoke theorem that refers to stability of homogeneous and heterogeneous linear systems.

The theorem of heterogeneous and homogeneous systems

The necessary and sufficient condition of stability of a heterogeneous linear system is stability of a zero solution of a corresponding homogeneous system. Stability of system $\dot{\mathbf{x}} = \mathbf{A}(t) \cdot \mathbf{x}(t) + \mathbf{f}(t)$, $\mathbf{x}(t_0) = \mathbf{x}_0$, with a heterogeneity element $\mathbf{f}(t)$, may be found through the study of a corresponding homogeneous system $\dot{\mathbf{x}} = \mathbf{A}(t) \cdot \mathbf{x}(t)$, $\mathbf{x}(t_0) = \mathbf{x}_0$.

We shall also invoke another important theorem that embraces the relation of stability in Lyapunov's sense and boundedness, which is relevant only for homogeneous linear systems.

The theorem of the relation of stability in Lyapunov's sense and limitedness

A homogeneous linear system is stable only when every solution of this system is limited for $t \in [t_0, \infty)$.

Stability of the point of equilibrium

When the theoretical considerations on stability are limited to the equilibrium of the system, then, analogously to the already quoted terms, the so-called point of equilibrium may be chosen for the subject of the definition. Thus, if there is a system of differential equations (2.1), that describes the system's movement in the neighbourhood of the point of equilibrium x_1, x_2, then the point of equilibrium will be called:

- stable, if for the given neighbourhood of the point - field ε, we find a neighbourhood of the point - field δ, such that the entire trajectory starting from field δ is enclosed within field ε;
- locally stable, if the size of field δ is limited;
- globally stable, if the size of field δ is unlimited and contains the whole state space;
- asymptotically stable, if field δ is infinitesimal and becomes the point of equilibrium;
- globally asymptotically stable, if field ε is unlimited and field δ is infinitesimal.

Study of stability in Lyapunov's sense

In every defined system, various methods may be used to study stability in Lyapunov's sense. The most popular are: the direct method (see Examples 2.1 and 2.2), and the indirect method that employs Lyapunov's function. Let us present the theorem that refers to the study of stability with the use of Lyapunov's function and illustrate it with an example.

Lyapunov's theorem

If in a field **D**, that contains the system's point of equilibrium, which serves as the origin of the co-ordinates system, there is a positive definite scalar function $V(x_1, x_2, \ldots, x_n)$, i.e.:

1 $\bigwedge_{\mathbf{x}\in\mathbf{D}} V(\mathbf{x}) \in C^1$ that is function V is continuous with first partial derivatives in relation to every variable,

2 $V(\mathbf{x}_0 = 0) = 0$,

3 $\bigwedge_{\mathbf{x}\in\mathbf{D},\mathbf{x}\notin 0} V(\mathbf{x}) > 0$ - for every non-zero variable from field **D** the function $V(x) > 0$, while its derivative along the solutions of system (2.1) is negative definite, i.e.: $\bigwedge_{\mathbf{x}\in\mathbf{D}} \dot{V}(\mathbf{x}) \in C^1$, that is function $\dot{V}$ is continuous with its

first partial derivatives in relation to every variable $\dot{V}(\mathbf{x}_0 = 0) = 0$ and $\bigwedge_{\mathbf{x}\in\mathbf{D},\mathbf{x}\notin 0} \dot{V}(\mathbf{x}) < 0$ for every non-zero variable from field **D** the function $\dot{V}(x)$ is smaller than zero, then the system defined by the system of equations (2.1) is asymptotically stable in field **D**.

If $\dot{V}(\mathbf{x})$ is negatively half-definite in field **D**, then the system is stable in that field, though not necessarily asymptotically.

If Lyapunov's function and its derivative do not change their signs in field **D** and $\bigwedge_{\mathbf{x}\in\mathbf{D},\mathbf{x}\notin 0} V(\mathbf{x}) > 0$ and $\bigwedge_{\mathbf{x}\in\mathbf{D},\mathbf{x}\notin 0} \dot{V}(\mathbf{x}) > 0$, or $\bigwedge_{\mathbf{x}\in\mathbf{D},\mathbf{x}\notin 0} V(\mathbf{x}) < 0$ and $\bigwedge_{\mathbf{x}\in\mathbf{D},\mathbf{x}\notin 0} \dot{V}(\mathbf{x}) < 0$ occur, then the system is unstable in field **D**.

The interpretation of Lyapunov's function for a two-dimensional state space

Let us assume the existence of function $V(x_i)$ in the form of $V = x_1^2 + x_2^2$, which geometrically corresponds to an equation of a circle. We can observe that it meets the three conditions mentioned above: $1^o, 2^o$ and 3°: it is positive definite, its value equals zero for $x_1 = 0$ and $x_2 = 0$ and it approaches infinity when $\|x\| \to \infty$. Figure 2.6 presents the trajectory of the state space of co-ordinates x_1 and x_2 and the assumed function $V = x_1^2 + x_2^2$ - a circle. The study of stability in this case consists in the study of the sign of the derivative of function V along the solutions of system (2.1) $\dot{V}(x_1, x_2)$, which will allow us to discover the tendency of the radius's change $V = x_1^2 + x_2^2$ in point T. If derivative $\dot{V}(x_1, x_2) < 0$, then the circle's radius in point T tends to shorten, and the value of $\dot{V}(x_1, x_2)$ is the velocity of changes of the radius's square related to time. If condition $\dot{V}(x_1, x_2) <)$ occurs for all $(x_1^2 + x_2^2)$, then function $V = x_1^2 + x_2^2$ is a Lyapunov function for system (2.1). If derivative $\dot{V}(x_1, x_2) = 0$, then trajectory $f_i(x_1, x_2)$, is tangent to circle $V = x_1^2 = x_2^2$, i.e. it creates a stable limit cycle. If derivative $\dot{V}(x_1, x_2) \leq 0$, then trajectory $f_i(x_1, x_2)$ does not approach infinity, only runs tangentially to circle $V = x_1^2 + x_2^2$ and system (2.1) is stable in Lyapunov's sense according to the definition.

EXAMPLE 2.4 *The movement of a system is described by differential equation $\dot{x} = -A \cdot x - B \cdot sign(x)$, where constants A and $B > 0$. Lyapunov's function will take the form $V(x) = 0.5 \cdot x^2$. Function V is positive definite in field R^n. The function's derivative along the solutions of the system's equations is described by the dependence:*

$$\dot{V}(x) = \frac{\partial V}{\partial x} \cdot \dot{x} = -A \cdot x^2 - B \cdot x \cdot sign(x) .$$

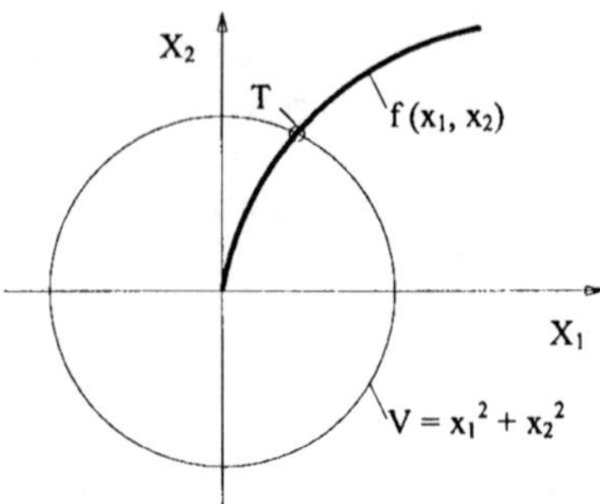

Figure 2.6. A trajectory in the state space of co-ordinates x_1 and x_2 and the assumed Lyapunov's function V.

As we can observe, function $\dot{V}$ is negatively definite in the whole field R^n. The studied system, with point of equilibrium $x_0 = 0$, is globally asymptotically stable in Lyapunov's sense.

EXAMPLE 2.5 *Let us study stability of a system composed of inertial linear term L and non-linear term N, according to the block diagram in Figure 2.7. The system to be considered may perform the role of a model of a vehicle (L) the radial motion of which – the angle of rotation ψ – is controlled by the driver by rotation of the steering wheel at angle β. Let us assume that the driver's action is non-linear (N) and is not delayed in time.*

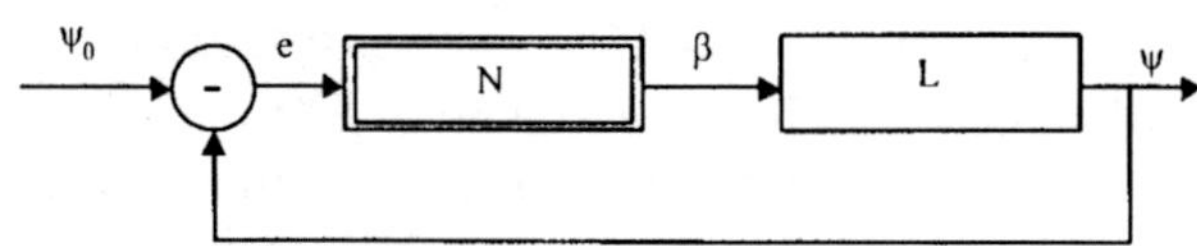

Figure 2.7. A block diagram of a system composed of inertial linear term L and non-linear term N.

The system's transmittance is described by the dependence:

$$G(s) = \frac{k}{s^2 + A \cdot s + B}, \quad \textit{where: } A, B, k > 0.$$

We will assume that the non-linear term (the model of the driver) is described by function

$$\beta = g(e) = \begin{cases} \sin(w \cdot e) & e \in \Omega = (-0,5\pi/w, \;\; 0,5\pi/w) \\ sign(w) & e \notin \Omega\,. \end{cases}$$

The graphic representation of function $g(e)$ is shown in Figure 2.8. The system's differential equation takes the form of $\ddot{\psi} + A \cdot \dot{\psi} + \psi = k \cdot \beta$. The issue being studied refers to the system's equilibrium at $\psi_0 = 0$. In this case error $e = -x$. Assuming that: $x_1 = e$, $x_2 = \dot{e}$, the differential equations of the second order will be reduced to two equations of the first order:

$$\dot{x}_1 = x_2; \quad \dot{x}_2 = -A \cdot x_2 - B \cdot x_1 - k \cdot g(x_1).$$

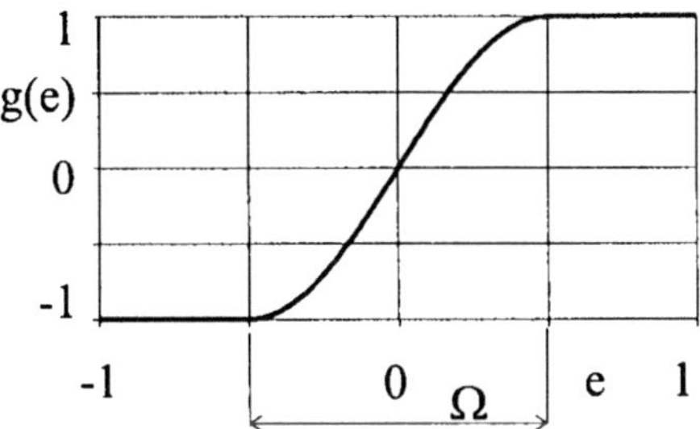

Figure 2.8. Function $\beta = g(e)$ – the model of the driver's actions.

Lyapunov's function will take the form[3] of:

$$V(x_1, x_2) = x_1^2 + \frac{1}{B} \cdot x_2^2 + \frac{2k}{B} \cdot \int_0^{x_1} g(x_1) dx_1,$$

$$\int_0^{x_1} g(x_1) dx_1 = \begin{cases} \int_0^{0,5\pi/w} \sin(w \cdot x_1)\, dx_1 & x_1 \in \Omega \\ \int_{0,5\pi/w}^{x_1} sign(x_1)\, dx_1 & x_1 \notin \Omega \end{cases}$$

$$= \begin{cases} \frac{1}{w} \cdot \left(1 - \cos\left(\frac{\pi}{2} \cdot x_1\right)\right) & x_1 \in \Omega \\ x_1 \cdot sign(x_1) - 0,5\pi/w & x_1 \notin \Omega \end{cases},$$

where: $\Omega = (-0.5\pi/w, 0.5\pi/w)$.

Lyapunov's function's derivative along the solutions of the system being studied takes the form of:

$$\dot{V} = \frac{\partial V}{\partial x_1} \cdot \dot{x}_1 + \frac{\partial V}{\partial x_2} \cdot \dot{x}_2$$

[3] In fact, this function may be called "Lyapunov's function", whose properties we studied first – see: "Lyapunov's theorem".

$$= 2x_1 \cdot \dot{x}_1 + 0 + \frac{2k}{B} \cdot g(x_1) \cdot \dot{x}_1 + 0 + \frac{2}{B} x_2 \cdot \dot{x}_2 + 0 = -\frac{2A}{B} \cdot x_2^2.$$

Having studied the properties of function $V(x_1, x_2)$ for the assumed function $g(x_1)$ and its variability periods for $x_1 \in \Omega$ and for $x_1 \notin \Omega$, we shall observe, that:

1. *in field $x_1, x_2 \in R^n$ function $V(x_1, x_2) \in C^1$,*
2. *for $x_1 = 0$ (point $x_1 = 0 \in \Omega$) and for $x_2 = 0$ function $V = 0$,*
3. *for $x_1 \in \Omega$ and $x_2 \in R^n$ and for $x_1 \neq 0$ and $x_2 \neq 0$ function $V > 0$, for $x_1 \neq \Omega$ the value of $\sup(x_1 \cdot sign(x_1)) > \frac{\pi}{2w}$, i.e. for $x_1 \notin \Omega$ ($x_1 \in (R^n - \Omega)$) function $V > 0$, finally: for $x_1 \in (R^n - 0)$ and $x_2 \in (R^n - 0)$ function $V > 0$.*

If all the three conditions are met, the assumed function $V(x_1, x_2)$ is positively definite in the whole state space.

Function $\dot{V} = -2A/B \cdot x_2^2$ is negatively definite in the whole state space:

1. $\bigwedge_{x \in R^n} \dot{V}(x_1, x_2) \in C^1$,
2. $\dot{V}(x_1 = 0,\ x_2 = 0) = 0$,
3. $\bigwedge_{x \in R^n, x \in 0} \dot{V}(x_1, x_2) < 0$.

Considering the fact, that we have found function V which meets all the requirements to be Lyapunov's function for the whole state space at $x_1, x_2 \in R^n$, we can confirm that the system's point of equilibrium $(x_1 = 0, x_2 = 0)$ is globally asymptotically stable in Lyapunov's sense.

There are not many differential equations with analytical solutions and most of them can be solved numerically. Due to the use of various numerical methods of calculation it is necessary to introduce the notion of discrete time. In such cases, the notion of a dynamic system with discrete time (a cascade) is introduced. Thus, every dynamic system (R^n, f) can form a cascade (discrete time) or a flow (continuous time).

In case of a cascade, the sequence of subsequent values $\{\Phi^n(x)\}, n = 0, 1, 2, ...$ is called the trajectory of point x. If there is such a natural number $k \geq 2$ and such a point x_0 that fulfils dependencies $x_0 = \Phi^k(x_0)$, and $x_0 \neq \Phi^l(x_0)$ for $0 < l < k$, then x_0 is called a periodical point with period k. Such a point is related to a trajectory (a periodical sequence), the k *-element set* $\{x_0, \Phi(x_0), \Phi^2(x_0)...\Phi^{k-1}(x_0)\}$ of which is called a periodical orbit related to point (x_0). Every point of that orbit is a periodical point with period k.

Point x^* is called the ω-*limit* point of trajectory $\{\Phi^n(x)\}$, if there is such a sub-sequence of it, that $\lim_{n^* \to \infty} \Phi^{n^*}(x) = x^*$. The set of all ω-*limit* points is called an Ω-limit set of trajectory $\{\Phi^n(x)\}$.

Compact subset x^* of space R^n is called an invariant set of cascade (R^n, Φ) if $\Phi(x^*) = x^*$.

The constant points, singular points and Ω-limit sets are examples of invariant sets. Moreover, we need to add here the quasi-periodical sets that belong to particular Ω-limit sets. They may appear in a system of at least third order, and their simplest example, most often observed in technical problems, is a harmonically forced linear oscillator (the frequency of proper vibrations and force excitations are not commensurable, e.g.:1 and $\sqrt{2}$). In this case, the observation and recording of the system's fixed state at every period of the excited force allow studying the quasi-periodical orbit that is a set of points constituting a closed curve.

On the basis of the previously introduced notions we can specify the notions of attractors and repellers. All of the sets listed so far, i.e. fixed points, periodical points and quasi-periodical solutions, may be attractors or repellers.

An attractor of a dynamic system (R^n, Φ) *is a closed and limited invariant system A, if there is a neighbourhood* $O(A)$ *of it, such that for any given* $x \in O(A)$ *trajectory* $\{\Phi^n(x)\}$ *remains in* $O(A)$ *and also approaches it asymptotically at* $n \to \infty$. *Moreover, the set of all values* x, *for which sequence* $\{\Phi^n(x)\}$ *approaches set A, is called its set (pool) of attraction.*

A repeller of a dynamic system (R^n, Φ) *is a closed and limited invariant system* $\bar{A} \subset R^n$, *if there is a neighbourhood* $O(\bar{A})$, *such that if* $x \notin \bar{A}$ *and* $x \in O(\bar{A})$, *then for every* ℓ *occurs* $\Phi^k(x) \notin O(\bar{A})$ *for* $k > \ell$, *regardless of the chosen neighbourhood.*

Recent literature contains many examples of peculiar types of both attractors and repellers with their sets of attraction or repulsion that may have very complex characters and properties [20].

Another step in adjusting the definition for the needs of real dynamic systems is the introduction of the notion of technical stability in definite time.

2.1 Lyapunov's functions and Lyapunov's second method

If

$$x_S(t) + y_S^*(t) = y_S(t), \tag{2.3}$$

then the substitution into (1.1) results in

$$\frac{dx_S}{dt} = X_S(t, x_1, ..., x_n), \quad s = 1, ..., n, \tag{2.4}$$

where:

$$X_S(t, x_1, ..., x_n) = F_S(t, x_1 + y_1^*, ..., x_n + y_n^*) - F_S(t, y_1^*, ..., y_n^*). \quad (2.5)$$

Values $x_S(t)$ can be understood as the disturbances of the analysed (undisturbed) solution $y_S^*(t)$. It is worth noticing, that solution $x_S \equiv 0$ of equations (2.4), called from now on the equations of disturbances, corresponds to solution $y_S^*(t)$. Therefore, the analysis of stability of solution $y_S^*(t)$ can be brought to the analysis of the trivial solution (the state of equilibrium) of the disturbed system's equations, also-called the reduced equations.

Now, we shall introduce certain notions and definitions that are necessary for further considerations.

DEFINITION 2.1 *Scalar function $V(t, x_1, ..., x_n)$ that is real and continuous, is called a positively or negatively slightly definite function, if respectively $V(t, x_1, ..., x_n) \geq 0$ lub $V(t, x_1, ..., x_n) \leq 0$.*

DEFINITION 2.2 *Function $V(t, x_1, ..., x_n)$ is called positively definite, if there is a scalar function $W(x_1, ..., x_n)$ such that*

$$V(t, x_1, ..., x_n) \geq W(x_1, ..., x_n) > 0, \quad (2.6)$$

$$V(t, 0, ..., 0) = W(0, ..., 0) = 0. \quad (2.7)$$

In a similar way we can define a negatively definite function V.

DEFINITION 2.3 *If there is $W(x_1, ..., x_n) > 0$ such that*

$$V(t, x_1, ..., x_n) \leq -W(x_1, ..., x_n) < 0, \quad (2.8)$$

$$V(t, 0, ..., 0) = W(0, ..., 0) = 0, \quad (2.9)$$

then function $V(t, x_1, ..., x_n)$ is called negatively definite.

The positively or negatively definite functions are called functions of the definite sign.

It can be proved, that if $V(t, x_1, ..., x_n) = V(x_1, ..., x_n)$ is a function of the definite sign, then there is such a number $H > 0$, for which every surface $V(x_1, ..., x_n) = C$ is closed around point $(0, ..., 0)$ for $|C| < H$. If function V includes time, then although surfaces $V(t, x_1, ..., x_n)$ change in the course of time, the appropriate choice of C makes surface $V(t, x_1, ..., x_n)$ stay all the

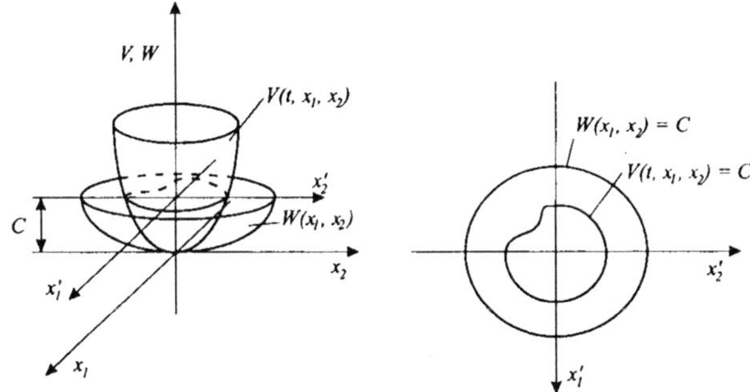

Figure 2.9. The illustration of a positively definite function for a two-dimensional system.

time within the field limited by the closed surface $W(x_1, ..., x_n)$. A geometric illustration for the system of two variables is presented in Figure 2.9.

During the study of stability with Lyapunov's method it is necessary to combine function $V(t, x_1, ..., x_n)$ with reduced system (2.4). If variable $(x_1, ..., x_n)$ of that system is the same variable that appears in function $V(t, x_1, ..., x_n)$, then function

$$\dot{V}(t, x_1, ..., x_n) = \frac{\partial V}{\partial t} + \sum_{i=1}^{n} \frac{\partial V}{\partial x_i} X_i(t, x_i, ..., x_n), \tag{2.10}$$

is called a complete derivative of function $V(t, x_1, ..., x_n)$ in relation to time t, matched with a reduced system.

THEOREM 2.1 *(Lyapunov's first theorem) If for a reduced system* (2.4) *there is a scalar function of the definite sign* $V(t, x_1, ..., x_n)$, *with derivative* $\dot{V}(t, x_1, ..., x_n)$ *in relation to time that is matched with system* (2.4), *which is a function of a constant sign opposite to V or is identically equal to zero, then the system's trivial solution* $(x_1, ..., x_n) = (0, ..., 0)$ *is stable in Lyapunov's sense for* $t \to +\infty$.

THEOREM 2.2 *(Lyapunov's second theorem) If for a reduced system* (2.4) *there is a function of the definite sign* $V(t, x_1, ..., x_n)$ *with a derivative of the definite sign* $\dot{V}(t, x_1, ..., x_n)$ *in relation to time that is matched with that system, and its sign is opposite to V, then the system's trivial solution* $(x_1, ..., x_n) = (0, ..., 0)$ *is asymptotically stable in Lyapunov's sense for* $t \to +\infty$.

THEOREM 2.3 *(the first theorem of instability) If for a reduced system* (2.4) *there is such a function* $V(t, x_1, ..., x_n)$ *that its derivative* $\dot{V}(t, x_1, ..., x_n)$ *in relation to t that is matched with the system is definite and in zero's neighbourhood may assume the values of the same sign as function* $V(t, x_1, ..., x_n)$*, then the trivial solution (and the undisturbed solution at the same time) is unstable.*

THEOREM 2.4 *(the second theorem of instability) If for a reduced system* (2.4) *we may find a bounded function* $V(t, x_1, ..., x_n)$*, the derivative of which is*

$$\dot{V} = \lambda V + W, \tag{2.11}$$

where λ *is a certain constant positive number, and* $W \equiv 0$ *or is a slightly definite function, and if in this case the derived function* $V(t, x_1, ..., x_n)$ *is not a slightly definite one with the opposite sign to function W, then the trivial solution, and the undisurbed solution at the same time, is unstable.*

The first two Lyapunov's theorems and the first theorem of instability enable us to present a graphic illustration. If function $V(t, x_1, ..., x_n)$ and its derivative $\dot{V}(x_1, ..., x_n)$ are functions of the definite and the opposite signs, then the phase point, which moves according to the increase of time along the phase trajectory, intersects every surface $V(x_1, ..., x_n) = C$ from outside to inside because the values of function V are supposed to decrease. In this case, phase trajectories should unlimitedly approach the origin (Figure 2.10a).

Finally, when the conditions of the first theorem of instability are fulfilled, the phase point intersects surface $V(t, x_1, ..., x_n) = C$, moving from the inside of the curve to the outside (Figure 2.10c). The case presented in Fig. 2.10b is special, because $\dot{V} = 0$.

As it may be observed, while formulating the theorem of instability we assumed, that the derivative $\dot{V}(t, x_1, ..., x_n)$ matched with the system, the stability of which we want to determine, is slightly definite in a certain complete neighbourhood of the origin. In order to show that a dynamic system may be unstable, it is enough to show that there is at least one derivative with that begins in the neighbourhood of the state of equilibrium and tends to move away from it.

Therefore, it is not necessary to consider the entire neighbourhood of the state of equilibrium. It is formulated in the following Chetayev's theorem:

THEOREM 2.5 *(Chetayev's theorem) Let a reduced system* (2.4) *have function* $V(t, x_1, ..., x_n)$ *which has constant partial derivatives the first order and the projection of the function's intersection onto the half-plane* $0, x_1, ..., x_n$ *is not an empty set; further it is open and its boundary contains the origin and, moreover, on the boundary* $V(t, x_1, ..., x_n) = 0$.

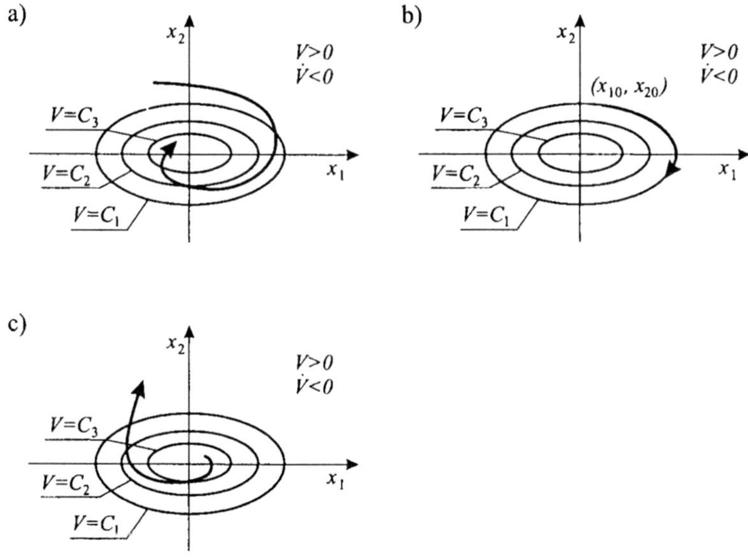

Figure 2.10. Graphic illustrations of the theorems of stability and instability.

Then, if function $V(t, x_1, ..., x_n)$ is limited in the area of the projection where it has its derivative $\dot{V}(t, x_1, ..., x_n)$ that is matched with the system (2.4), and in every sub-area of the area of projection for which $V(t, x_1, ..., x_n) \geq \alpha > 0$ the inequality $\dot{V}(t, x_1, ..., x_n) \geq \beta > 0$ is fulfilled, where $\beta = \beta(\alpha)$ is a certain positive number dependent on positive number α, then the trivial solution of system (2.4) is unstable in Lyapunov's sense for $t \to +\infty$.

Figure 2.11 presents the projection of the section of function $V(t, x_1, ..., x_n)$ on the half-plane $0, x_1, ..., x_n$ and the origin belongs to the projection's boundary. If there is such an interior point $(t_0, x_1^{(0)}, ..., x_n^{(0)})$ that $0 < \|\{x^{(0)}\}\| < \delta$, and $V(t_0, x_1^{(0)}, ..., x_n^{(0)}) = \alpha > 0$, and $\dot{V}(t, x_1(t), ..., x_n(t)) > 0$, then the projection of the trajectory that originates in the shadowed area will leave that area.

Let us consider again the reduced system (2.4). If the system's trivial solution is globally asymptotically stable, then it is asymptotically stable in Lyapunov's sense and its set of attraction is the entire considered space.

Function $V(t, x_1, ..., x_n)$ is called infinitely large, if for any given positive number α there is such a positive number R, that inequality $V(x_1, ..., x_n) > \alpha$ occurs for every $x_1, ..., x_n$ within a sphere with radius R.

THEOREM 2.6 *(Barbashin–Krasowski's theorem) If there is a positively definite infinitely large function $V(x_1, ..., x_n)$ and its derivative, which is matched*

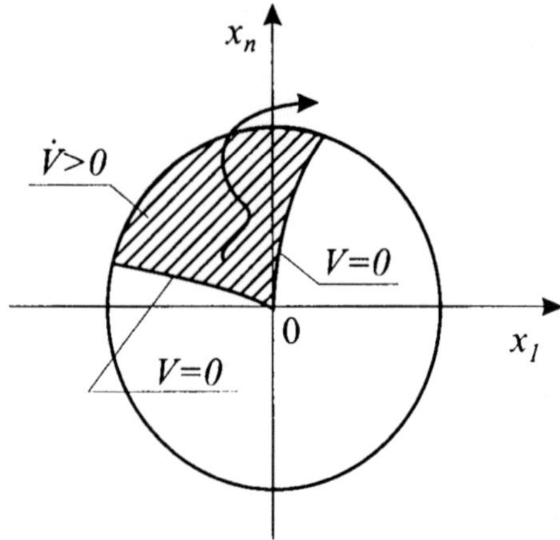

Figure 2.11. The graphic representation of Chetayev's theorem.

with system (2.4), *is a negatively definite function (and* $dV/dt = 0$ *occurs for* $(x_1, ..., x_n) = (0, ..., 0)$*), then the trivial solution of system* (2.4) *is globally asymptotically stable.*

The following theorem refers to the necessary and sufficient conditions of the system's stability in Lagrange's sense.

THEOREM 2.7 *The necessary and sufficient condition for stability of system (1.1) in Lagrange's sense is the existence of function* $V(t, y_1, ..., y_n)$ *that meets the following conditions:*

a) $V(t, y_1, ..., y_n) \geq W(y_1, ..., y_n)$, *and* $W(y_1, ..., y_n) \to \infty$ *at* $\|(y_1, ..., y_n)\| \to \infty$*;*

b) for every solution $y_s(t; t_0, y_{10}, ..., y_{n0}), s = 1, ..., n$ *function* $V(t, y_s)$ *is non-increasing in relation to t.*

The latter condition is equivalent to condition $\dot{V}(t, y_1, ..., y_n) \leq 0$, where $\dot{V}$ is a derivative of a function matched to the analysed system.

Let us consider a very general non-linear oscillator governed by equation [47],

$$\ddot{x} + p(t)\dot{x} + q(t)f(x) = 0, \tag{2.12}$$

with the following conditions imposed on functions $p(t), q(t)$ and $f(x) : 0 < q(t) < M, p(t) \geq -\dot{q}/2q$ and $\int_0^{\pm\infty} f(x)dx = +\infty$.

It may be shown that the equation's solutions (and their derivatives) are bounded for $t \in [t_0, \infty]$. Let us make the transformation of system (2.12) into two equations of the first order:

$$\frac{dx}{dt} = y, \tag{2.13}$$

$$\frac{dy}{dt} = -p(t)y - q(t)f(x). \tag{2.14}$$

Function V will assume the form of:

$$V(t, x(t), y(t)) = \int_0^{x(t)} f(\zeta)d\zeta + \frac{y^2(t)}{2q(t)}, \tag{2.15}$$

and because $q(t) \leq M$ then we assume

$$W(x, y) = \int_0^{x} f(\zeta)d\zeta + \frac{y^2}{2M}, \tag{2.16}$$

where: $V(t, x, y) \geq W(x, y)$ and $W(x, y) \to +\infty$, when $x^2 + y^2 \to \infty$.

Then, we calculate the derivative matched with system (2.12)

$$\frac{dV(t, x, y)}{dt} = f(x)y - \frac{y}{q}[py + qf] - \frac{y^2\dot{q}}{2q^2}, \tag{2.17}$$

which, after the transformations, leads to equality

$$\dot{V} = -\frac{y^2}{q}(p + \frac{\dot{q}}{2q}). \tag{2.18}$$

According to the assumptions $p(t) \geq -q/(2q)$ and $q < 0$, we obtain $\dot{V}[t, x(t), y(t)] \leq 0$, which proves boundedness of $x(t)$ and $\dot{x}(t)$ within $(0, \infty)$.

Now, we shall state several theorems referring to characteristic exponents, the proofs of which can be found in monograph [47].

THEOREM 2.8 *A characteristic exponent of the sum of a finite number of functions $f_m(t)(m = 1, ..., M)$ is not larger than the largest characteristic exponent of the sum's every element.*

THEOREM 2.9 *A characteristic exponent of the product of a finite number of functions $f_m(t)(m = 1, ..., M)$ is not larger than the sum of the functions' characteristic exponents.*

A characteristic exponent of matrix $[F_{jk}(t)]$ is the number or the symbol $(\pm\infty)$ described by the following formula:

$$\chi[F] = \max_{j,k} \chi[F_{j,k}(t)], \tag{2.19}$$

THEOREM 2.10 *A characteristic exponent of the sum of a finite number of matrices does not exceed the largest among the matrices' characteristic exponents.*

THEOREM 2.11 *A characteristic exponent of the product of a finite number of matrices is not larger than the sum of the matrices' characteristic exponents.*

Solution $y(t)$ which is not a trivial solution of system (2.4) is exponentially stable, if this solution and the solution $x(t)$ close to it for $t = t_0$ fulfil the following inequality:

$$\|x(t) - y(t)\| \leq L\|x(t_0) - y(t_0)\|e^{-\beta(t-t_0)}, \tag{2.20}$$

for $t \geq t_0$ and for certain positive constants L and β.

For a homogeneous linear system with constant coefficients, its exponential stability results from its trivial solution's asymptotic stability. However, it is not a generally true statement, if the coefficients of a linear system are variable.

THEOREM 2.12 *If there is a positively definite quadratic form*

$$V(x_1, ..., x_n) = ([A]x, x), \tag{2.21}$$

the derivative $\dot{V}(x_1, ..., x_n)$ *of which, matched with the reduced system* (2.4), *fulfils the inequality*

$$\dot{V}(x_1, ..., x_n) \leq W(x_1, ..., x_n), \tag{2.22}$$

and

$$W(x_1, ..., x_n) = -([B]x, x), \tag{2.23}$$

is a negatively definite quadratic form, and matrices [A] and [B] are constant and symmetric, then the trivial solution of system (2.4) *is exponentially stable for t* $\rightarrow \infty$.

Now, we shall study stability with the use of Lyapunov's function.

EXAMPLE 2.6 *Study stability of a pendulum's states of equilibrium described by the following equations:*

$$\ddot{\phi} + \alpha^2 sin\phi = 0. \tag{2.24}$$

We reduce (2.24) *to the system of two equations of the first order:*

$$\frac{d\phi}{dt} = \nu,$$

$$\frac{d\nu}{dt} = -\alpha^2 \sin\phi. \tag{2.25}$$

However, first we shall consider a more general problem, i.e.: we shall study the following dynamic system:

$$\frac{d\phi}{dt} = \nu,$$

$$\frac{d\nu}{dt} = -F(\phi). \tag{2.26}$$

We shall prove, that if $F(\phi)$ is a function that meets the two conditions:

$$F(0) = 0, \tag{2.27}$$

$$\phi F(\phi) > 0 \quad for \quad \phi \neq 0, \tag{2.28}$$

then the state of equilibrium $\phi = \nu = 0$ is stable. It may be observed, that system (2.26) includes the first integral in the form of:

$$\frac{\nu^2}{2} + \int_0^\phi F(\zeta)d\zeta = C. \tag{2.29}$$

By differentiating (2.29) we obtain

$$\nu d\nu + F(\phi)d\phi = 0, \tag{2.30}$$

which results from equations (2.26). Within range $(0, \phi)$ function $F(\phi) > 0$, which means, that

$$\int_0^\phi F(\zeta)d\zeta > 0 \tag{2.31}$$

for $\phi \neq 0$. Thus, function

$$V(\phi, \nu) = \frac{\nu^2}{2} + \int_0^\phi F(\zeta)d\zeta, \tag{2.32}$$

is a positively definite function. Its derivative equals

$$\frac{dV}{dt} = \frac{\partial V}{\partial \phi}\nu - \frac{\partial V}{\partial \nu}F(\phi). \tag{2.33}$$

Having included (2.32) in (2.33) we obtain

$$\frac{dV}{dt} = F(\phi)\nu - \nu F(\phi). \tag{2.34}$$

According to Lyapunov's first theorem (Theorem 2.1), the state of equilibrium $\phi = \nu = 0$ is stable, because function $\dot{V}$ is identically equal to zero.

Now, let us return to system (2.25). *This system has an infinitely large number of the states of equilibrium in the form of* $(\phi, \nu) = (k\pi, 0)$, *for* $k = 0, 1, 2, ...$ *Due to periodicity of function* $sin\phi$, *it is enough to study only one lowest and highest position of the pendulum, i.e.:* $(0, 0)$ *and* $(\pi, 0)$. *Let* (ϕ_0, ν_0) *be the particular solution that we want to study. Then* (2.25) *results in*

$$\frac{d(\phi_0 + \phi)}{dt} = \nu_0 + \nu,$$

$$\frac{d(\nu_0 + \nu)}{dt} = -\alpha^2(sin\phi_0 cos\phi + cos\phi_0 sin\phi), \tag{2.35}$$

because for small ϕ *we have* $sin\phi \approx \phi$ *and* $cos\phi \approx 1$.

After the linearization we obtain

$$\frac{d\phi}{dt} = \nu,$$

$$\frac{d\nu}{dt} = -\alpha^2 \phi cos\phi_0. \tag{2.36}$$

The solutions of (2.36) *that we are looking for should have the form of* $\phi = Ae^{\lambda t}, \nu = Be^{\lambda t}$, *and we obtain the following characteristic equations:*

$$\begin{vmatrix} \lambda & -1 \\ \alpha^2 cos\phi_0 & \lambda \end{vmatrix} = \lambda^2 + \alpha^2 cos\phi_0 = 0. \tag{2.37}$$

For the highest position of the pendulum, i.e: $(\pi, 0)$ *we have* $\lambda = \pm\alpha$ *and because* $\alpha > 0$ *then the state of equilibrium is unstable. For the lowest position of the pendulum, i.e.: (0, 0) we have* $\lambda = \pm i\alpha$ *and stability of that state of equilibrium may be studied with the use of Lyapunov's function.*

According to (2.32) *we have*

$$V(\phi, \nu) = \frac{\nu^2}{2} + \int_0^{\phi} \alpha^2 sin\varsigma d\varsigma = \frac{\nu^2}{2} + \alpha^2(1 - cos\phi). \tag{2.38}$$

Function $V(\phi, \nu)$ *is positively definite. Let us study its derivative*

$$\frac{dV(\phi, \nu)}{dt} = -\nu\alpha^2 sin\phi + sin\phi\alpha^2\nu \equiv 0. \tag{2.39}$$

Thus, according to Theorem 2.1, the equilibrium state $\phi = \nu = 0$ *is stable.*

EXAMPLE 2.7 *Study stability of solution (0,0) for a dynamic system*

$$\frac{dx}{dt} = ax^5 + by,$$

$$\frac{dy}{dt} = cx + dy^3, \tag{2.40}$$

if we know that $a < 0$ and $d < 0$.

On the basis of the linearized equation we obtain

$$\frac{dx}{dt} = by,$$

$$\frac{dy}{dt} = cx, \tag{2.41}$$

which yields the following characteristic equation:

$$\begin{vmatrix} \lambda & b \\ c & \lambda \end{vmatrix} = \lambda^2 - bc = 0. \tag{2.42}$$

If $bc > 0$, then (as it can be easily observed) the studied state of equilibrium is unstable. However, if $bc < 0$, then equation (2.43) results in purely imaginary conjugate roots $\lambda_{1,2} = \pm i\sqrt{bc}$, which corresponds to a critical case. The study of stability of the state of equilibrium must be performed with the use of Lyapunov's function.

We are looking for Lyapunov's function in the form of:

$$V(x, y) = A_1(x) + A_2(y) \tag{2.43}$$

where $A_1(0) = A_2(0) = 0$. The function's derivative equals:

$$\frac{dV}{dt} = \frac{dA_1}{dx}(ax^5 + by) + \frac{dA_2}{dy}(cx + dy^3). \tag{2.44}$$

We want function (2.45) to have the following form:

$$\frac{dA_1}{dx}by + \frac{dA_2}{dy}cx = 0. \tag{2.45}$$

Equation (2.45) results in

$$\frac{by}{\frac{dA_2}{dy}} = -\frac{cx}{\frac{dA_1}{dx}} = \frac{1}{2} \tag{2.46}$$

where, as it has been assumed, the ratio is assumed to be 1/2. Equations (2.46) result in

$$\frac{dA_2}{dy} = 2by, \tag{2.47}$$

$$\frac{dA_1}{dx} = -2cx, \tag{2.48}$$

and after the integration:

$$A_2 = by^2, \tag{2.49}$$

$$A_1 = cx^2. \tag{2.50}$$

If we include those expressions in formula (2.43), *we shall obtain the function we have been looking for*

$$V(x, y) = by^2 - cx^2. \tag{2.51}$$

The function's derivative equals

$$\frac{dV}{dt} = 2bdy^4 - 2acx^6. \tag{2.52}$$

If parameters b and c have opposite signs, then $V(x, y)$ *is a function of the definite sign, for it is positive or negative in every place, except for* $x = y = 0$. *Function* $\dot{V}$ *has its zero points different from* $x = y = 0$, *if the following conditions are fulfilled:*

$$bdy^4 = acx^6 \tag{2.53}$$

If $a < 0, d < 0$ *and* $bc < 0$, *then trivial solution (0,0) is asymptotically stable.*

3. Stability in Lagrange's[4] Sense

There is a system of differential equations:

$$\dot{\mathbf{x}} = f_i(t, x_1, x_2, \ldots, x_n), \;\; i = 1, 2, \ldots, n, \;\; \mathbf{x}(t_0) = \mathbf{x}_0, \tag{2.54}$$

$t_0 \in \langle 0, +\infty)$. Let us assume, that function f_i meets the conditions of existence and uniqueness of the solutions. The general solution of equation (2.54) will be presented in Cauchy's form:

$$\mathbf{x} = \mathbf{x}(t, t_0, \mathbf{x}_0).$$

Definition — Stability in Lagrange's sense

Solutions $\mathbf{x}(t, t_0, \mathbf{x}_0)$ of the system (2.54) will be called stable in Lagrange's sense, if they are definite in future, in period $[t_0, \infty)$ and if the norm of every solution is bounded

$$\|\mathbf{x}(t, t_0, \mathbf{x}_0)\| = \sum_{i=1}^{n} \left\{ [x_i(t, t_0, x_{0i})]^2 \right\} < M^2 = const < \infty, \quad \text{for } t \geq t_0.$$

[4]de Lagrange Joseph Louis, a French mathematician: 1736-1813

In other words, solutions $x_1(t), x_2(t), \ldots, x_n(t)$ of system (2.54) are called stable in Lagrange's sense if they all are bounded.

Comparison of the definitions of stability in Lyapunov's sense and Lagrange's sense reveal essential differences between them. Stability in Lyapunov's sense refers to individual solutions and stability in Lagrange's sense – to all solutions; it refers to the system's stability. Stability in Lyapunov's sense is not related to the boundedness of solutions, but to the proximity of their trajectories, whereas stability in Lagrange's sense does not require the proximity of trajectories, only their boundedness.

EXAMPLE 2.8 — Stability in Lyapunov's sense vs. stability in Lagrange's sense
There is an equation of a non-linear system's movement: $\dot{x} = \sin^2(x)$, $x(0) = x_0$. *The solution of this equation takes the form of:*

$$x = arcctg\left[ctg\left(x_0 - t\right)\right], x_0 \neq k \cdot \pi, k = 0, \pm 1, \pm 2, \ldots \,.$$

The solutions for various initial conditions are shown in Figure 2.12. All solutions of the system are bounded, thus they are all stable in Lagrange's sense. The zero solution of the system is unstable in Lyapunov's sense; there can be no solution x_0 *approaching zero, that would be found in zero's proximity for* $t > t_0$ *– see Figure 2.12.*

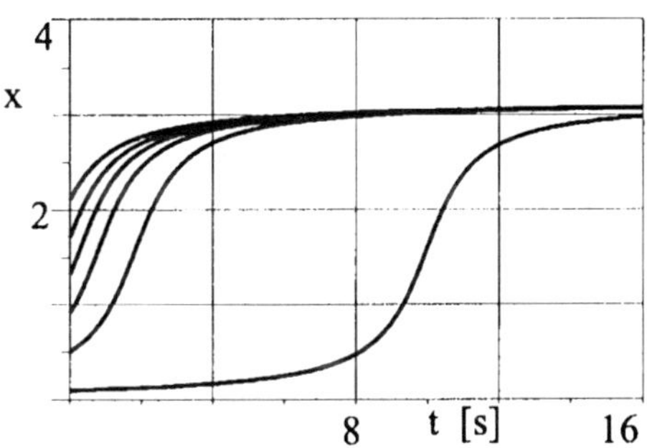

Figure 2.12. The solutions of system $\dot{x} = \sin^2(x)$ for various initial conditions.

4. Stability in Poincaré's[5] Sense-Orbital Stability

There is a system of differential equations:

$$\frac{dx}{dt} = f_i(t, x_1, x_2, \ldots, x_n), \;\; i = 1, 2, \ldots, n \tag{2.55}$$

[5]Poincaré Henri (1854-1912): a French mathematician, physicist and philosopher, the pioneer of the theory of relativity.

with initial conditions: $\mathbf{x}(t_0) = \mathbf{x}_0$, time $t_0 \in \langle 0, +\infty)$. We assume, that function f_i meets the conditions of existence and uniqueness of the solutions. The solution of equation (2.55) will be presented in the form of $\mathbf{x} = \mathbf{x}(t)$, and $\mathbf{L}$ will denote the trajectory of this solution in range $t \in [t_0, \infty)$.

Definition 1 — Stability in Poincaré's sense

Solution $\mathbf{v} = \mathbf{v}(t)$ of system (2.1) is called *stable in Poincaré's sense – orbitally stable*, if trajectories $\mathbf{L}$ of all solutions $\mathbf{x} = \mathbf{x}(t)$ in range $[t_0, \infty)$, which within time t_0 are sufficiently close to solution $\mathbf{v}(t)$, are throughout $t > t_0$ enclosed in an arbitrarily small neighbourhood, ε, of trajectory $\mathbf{L}$ of solution $\mathbf{v}(t)$.
In other words, for an arbitrarily small e there is a number $\mu(\varepsilon, t_0) > 0$, such that

$$\|\mathbf{x}(t_0) - \mathbf{v}(t_0)\| < \mu \Rightarrow \rho\left(\mathbf{x}(t), \mathbf{L}_0\right) < \varepsilon \text{ for every } t \geq t_0,$$

where: $\rho(x(t), \mathbf{L}_0)$ is the distance between the point of solution $\mathbf{x}(t)$ and trajectory $\mathbf{L}_0$.

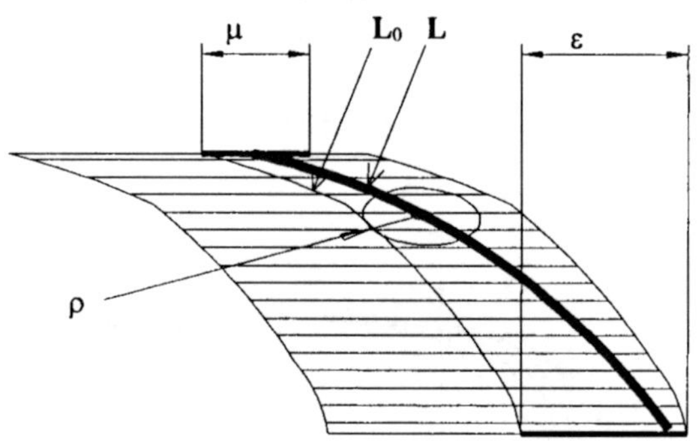

Figure 2.13. A stable solution in Poincaré's sense.

Solution $\mathbf{v}(t)$ which is stable in Lyapunov's sense is also stable in Poincaré's sense. There is no converse theorem; the condition of orbital stability is less restrictive than the stability condition in Lyapunov's sense.

EXAMPLE 2.9 *Let us consider the system of differential equations:*

$$\dot{r} = a \cdot r \cdot \left(1 - r^2\right); \; x_1 = r \cdot \cos\left(t + \theta_0\right); \; x_2 = r \cdot \sin\left(t + \theta_0\right), \; \textit{where } a > 0.$$

The graphic representation of solutions $x_2(x_1)$ on a phase plane (Figure 2.14) contains a limit cycle for periodical solution $r = 1$. All the other trajectories of solutions approach that limit cycle. Our system is a so-called self-excited system. It means that once it is out of balance (in points $x_1 = 0$ and $x_2 = 0$), the system cannot return to the previous state of equilibrium but for $t \to \infty$

tends towards periodic motion described by:

$$x_1 = \cos(t + \theta_0); \quad x_2 = \sin(t + \theta_0).$$

To sum up, the limit cycle that corresponds to the periodic solution (motion) is asymptotically stable in Poincaré's sense, i.e., it is orbitally stable.

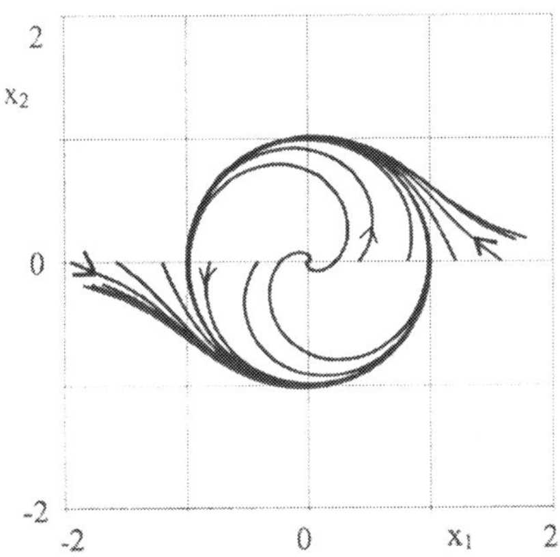

Figure 2.14. The graphic representation of solutions $x_2(x_1)$ with a limit cycle for periodical solution $r = 1$.

5. Stability in Poisson's[6] Sense

As far as periodical or almost periodical motion systems are concerned, the proximity of trajectories in the neighbourhood of the points of equilibrium, to which the trajectories infinitely return, may be especially significant. If trajectories T (Figure 2.15) that leave neighbourhood δ return every time to neighbourhood ε, then the point of equilibrium is stable in Poisson's sense. Phase trajectory $x(t)$ is stable in Poisson's sense, if, and only if it is bounded and when for every t_0 and $\delta > 0$ there are sufficiently large t_i, where $i = 1, 2, 3, \ldots, M$ and $t_0 < t_1 < t_2 < t_3 < \ldots < t_M$, for which $\|x(t_i) - x(t_0)\| < \delta$.

6. Technical Stability. Stability in Bogusz's Sense

The notion of stability in Lyapunov's sense and the notions that originate from it, although practically used in many fields of technology, may in many cases appear inadequate for a given phenomenon or an object of study. To illustrate the problem let us consider an inertial system of the second order that can be presented as the following equation: $\ddot{x} + C \cdot \dot{x} + K \cdot x = 0$. As we

[6]Poisson Simeon Denis (1781-1840): a French mathematician and physicist.

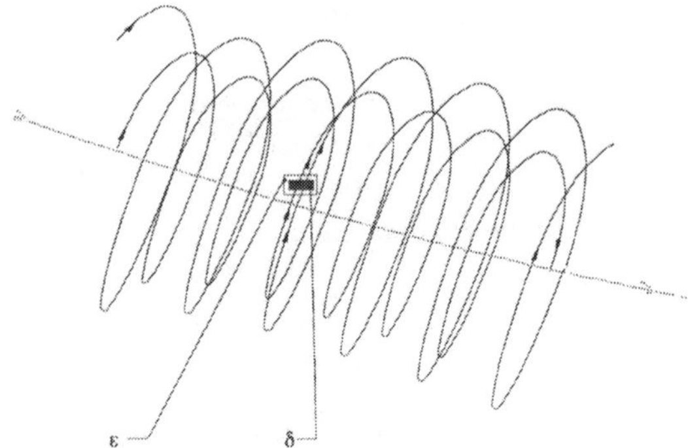

Figure 2.15. Stability in Poisson's sense – the periodical trajectories that leave neighbourhood δ return to neighbourhood ε every time.

already know, the condition of stability (in Lyapunov's sense) of this equation's solutions is positivity of coefficients C and K. Figure 2.16 presents graphic time representation $x(t)$ and phase representation $\dot{x}(x)$ for two essentially different cases of the equation's coefficients. Figure 2.16a shows the solutions for coefficient $C < 0$, i.e. for an unstable system, and Figure 2.16b – the solutions for coefficient $C > 0$, i.e. for a stable system.

If we study the system's behaviour for a sufficiently long time, we will see the difference in behaviour between the stable system and the unstable one in Lyapunov's sense. However, the observation time is very often limited and relatively short due to the type of the system or the nature of its motion (it could be landing of a plane, a car's braking or changing lanes). Such is the case illustrated in Figure 2.16: we are interested in the first 10 seconds of the system's motion. The study of stability in Lyapunov's sense of such a system for $t \rightarrow \infty$ is highly impractical.

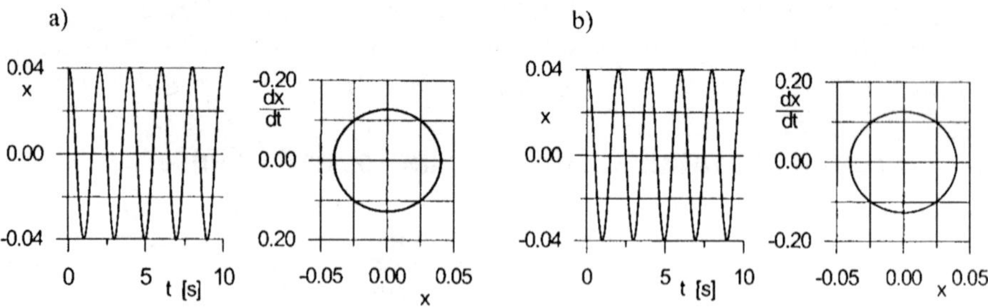

Figure 2.16. Motion of the inertial system of the second order; (a) - the unstable system in Lyapunov's sense $K = 10$, $C = -0.002$; (b) - the stable system in Lyapunov's sense $K = 10$, $C = 0.002$.

In a technical sense, we would be curious whether the system's solutions would be close to one another within a completed period of time, for example within 10 seconds. From the practical point of view, the possibility of defining proximity of solutions or movements would be extremely advantageous. The theoretical notions of technical stability [36, 37] seem to follow those tendencies. The use of the term "technical stability" depends on the conditions that affect the description of the studied system. There are four of them:

- when the initial conditions are determined a priori - the initial conditions field,
- when the requirements concerning movements' boundaries are determined a priori - the acceptable solutions field,
- when disturbances occur, and their maximum values are determined,
- when the time of the system's functioning is definite (the range is determined).

Definition 1 — Technical stability

Let us study the system of equations:

$$\dot{x}_i = f_i(t, x_1, x_2, \ldots, x_n) + R_i(t, x_1, x_2, \ldots, x_n),$$
$$i = 1, 2, \ldots, n, \quad x_i(t_0) = a_i, \tag{2.56}$$

where: $a_1, a_2, \ldots, a_n$ are given numbers, time $t_0 \in \langle 0, +\infty)$, functions $R_i(t, x_1, \ldots, x_n)$ are constantly occurring disturbances that for $t \geq 0$ fulfil inequality $\|\mathbf{R}(t, x_1, \ldots, x_n)\| < r^2$, and r is a sufficiently small number.

If all solutions $x_1(t), x_2(t), \ldots, x_n(t)$ of system (2.56) that meet the initial condition:

$$\|\mathbf{a} - \mathbf{b}\| = \sum_{i=1}^{n} \left[(a_i - b_i)^2\right] < \delta^2,$$

at limited disturbances $\|\mathbf{R}(t, x_1, \ldots, x_n)\| < r^2$, fulfil the condition:

$$\|\mathbf{x}(t) - \mathbf{x}^*(t)\| = \sum_{i=1}^{n} \left\{[x_i(t) - x_i^*(t)]^2\right\} < \varepsilon^2 \quad \text{for every } t \geq t_0,$$

then the solutions will be called technically stable in relation to the values of limits δ_2, ε_2 and r_2. The field determined by the number 2δ will be called the initial conditions field and notated as ω. The field determined by the number 2ε will be called the field of acceptable solutions and notated as Ω.

Comparison of the definition of technical stability and the definition of stability in Lyapunov's sense (at constantly occurring disturbances) reveals the following differences:

- In Lyapunov's definition every field of acceptable solutions Ω must contain such an initial conditions field ω, so that the solution that starts from it could remain in field Ω all the time;
- There are no such requirements in the definition of technical stability. Initial conditions field ω may be assumed independently of the field of acceptable solutions Ω. The assumption of the fields' values may include various technical conditions imposed upon the system, quality standards, the standards of current level of technological development or the imagination of decision-makers.

The following definition – the definition of technical stability within definite time – is extremely useful for practical purposes. In order to show appreciation of W. Bogusz's outstanding creative input in the development of the theory of technical stability (see [36, 37]), the notion of technical stability within definite time was called stability in Bogusz's sense.

Definition 2 — Technical stability within definite time in Bogusz's sense [37]
Let the system be described by the following equations of motion:

$$\dot{x} = f(x,t) + R(x,t),$$

where:

$$f(x,t) = [f_1(x,t), ..., f_n(x,t)],$$

$$R(x,t) = [R_1(x,t), ..., R_n(x,t)], \quad x = [x_1, ..., x_n].$$

There are initial conditions $x(t_0) = x_0$, $x_0 \in \omega$.

Let $\omega \subset \Omega \subset E_n$, where set Ω is limited and closed, whereas set w meets the following condition:
the neighbourhood of every point $x \in \omega$ is embraced by set Ω, i.e.:

$$\bigwedge_{x\in\omega} \bigvee_{R>0} B(x,R) \subset \Omega,$$

where: $B(x,R)$ denotes a sphere with centrepoint x and radius R in E_n.

The system is called technically stable in relation to (ω, Ω) within time $T > 0$ (see Figure 2.17), if every trajectory that starts at t_0 in $x_0 \in \omega$ does not exceed the limits of set Ω within the time shorter than $t_0 + T$, i.e.:

$$\bigwedge_{x_0\in\omega, x(t_0)=x_0} x(t) \in \Omega \text{ for } t_0 \leqslant t \leqslant t_0 + T.$$

THEOREM 2.13 *Let us assume that there is a positively definite function $V(x)$ of C^1 class on E_n. If the following conditions:*

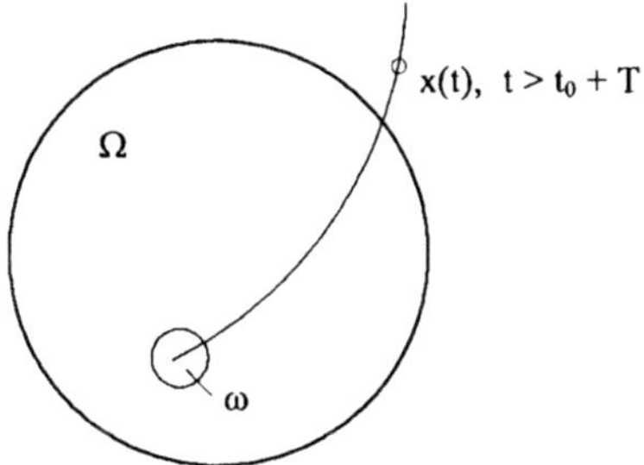

Figure 2.17. Stability in the technical sense – trajectory $x(t)$ does not exceed the limits of set Ω within time $t_0 + T$.

1. $\frac{dV}{dt}(x(t)) < 0$ *for every* $x(t) \in \Omega/\omega$ *and* $0 < t - t_0 \leq T$;

2. $V(x_1, t_1) \leq V(x_2, t_2)$ *for every* $x_1 \in \omega$, $x_2 \in E/\Omega$ *and* $0 < t_2 - t_1 < T$ *are fulfilled, then system* (2.56) *is technically stable within time* T.

Proof: Let $x(t)$ be a trajectory "starting" from a given point $x_0 \in \omega$ at t_0. Let us suppose, that at t_3, $0 < t_3 - t_0 < T$, the trajectory "left" set Ω (Figure 2.18). Since set $\omega \subset \Omega$, then while being in "motion" the trajectory had to "stay" in set Ω. Let t_2 denote the moment when the trajectory left set Ω, and t_1 the moment it left set ω. Obviously, $0 < t_2 - t_1 < T$ or from 2): $V(x(t_1), t_1) \leq V(x(t_2), t_2)$. At time $t_1 < t < t_2$ $(0 < t - t_0 < t_2 - t_0 < T)$ the trajectory was in set Ω/ω, however the energy there grows smaller along the trajectory, i.e.:

$$V\left(x\left(t_2\right), t_2\right) - V\left(x\left(t_1\right), t_1\right) = \int_{t_1}^{t_2} \frac{\partial V}{\partial t}\left(x\left(t\right)\right) dt < 0,$$

contradicts condition 2.

THEOREM 2.14 *Let us assume that there is a positively definite function* $V(x)$ *of* C^1 *class on* E_n. *If the following conditions:*

1. $V(x_1, t_1) \leq C_0$ *for every* $x_1 \in \omega$, $0 < t_1 - t_0 < T$,

2. $V(x_2, t_2) \geq C_1$ *for every* $x_2 \in E_n/\Omega$, $0 < t_2 - t_0 < T$, *where* $C_0 < C_1$,

3. $\frac{dV}{dt}(x(t)) < \frac{C_1 - C_0}{T}$ *for* $x(t) \in \Omega/\omega$ *and* $0 < t - t_0 < T$,

are fulfilled, then system (2.56) *is technically stable within time* T.

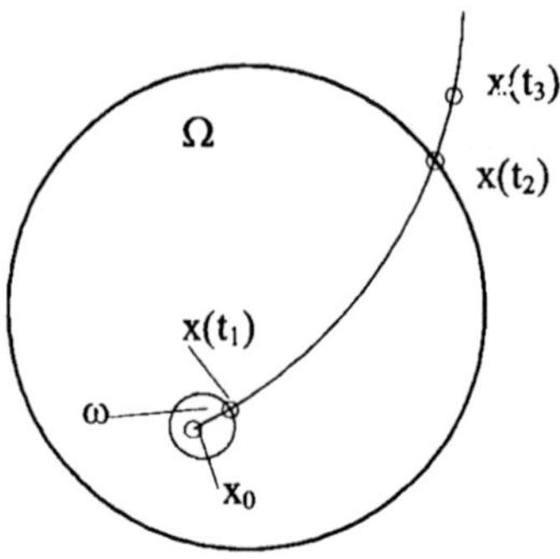

Figure 2.18. The moments of leaving fields ω and Ω by trajectory $x(t)$.

Proof

Analogous to the proof of Theorem 2.13.

EXAMPLE 2.10 — Comparison of requirements resulting from Lyapunov and Bogusz's definitions of stability

Let us explore the problem by studying an example of a system defined by the following equation of motion:

$$\ddot{x} + a \cdot \dot{x} + b \cdot x = 0.$$

The system of equations of the first order that corresponds to the equation above will be composed by means of the following substitutions $x = x_1$, $\dot{x} = x_2$:

$$\dot{x}_1 = x_2; \quad \dot{x}_2 = -b \cdot x_1 - a \cdot x_2.$$

a) The conditions of stability in Lyapunov's sense

For the linear homogeneous differential equation of the second order with constant coefficients, the sufficient and necessary condition of asymptotic stability of the stationary state ($\ddot{x} = 0, \dot{x} = 0$) meets the assumptions concerning the constant coefficients: the condition of stability in Lyapunov's sense: $a > 0$ and $b > 0$.

b) The conditions of stability in Bogusz's sense

Let us assume the initial conditions field ω for $t = t_0 = 0$,

$$\omega \equiv \left\{ x_1^2 + x_2^2 < r_0^2 \right\}.$$

We will assume the field of acceptable solutions Ω for $t \leq T$,

$$\Omega \equiv \left\{ x_1^2 + x_2^2 \leqslant R_{dop}^2 \right\},$$

while the values $r_0 < R_{dop}$, so that condition $\omega \subset \Omega$ is fulfilled.

Bogusz's function will take the form of:

$$V(x_1, x_2) = \frac{1}{2} \cdot b \cdot x_1^2 + \frac{1}{2} \cdot x_2^2,$$

and as we can see, the function meets the requirements listed in Theorem 2.13 when $b > 0$. The maximum of Bogusz's function in field ω

$$C_0 = \inf_{x_1, x_2 \in \omega} [V(x_1, x_2)] = \frac{1}{2} \cdot b \cdot r_0^2 + \frac{1}{2} \cdot r_0^2 .$$

The minimum of Bogusz's function in field Ω

$$C_1 = \underset{x_1, x_2 \notin \Omega}{sub} [V(x_1, x_2)] = \frac{1}{2} \cdot b \cdot R_{dop}^2 + \frac{1}{2} \cdot R_{dop}^2 .$$

The derivative of Bogusz's function along the solutions of the studied system

$$\dot{V} = \frac{\partial V}{\partial x_1} \cdot \dot{x}_1 + \frac{\partial V}{\partial x_2} \cdot \dot{x}_2 = b \cdot x_1 \cdot \dot{x}_1 + 0 + 0 + x_2 \cdot \dot{x}_2 = -a \cdot x_2^2 .$$

The condition of the system's stability at the stationary state fulfills the following inequality:

$$\dot{V}_{x_1, x_2 \notin \omega, 0 \leqslant t \leqslant T} < (C_1 - C_0)/T.$$

Let us find the maximum value of function $\dot{V}$ for $x_1, x_2 \notin \omega$
for $a > 0$ $\inf_{x_1, x_2 \notin \omega} \left[\dot{V}\right] = -a \cdot r_0^2$
for $a = 0$ $\inf_{x_1, x_2 \notin \omega} \left[\dot{V}\right] = 0$
for $a < 0$ $\inf_{x_1, x_2 \notin \omega} \left[\dot{V}\right] = \left|a \cdot R_{dop}^2\right|.$

The substitution of $\dot{V}$, C_1, C_0 with the received dependencies will yield:
for $a > 0$: $-|a \cdot r_0^2| < 0.5(b+1) \cdot (R_{dop}^2 - r_0^2)/T$, *which occurs for* $\omega \subset \Omega$, *so* $R_{dop} > r_0$,
for $a = 0$: $0 < 0.5(b+1) \cdot (R_{dop}^2 - r_0^2)/T$, *which occurs for* $\omega \subset \Omega$, *so* $R_{dop} > r_0$,
for $a < 0$: $|a \cdot R_{dop}^2| < 0.5(b+1) \cdot (R_{dop}^2 - r_0^2)/T$.

Thus the conditions of stability in Bogusz's sense are as follows:

$$b > 0 \text{ and for } a < 0 \ |a| < 0.5(b+1) \cdot (1 - (r_0/R_{dop})^2)/T.$$

Considering the studied system as a mechanical system, we may give the following facts about the derived stability conditions:

- *the condition of stability in Lyapunov and Bogusz's sense is $b > 0$,*

- *the condition of stability in Lyapunov's sense is $a > 0$,*
- *the condition of stability in Bogusz's sense is*

$$a > 0.5 \cdot (b+1) \cdot \left((r_0/R_{dop})^2 - 1 \right) / T,$$

which means that this coefficient may be smaller than zero, i.e.: "the negative damping" of the system's motion is acceptable.

Figure 2.19 presents the solutions of two systems defined by differential equation $\ddot{x} + a \cdot \dot{x} + b \cdot x = 0$.

System L with coefficients $a > 0$ and $b > 0$ is asymptotically stable in Lyapunov's sense. All of its solutions with any given initial condition x_0 approach the stationary state (zero) for $t \to \infty$.

System B with coefficients $a < 0$ and $b > 0$ is stable in Bogusz's sense. None of its solutions with initial condition $x_0 \in \omega$ exceed the limits of the assumed field of acceptable solutions Ω for $t < T$.

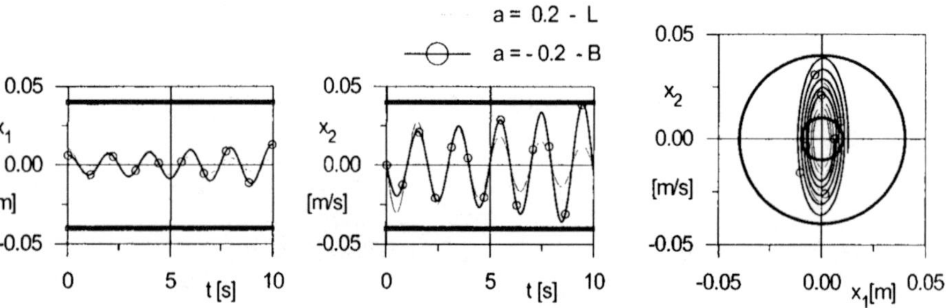

Figure 2.19. Trajectories (x_2, x_1) and $x(t)$ within the assumed fields of initial conditions $\omega\{r_0 = \pm 0.01$ and acceptable conditions $\Omega\{r_0 = \pm 0.04$ for observation time $T = 10s$. L - the solution for $a = 0.2$ (the trajectory): stable in Lyapunov's sense; B - the solution for $a = -0.2$: stable in Bogusz's sense.

System L, the solutions of which are stable in Lyapunov's sense, is stable in Lagrange's sense, since for $a > 0$ and $b > 0$ all the solutions of the system are bounded.

7. Stability in Szpunar's Sense

The definition of technical stability presented in Sect. 2.6 imposes a serious condition: the field of acceptable solutions contains the field of initial conditions – $\omega \subset \Omega$. This assumption specifies the notion of technical stability in such a way that for many objects it may turn out to be inadequate. In case of certain objects or processes, if we determine the field of initial conditions, we can immediately determine the field of acceptable solutions but it will only be the kind that does not contain the field of initial conditions - $\omega \not\subset \Omega$. It is the nature of our object that its trajectories that leave the field of initial conditions $\omega \not\subset \Omega$ enter field Ω and stay in it forever. Such behaviour is considered to be "stable"; the object that leaves the point of equilibrium in the field of initial conditions moves to another point of equilibrium. It corresponds with the widely understood notions of stability, and the definition of technical stability that contains the condition $\omega \subset \Omega$ is inapplicable in this case.

In order to solve the problems discussed above, K. Szpunar presented a modified definition of technical stability. In this book, this definition was named after its author.

Definition — Technical stability in Szpunar's sense [106]

We shall study the following system of equations:

$$\dot{\mathbf{x}} = \mathbf{f}\,(\mathrm{t},\mathbf{x}) + \mathbf{R}\,(\mathrm{t},\mathbf{x}) \tag{2.57}$$

with initial conditions: $\mathbf{x}(0)$, function $\mathbf{R}(t,\mathbf{x})$ that represents constant disturbances and fulfils inequality $\|\mathbf{R}(t,\mathbf{x})\| < r$ for $t \geq 0$, where r is a sufficiently small number.

We shall determine two limited and closed fields: field $\omega \subset E^n$ and field $\Omega \subset E^n$, in such a way, that dependence $\omega \not\subset \Omega$ will be satisfied. The system will be called technically stable in relation to (ω, Ω) within $t > T$, if every trajectory that starts at $t = 0$ in $x(0) \in \omega$ does not exceed the limits of field Ω within a time shorter than $t \geq T$, that is: if $\mathbf{x}(0) \in \omega \Rightarrow \bigwedge_{t \geqslant T} \mathbf{x}(t) \in \Omega$.

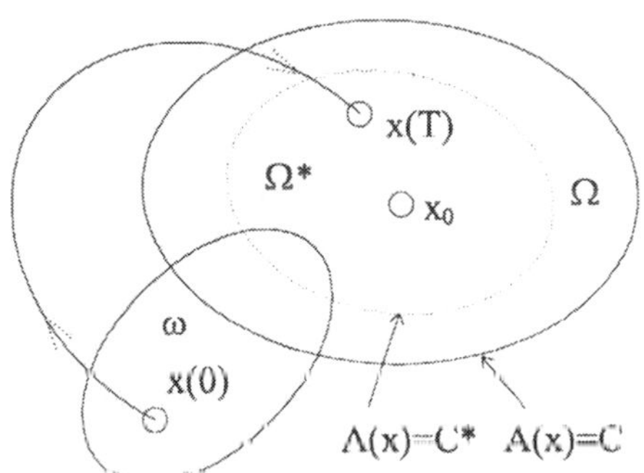

Figure 2.20. Trajectory $x(t)$ within fields ω and Ω.

For effective study of technical stability in Szpunar's sense we may use the theorem presented in [106].

THEOREM 2.15 *Let us assume, that there is a positively definite function $H(t - T, x)$ of class C^1 on E^n. If the following conditions: $H(0, x) = A(x)$,*

$H(t-T,x) \to \infty$ for $|x| \to \infty$ within range $0 \le t \le T$ are satisfied, the derivative of function H along the solutions of system (2.57) fulfils

$$\frac{dH}{dt} < -\frac{M - C^*}{T} \quad \textit{for } t \geqslant 0, x \notin \Omega^* = \{x; A(x) \leqslant C^*\},$$

where: $M = \max_{x(0)\in\Omega_0} H[-T, x(0)]$ and C^* the following inequality is satisfied

$$M - m \leqslant C^* \leqslant M \quad \textit{and} \quad C^* < C,$$

where: $m = \min_{x(0)\in\omega} H[-T, x(0)]$ and the following inequality is satisfied $H(t_1 - T, x_1) < H(t_2 - T, x_2)$ for $t_2 > t_1$, $x_1 \in \Omega^*$, $x_2 \not\subset \Omega$, then system (2.57) is technically stable in Szpunar's sense.

8. General Stability Estimation[7]

8.1 Introduction

Among different methods of stability investigation, Lyapunov's second method is the most powerful method. It was proved by qualitative analysis (see [88, 89, 90, 97, 98]) that the Lyapunov method needs the weakest assumptions to analyze a system and gives the best estimations of stability. However the problem of finding a reasonable Lyapunov function that will guarantee good stability estimation is not solved as of today in the general case. Usually a selected class of the Lyapunov function is considered. Very often (see for example [88, 89, 90, 97, 98, 74, 99, 108, 85]) the Lyapunov function is assumed to be a function in the class of the generalised Euclidean norms, i.e. in the form $V(v) = \|x\|_S = (x, Sx)^{1/2}$, where S is a symmetric, positive-defined matrix. The problem of defining the matrix S is important in determining stability, especially in estimation of the stability regions or other stability indices (for example the time for a transient process to decrease), but it is difficult to find in published papers a useful criterion for construction of the optimal matrix S. Earlier works have used the identity or diagonal matrix as a matrix S. This approach does not always allow one to obtain a stability estimation. It is known (see [98]) that the considered problem leads to a min-max optimisation problem with respect to the elements of the matrix S and the points of a region of the system phase space. This method gives the best results but the problem is rather complicated, particularly for systems of many degrees-of-freedom, where many local extremes may exist.

In the paper [108] the stability domain estimation via a quadratic Lyapunov function for polynomial systems of small degree was considered. By a kind of

[7]Courtesy of Professor B. Radziszewski

parameterisation the optimisation problem with respect to the space parameters was reduced to a problem with one extreme which makes the solution easier, but the problem of the optimal matrix S remained open as before. The authors of [108] proposed to consider the sum of all estimations of the stability region obtained for all matrices S. For many degrees-of-freedom systems that is a very time-consuming process.

The aim of the present section is to determine the best matrix S for the linearized system and to apply it to the stability estimation of the non-linear system. The results obtained by this way are not optimal for non-linear systems but the method requires a moderate number of numerical calculations. It gives also possibilities for estimation of the stability region and the times of the transient process decreasing for many degrees-of-freedom systems even when the considered equilibrium point of the system is exponentially stable. The first results of the method were given in the paper [90].

The construction of the matrix S based on the eigenvectors of the linear system matrix is presented in Section 2.8.2. It was assumed that the problem of eigenvalues and the eigenvectors of a matrix is the standard problem. The estimation of the exponential stability domain for a non-linear system applying the best matrix S for a linear system is given in Section 2.8.3. The general considerations are illustrated by two analytical and two numerical examples.

8.2 Linear System

Let us consider the linear system in the form

$$\dot{x} = Ax\,, \tag{2.58}$$

where: $x : R \to R^n$, $A \in M_n(R)$, $M_n(R)$ – the set of $n \times n$ real matrices, and a Lyapunov function in the form of the generalised Euclidean norm

$$V(x) = \|x\|_S = (x, Sx)^{1/2}, \tag{2.59}$$

where: $S \in M_n^+(R)$, $M_n^+(R)$ is a set of symmetric $n \times n$ positive-definite real matrices. The right upper derivative of (2.59) along the solutions of (2.58)

$$\dot{V}_{(1)}(x) = \max_{h \to 0^+} \frac{\|x + hAx\| - \|x\|}{h} \tag{2.60}$$

is equal to

$$\dot{V}_{(1)}(x) = \frac{(x, SAx)}{\|x\|_S}. \tag{2.61}$$

The system (2.58) is exponentially stable if and only if there exists $\gamma < 0$ such that

$$\dot{V}_{(1)}(x) \leq \gamma V(x) \tag{2.62}$$

for every $x \in R^n$. Let us assume

$$\gamma_s = \sup_{x \in R^n} \frac{\dot{V}_{(1)}(x)}{V(x)} = \sup_{x \in R^n} \frac{(x, SAx)}{(x, Sx)}. \tag{2.63}$$

The system (2.58) is exponentially stable if and only if there exists a symmetric positive-definite matrix S such that $\gamma < 0$. For the same systems (2.58) and different matrices S the values of γ_s may be non-negative. Thus it is interesting to look for the minimal value of γ_s with respect to the matrix S, i.e. to solve the problem

$$\tilde{\gamma} = \inf_{S \in M_n^+(R)} \sup_{x \in R^n} \frac{(x, SAx)}{(x, Sx)}. \tag{2.64}$$

The matrix S_0 for which $\tilde{\gamma} = \sup_{x \in R^n} \frac{(x, S_0 Ax)}{(x, S_0 x)}$ will be called *the best matrix* S for the linear system (2.58). The matrix $S_0 \in M_n^+(R)$ may not exist. If the matrix A has not a simple structure, the matrix S_0 is singular and $S_0 \notin M_n^+(R)$. It is easy to prove that

$$\tilde{\gamma} = \max_j \mathrm{Re}\lambda_j(A), \tag{2.65}$$

where $\lambda_j(A)$ is an eigenvalue of A. The condition

$$\tilde{\gamma} < 0 \tag{2.66}$$

is the known necessary and sufficient condition for the exponential stability of (2.58). The best matrix S for a linear system is given by the following lemma:

LEMMA 2.1 *The best matrix S_0 for the linear system* (2.58) *has the form $S_0 = X^* X$, where X is the set of matrices of eigenvectors of A.*

Proof. From the matrix properties for arbitrary matrices A and S we have

$$\sup_{x \in R^n} \frac{(x, SAx)}{(x, Sx)} \geq \max_j \mathrm{Re}\lambda_j(A).$$

Let Λ and X denote the diagonal matrix of eigenvalues and the matrix of eigenvectors of the matrix A respectively, i.e. $XA = \Lambda X$. Let $S_0 = X^* X$, where $X^* = \bar{X}^T$. In case we have

$$\sup_{x \in R^n} \frac{(x, S_0 Ax)}{(x, S_0 x)} = \sup_{x \in R^n} \frac{(Xx, \Lambda Xx)}{(Xx, Xx)} = \sup_{x \in R^n} \frac{(Xx, \Pi Xx)}{(Xx, Xx)}$$

$$\leq \sup_{y \in C^n} \frac{(y, \Pi y)}{(y, y)} = \sup_{y \in R^{2n}} \frac{(y_1, \Pi_1 y_1)}{(y_1, y_1)} = \max_{j=1,\dots,n} \mathrm{Re}\lambda_j(A),$$

where: $y = Xx$, $y = u + iv$, $u, v \in R^n$, $y_1 = [u\ v]^T$ and $\Pi = \frac{1}{2}(\Lambda + \Lambda^*)$, $\Pi \in M_n(R)$, $\Pi_1 = \begin{bmatrix} \Pi\ 0 \\ 0\ \Pi \end{bmatrix}$. By comparison of the inequalities we obtain $\tilde{\gamma} = \sup\limits_{x \in R^n} \dfrac{(x, S_0 Ax)}{(x, S_0 x)}$.

EXAMPLE 2.11 *For the linear system* (2.58) *with the matrix* $A = \begin{bmatrix} 0 & 1 \\ -1 & 2a \end{bmatrix}$, $a \in R$ *we have* $\lambda_{1/2}(A) = a \pm \sqrt{a^2 - 1}$ *and* $X = k \begin{bmatrix} 1 & -\lambda_1(A) \\ 1 & -\lambda_2(A) \end{bmatrix}$ *or* $X = k \begin{bmatrix} 1 & -\lambda_1(A) \\ -\lambda_1(A) & 1 \end{bmatrix}$, *where k is an arbitrary non-zero real number. The index* $\tilde{\gamma}$ *and the optimal matrix* S_0 *depend on the values of the parameter* a. *If* $|a| \leq 1$, *then* $\tilde{\gamma} = a$ *and* $S = X^*X = 2k^2 \begin{bmatrix} 1 & -a \\ -a & 1 \end{bmatrix}$. *If* $|a| > 1$ *then* $\tilde{\gamma} = a + \sqrt{a^2 - 1}$ *and* $S = 2k^2 \begin{bmatrix} 1 & -a \\ -a & 2a^2 - 1 \end{bmatrix}$ *or* $S = 2k^2 \begin{bmatrix} 1 & -\frac{1}{a} \\ -\frac{1}{a} & 1 \end{bmatrix}$.

8.3 Non-Linear System

Let us consider a non-linear system

$$\dot{x} = f(t, x), \tag{2.67}$$

where: $f : I \times \Omega$ *to* R^n, $I = (\tau, \infty)$, $\tau \geq -\infty$, Ω - domain of R^n containing the origin. We assume that $f(t, 0) = 0$ for all $t \in I$, so that (2.67) has the zero solution. We will study the problem of the exponential stability of the zero solution of (2.67), interior estimation of the exponential stability domain (ESD) and the exponential convergence index (ECI) valuations.

The exponential stability domain (ESD) is defined in the form

$$\Omega_e(t_0) = \left\{ x_0 \in \Omega : \ (\exists \eta \geq 1)(\exists \sigma > 0)(\forall t \geq t_0)\ \|x(t, t_0, x_0) < \eta \|x_0\| e^{-\sigma(t - t_0)} \right\}, \tag{2.68}$$

where $\|.\|$ denotes an arbitrary vector norm in R^n. The exponential convergence index (ECI) is defined as a boundary value of a Lyapunov function derivative. The right upper derivative of (2.59) along the solutions of (2.67) has the form

$$\dot{V}_{(9)}(t, x) = \frac{(x, Sf(t, x))}{\|x\|_S}. \tag{2.69}$$

We will consider ECI in the form

$$\gamma_s(\rho) = \sup_{t \in I} \sup_{x \in B^S_\rho} \frac{(x, Sf(t, x))}{(x, Sx)}, \tag{2.70}$$

where: $B_\rho^S = \{x \in R^n : \|x\|_S < \rho\}$.

The knowledge of ECI gives many advantages. Krasowski's theorem for the exponential stability follows (see [65]) that if there exists $S \in M_n^+(R)$ and $\rho > 0$ such that

$$\gamma_S(\rho) < 0, \tag{2.71}$$

than the zero solution of (2.67) is exponentially stable and $B_\rho^S \in \Omega_e(t_0)$ for every $t_0 \in I$. Also if there exist $S \in M_n^+(R)$ such that

$$\tilde{\gamma}_S(\rho) = \lim_{\rho \to 0} \gamma_S(\rho) < 0, \tag{2.72}$$

then the zero solution of (2.67) is exponentially stable.

In the paper [65] the time of the transient processes decreases as the time of convergence to some set in the vicinity of the equilibrium point is defined. For two arbitrary sets $\Omega_i = \{x : \|x\|_S = \rho_i\}$, $i = 1, 2$, such that $\rho_2 < \rho_1 < \rho$, if (2.71) is fulfilled, the time of convergence may be directly estimated by ECI in the following form:

$$T_{S,\rho}(\rho_1, \rho_2) \leq \frac{1}{\gamma_s(\rho)} \ln \frac{\rho_2}{\rho_1}. \tag{2.73}$$

For all $x \in B_\rho^S$ the times of k-times decreasing $(k > 1)$ of the distance between a representative point and the equilibrium point in the sense of the norm $\|, \|_S$ is estimated

$$T_{S,\rho}^k \leq -\frac{\ln k}{\gamma_S(\rho)}.$$

The value of ECI depends on the matrix S. If $\gamma_S(\rho) < 0$, the smaller $\gamma_S(\rho)$, the better estimations of $\Omega_e(t_0)$ and $T_{S,\rho}^k$. The minimal value of ECI with respect to the matrix $S \in M_n^+(R)$,

$$\tilde{\gamma}_\rho = \lim_{S \in M_n^+} \gamma_S(\rho), \tag{2.74}$$

is desirable. The analytical solution of this problem may exist only for some particular cases of the system with small-degree-of-freedom. For the non-linear systems in the general case the problem becomes the numerical evaluation of the extremal values of the quotient $g(t, x) = \dfrac{(x, Sf(t, x))}{(x, Sx)}$. The dimension of the full optimisation problem (2.74) is equal to $(n^2 + 3n)/2$, where n denotes the dimension of the phase space of the system. It grows with the square of n. If the matrix S is given, the dimension of the problem is equal to $n + 1$ and is a linear function of n. Thus to reduce the problem dimension we will assume the best matrix S_0 for the linear part of the non-linear equation (2.67) as the matrix S in (2.70).

We will estimate ESD and ECI using the linear part of the non-linear equation (2.67). By the linearization of the function f we obtain

$$f(t,x) = Ax + h(t,x), \tag{2.75}$$

where: $A \in M_n(R)$ and $h: I \times R^n \to R^n$ is a continuous function at the point $(t,0)$ and $\lim\limits_{x\to 0} \dfrac{h(t,x)}{\|x\|} = 0$ uniformly with respect to t, where $\|.\|$ is an arbitrary vector norm. By (2.70) and (2.75) we obtain

$$\tilde{\gamma}_S = \lim_{\rho\to 0} \sup_{t\in I} \sup_{x\in B_\rho^S} \frac{(x, SAx) + (x, Sh(t,x))}{(x, Sx)} = \max_i \mathrm{Re}\lambda_i \left[\frac{1}{2}\left(A_S^T + A_S\right)\right], \tag{2.76}$$

where: $A_S = S^{1/2}AS^{-1/2}$. If $S = S_0 \equiv X^*X$, where X fulfils $XA = \Lambda X$, then

$$\tilde{\gamma}_{S_0} = \lim_i \mathrm{Re}\lambda_i(A). \tag{2.77}$$

The exponential convergence index $\gamma_{S_0,\rho}$ is a non-decreasing function of ρ. By the condition (2.71) and the consideration above we have the following theorem:

THEOREM 2.16 *If* $\max\limits_i Re\lambda_i(A) < 0$ *and* $\lim\limits_{x\to 0} \dfrac{h(t,x)}{\|x\|} = 0$ *uniformly with respect to t, then the zero solution of* (2.67) *is exponentially stable and* $B_{\rho_0}^{S_0} \in \Omega_e(r_0)$, *where* $S_0 \equiv X^*X$ *and*

$$\rho_0 = \sup\{\rho : \gamma_S(\rho) < 0\}. \tag{2.78}$$

The theorem above gives us a useful tool for estimating the ESD and ECI for a non-linear system.

EXAMPLE 2.12 *Let us consider a Van der Pol equation of the form*

$$\dot{x} = Ax + h(x), \tag{2.79}$$

where: $A = \begin{bmatrix} 0 & 1 \\ -1 & -1 \end{bmatrix}$ *and* $h(x) = \begin{bmatrix} 0 \\ x_1^2 x_2 \end{bmatrix}$. *The conditions of Theorem 2.16 are fulfilled because* $\max\limits_i Re\lambda_i(A) = -\dfrac{1}{2}$ *and* $\lim\limits_{x\to 0} \dfrac{x_1^2 x_2}{\|x\|} = 0$. *In this case the matrix* $S_0 = \begin{bmatrix} 1 & \frac{1}{2} \\ \frac{1}{2} & 1 \end{bmatrix}$.

Figure 2.21 presents the function $\gamma_{S_0}(\rho)$ for this case. The numerical calculations follow $\gamma_{S_0}(\rho) < 0$ for $\rho \le \rho_0 \approx 0.957$. The ball $B_{\rho_0}^{S_0}$ is the best estimation of ESD of the zero solution obtained by application of the matrix S_0,

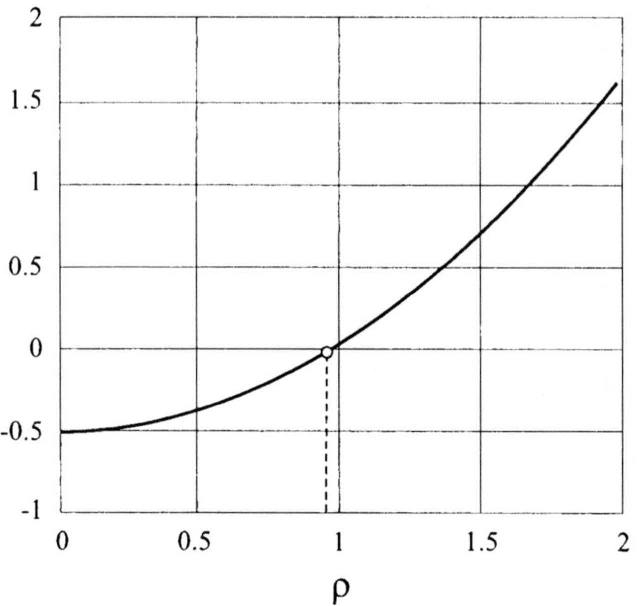

Figure 2.21. The function $\gamma_{S_0}(\rho)$.

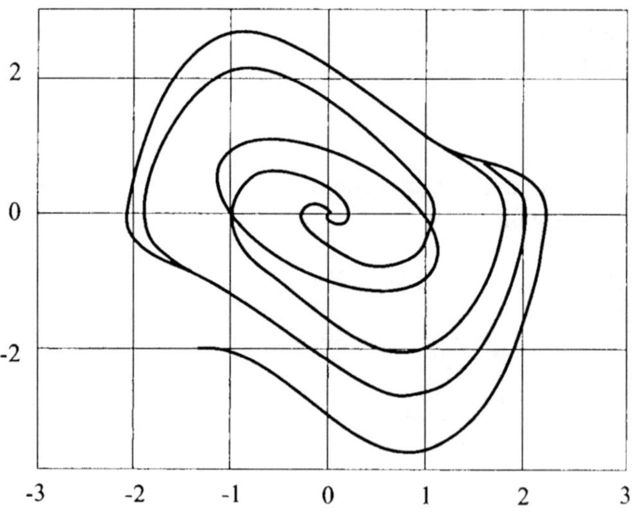

Figure 2.22. Phase-plane portrait of equation (2.79).

which is optimal for the linear system. This estimation with the phase-plane portrait of the equation (2.79) is shown in Figure 2.22.

EXAMPLE 2.13 *Let us consider the equation*

$$\dot{x} = Ax + h(t, x), \tag{2.80}$$

where: $A = \begin{bmatrix} 0 & 1 \\ -1 & 2a \end{bmatrix}$, $a \in (-1, 0)$ *and* $h(t, x) = \begin{bmatrix} 0 \\ -b(t)x_1^n \end{bmatrix}$, $b_0 = \sup_{t \geq t_0} b(t)$, $b : I \to R^+$, $n \to N$, $n \neq 1$. *In this case the matrix* S_0 *may be assumed as* $S_0 = \begin{bmatrix} 1 & -a \\ -a & 1 \end{bmatrix}$ *and the function* $\gamma_{S_0}(\rho)$ *has the following form:*

$$\gamma_{S_0,\rho} = a + c(n, a, b_0)\rho^{n-1}, \tag{2.81}$$

where:

$$c(n, a, b_0) = \frac{n^{n/2}}{(n+1)^{(n+1)/2}} \frac{b_0}{(1-a^2)^{n/2}}.$$

The ball $B_{\rho_0}^{S_0}$, *where* $\rho_0 = \sqrt[n-1]{-a/c(n, a, b_0)}$ *is the estimation of ESD of the zero solution obtained by application of the optimal matrix* S_0 *for the linear system. This estimation with the phase-plane portrait of the equation* (2.80) *for* $a = -\frac{1}{2}$, $n = 2$ *and two forms of the function* $b(t)$, *i.e.* $b(t) = b_0 = 1$ *and* $b(t) = \sin^2 5t$ *are shown in the Figs. 2.23 and 2.24.*

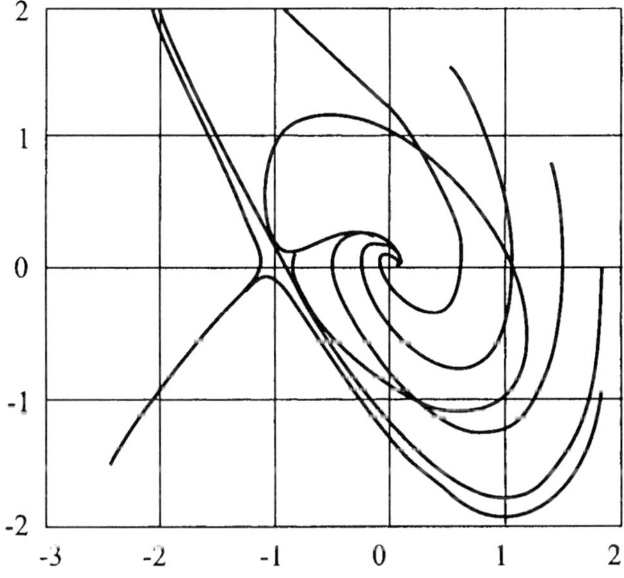

Figure 2.23. Phase-plane portrait of equation (2.80) for $b(t) = 1$.

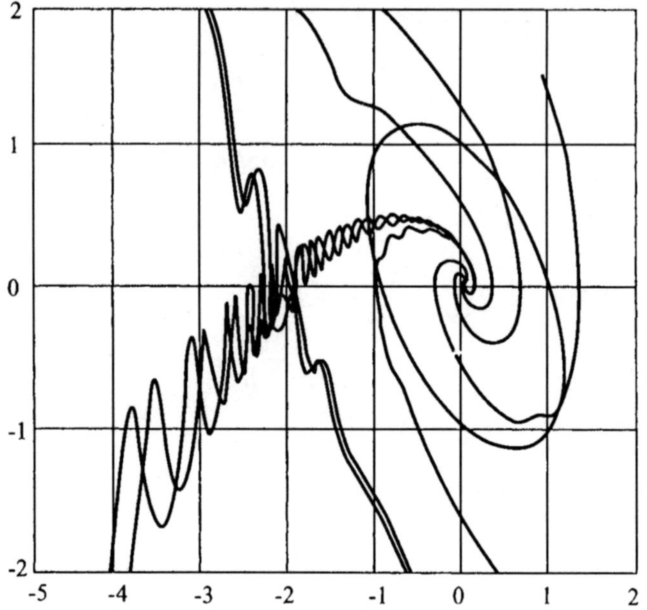

Figure 2.24. Phase plane portrait of equation (2.80) for $b(t) = \sin^s 5t$.

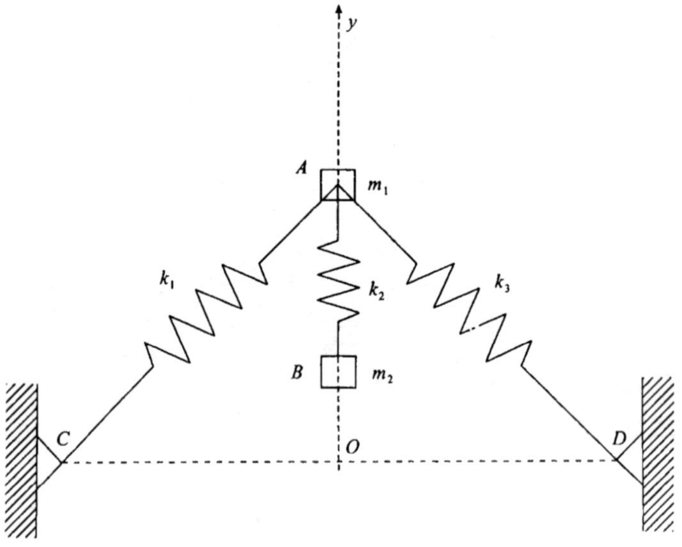

Figure 2.25. Two degrees-of-freedom system analysed in Example 2.14.

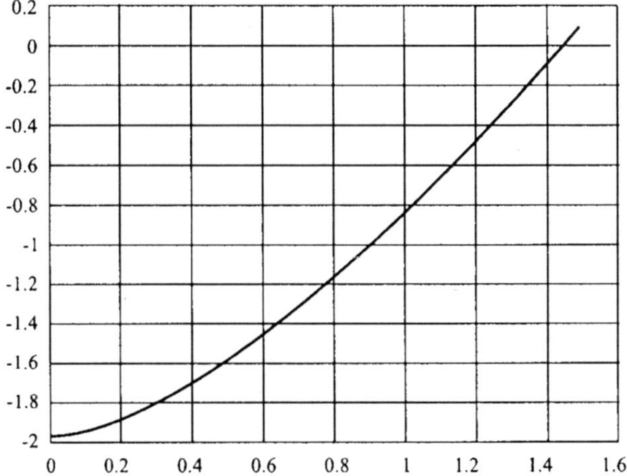

Figure 2.26. Dependence $ECI(\rho)$ (see Example 2.14).

EXAMPLE 2.14 *Let us consider the two degrees-of-freedom system shown in Fig. 2.25.*

Two masses A and B move according to constraints along the vertical axle y. The mass A is joined to the ground at C and D by two identical linear springs with revolute joints. The mass B is joined to the mass A by the other linear spring. It is assumed that the movements of A and B are linearly damped. Three points of the equilibrium are possible in the system. The motion equation of the system near the top equilibrium point has the form:

$$\dot{x} = Ax + h(x) \tag{2.82}$$

where:

$$A = \begin{bmatrix} 0 & 1 & 0 & 0 \\ a_{21} & -2c' & h' & 0 \\ 0 & 0 & 0 & 1 \\ k'' & 0 & -k'' & -2c'' \end{bmatrix},$$

$$h(x) = \begin{bmatrix} 0 \\ h_2(x_1) \\ 0 \\ 0 \end{bmatrix}, \quad x = \begin{bmatrix} z_1 - z_{10} \\ \dot{z}_2 \\ z_2 - z_{20} \\ \dot{z}_2 \end{bmatrix},$$

$$z_1 = \frac{y_1}{l}, \quad z_2 = \frac{y_2}{l} + \frac{a}{l}, \quad \alpha = \frac{\Delta l}{l}, \quad k = \frac{k_1}{m_1},$$

$$k' = \frac{k_2}{m_1}, \quad k'' = \frac{k_2}{m_2}, \quad c' = \frac{c_1}{m_1}, \quad c'' = \frac{c_2}{m_2},$$

$$a_{21} = -2k - k' + \frac{2k(1+\alpha)}{\left(1+z_{12}^2\right)^{3/2}},$$

$$h_2(x_1) = 2k(1+\alpha)\left[\frac{x_1+z_{12}}{\sqrt{1+(x_1+z_{10})^2}} - \frac{z_{10}}{\sqrt{1+z_{10}^2}} - \frac{x_1}{(1+z_{10}^2)^{3/2}}\right],$$

and $l = \frac{1}{2}CD$, $l + \Delta l$ *– length of undeformed springs* AC *and* AD, y_1, y_2 *– current positions of the masses* A *and* B, *parameter* z_{10} *corresponds to the top equilibrium point.*

By numerical calculation the estimation of the ESD and of the ECI for the top equilibrium point $z_{10} = 1.354$, $\dot{z}_{10} = 0$, $z_{20} = 1.157$, $\dot{z}_{20} = 0$ *in the case* $\alpha = 1m$, $l = 1m$, $a = \sqrt{3}m$, $k = 20\frac{1}{s^2}$, $k' = 2\frac{1}{s^2}$, $k'' = 50\frac{1}{s^2}$, $c' = 2\frac{1}{s}$, $c'' = 12.5\frac{1}{s}$, $\frac{m_1}{m_2} = 25$ *was determined. The matrix* S_0 *optimal for the linear part of the system was obtained in the form*

$$S_0 = \begin{bmatrix} 2.347 & 0.142 & -0.42 & -0.205 \\ 0.142 & 0.087 & 0.121 & 0.015 \\ -0.42 & 0.121 & 1.438 & 0.259 \\ -0.205 & 0.015 & 0.259 & 0.101 \end{bmatrix}. \tag{2.83}$$

The numerically derived dependence between ECI and the radius ρ *is shown in Fig. 2.26.*

The ball $B_{\rho_0}^{S_0}$, *where* $\rho_0 \cong 1.46$, *encompasses the estimations of ECI of the equilibrium point of the system obtained by the presented method.*

To conclude this section, a method for the estimation of the exponential stability domain and the exponential convergence of an arbitrary exponential stable equilibrium point of a non-linear system was presented. The method applies the best Lyapunov quadratic function obtained for the linearized system. This approach allows us for avoid the problems of determining the Lyapunov function for the non-linear system. The obtained results may by improved by optimisation the Lyapunov function for a non-linear system but in the case the time of the numerical calculations increases very quickly with the degree of the system.

Chapter 3

AN INTRODUCTION TO STABILITY OF A WHEELED VEHICLE

Wheeled car stability is addressed in section 3.1, pneumatic tire properties are illustrated and discussed in section 3.2. The road surface-car wheel-car body dynamics is considered in section 3.3, and a brief introduction to stability of a moving car is given in section 3.4.

In this chapter our attention is focused on the stability of a wheeled vehicle. In particular, the stability of various mathematical models of a road vehicle and its sub-assemblies will be analysed.

1. Introduction

Note that a requirement for being in operation of any wheeled vehicle (for instance, a car) consists of the following items:

(i) efficiency (a proper carrying out of the given tasks);

(ii) reliability and controllability (a possibility to achieve the defined states);

(iii) stability (a mutual closeness of solutions (motions));

(iv) boundedness and restrictions of solutions (motions).

Stability is without any doubt one of the most important properties of motion ..."stability characterizes a system ability to keep on conservation of the required functional property [...] a stability is closely related to sensitivity of solutions, for instance of the differential equations, on various changes of the initial conditions, structural parameters and disturbances".

The most celebrated stability in the Lyapunov sense [47] is that any solutions (motions) that are close to each other at a certain time instant t_0 become arbitrarily close during the whole observation time.

Stability in the sense of Poincaré (or orbital stability) denotes, roughly speaking, that each solution that is close to an investigated one remains arbitrarily close during the observed future (observe that contrary to the Lyapunov stability, now there is no requirement for solutions to be close to each other in the same time instant). Finally, a system is stable in Lagrange's sense if all its solutions are bounded (for more details see [47]).

First of all, it should be emphasized that in engineering practice particularly in respect to *car stability* concepts, the last one very often has no relation to stability either in a mathematical or in a control theory sense. Nowadays two clear paths can be distinguished while analysing wheeled cars:

- the Lyapunov concepts are widely used in *simulating investigations* of stability with respect to mathematical models of cars, in particular where an investigated motion exists for a relatively long time (a power transmission system or a straight line vehicle movement);

- the so-called *'technical stability'* concepts are very often used in *experimental investigations* while different kinds of motion are analysed during observation in a relatively short time interval. As an example one may refer to the standard stability studies of a braking vehicle moving on a stationary circular trajectory. In many laboratories, as a criterion the increment of rotational car velocity measured around a vertical axle after one second of the disturbance beginning (i.e. braking) is taken. The mentioned time interval of one second possesses an important practical meaning, since it corresponds strictly to beginning of the potential driver reaction to any dangerous parameter changes of the vehicle motion.

Stability concepts originating from Lyapunov stability are certainly important for the models governed by differential equations, but not necessarily for various other technical objects. In particular, in many engineering systems the motion duration is bounded and sometimes very short, and often a priori constraints are attached which describe a safe system (vehicle) behaviour (for instance, a width of a public road). Therefore, especially during investigation of a car's movement, much more important is the so-called technical stability definition [36, 37]. This point of view is applied to formulate stability conditions of a wheeled car and its subsystems for various types of motion.

2. A Wheeled Car Stability

2.1 Dynamics of a Stiff Tire and Stability of a Wheeled Car

In the beginning of motorization a car was equipped with wheels having relatively stiff solid tyres, coused by the tyre's construction and by application of a relatively large internal pressure (about $0.7MPa$). Transversal deformation of such a tyre was relatively small. The so-called *transversal car mechanics* was reduced drastically to the Ackermann–Jeantaud scheme (see Figure 3.1).

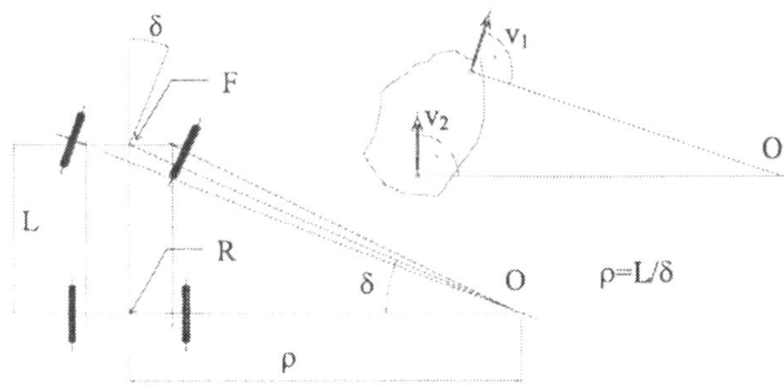

Figure 3.1. A scheme of transversal kinematics due to Ackermann–Jeantaud.

The Ackermann–Jeantaud scheme describes dependence of the position of an instantaneous rotation center car point on directions of velocity vectors of two car points associated with frontal and back axles (points F and R). It is assumed, that directions of these two vectors overlap with directions of frontal steered wheel planes (for example, these directions can be defined by both the averaged steer angle δ of these wheels - see Fig. 3.1) and by the direction of back wheel planes. In the case of circular car movement one may define the so-called *limiting car velocity* v_l governed by the following equation

$$(v)_l^2 = \frac{\mu \cdot g \cdot L}{\delta} \tag{3.1}$$

where: μ is the fiiction coefficient describing tire and road adherence, g is gravitational acceleration and δ is the angle of steering of front wheels (in fact it is averaged value of left and right angles).

2.2 Dynamics of an Elastic Tire and Stability of a Wheeled Car

The Ackermann–Jeantaud scheme was used until 1925. In the twentieth of XXth cenutry "low pressure" tires began to be used and an unexpected behaviour appeared. Namely, "wheel flutter" occurred, referred to further as *shimmy*, which is exhibited by an unpleasant wheel vibration around the backing-off axle. It was clear that an increase of tires elasticity caused this harmful behaviour. During analysis of the described phenomenon the so-called "tire wear out" was explained. Any transversal load acting on a rotating car wheel may change direction of its rotation. This force (for example, during a turn motion) moves a wheel, in a direction with deviation in respect to its plane, described by the so-called *slip* angle α. The value of a slip angle depends on the value of the lateral force.

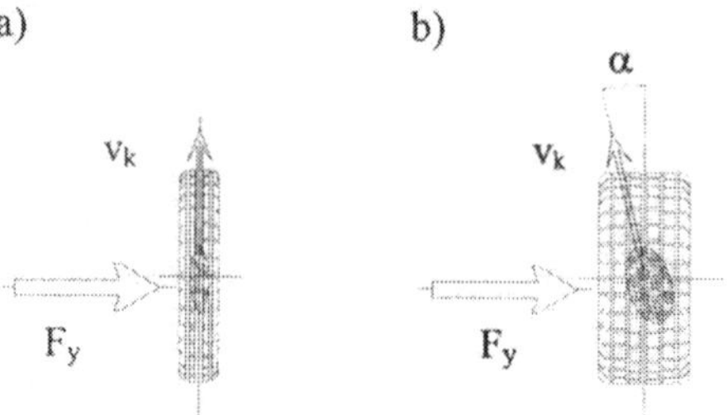

Figure 3.2. Car wheel mechanics for a stiff (a) and elastic (b) tire. In the first case $\alpha = 0$, whereas $\alpha \neq 0$ in the second case, since a transversal drift appears.

The mechanical scheme of a car wheel with depicted lateral force F_y, and slip angle α is reported. This figure incudes also the stiff tire case (a), where for $F_y \neq 0$ the angle $\alpha = 0$.

The mechanical scheme of a car motion having elastic tires is shown in Figure 3.3. A shift of an instantaneous rotation center movement depending on velocity directions of the flat car body associated with front and rear wheel axles is remarkable.

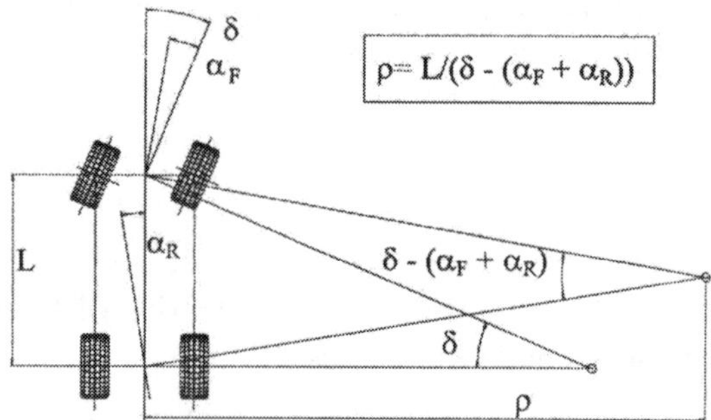

Figure 3.3. A transversal kinematics of a wheeled car taking into account a transversal tire drift.

The described shimmy phenomenon forced engineers to develop independent car suspensions. In addition, it effected development of both experimental investigation as well as stability and steerability theories applied to car dynamics. Among others, a *stability rule* of a car moving is a straight line has been formulated (for more details see chapter 5, section 1):

$$\mu \cdot g \cdot L/v^2 + \alpha_F - \alpha_R > 0. \tag{3.2}$$

The formula (3.2) yields three important relations between slip angles of front and rear tires:

(i) $\alpha_F > \alpha_R$ (stability is conserved for any car velocity);

(ii) $\alpha_F = \alpha_R$ (the motion is stable, but it is a critical case);

(iii) $\alpha_F < \alpha_R$ (when a car approaches the critical velocity value v_{cr}, then instability occurs).

The mentioned critical car velocity value is defined by the relation

$$v_{cr} = \sqrt{\frac{\mu \cdot g \cdot L}{\alpha_R - \alpha_F}} \quad \text{for} \quad \alpha_R > \alpha_F. \tag{3.3}$$

One may conclude from (3.3) that for a given difference $\alpha_R - \alpha_F$ a value of critical velocity increases owing to increase of tractive and road adhesion (μ) as well as a axle base track (L).

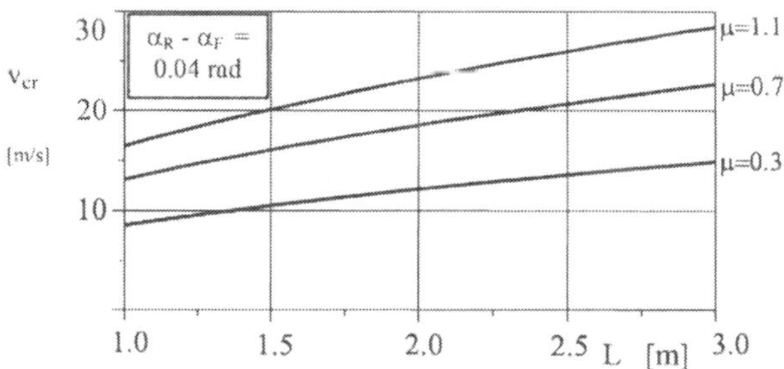

Figure 3.4. A critical car speed value as the function of axle base L for various adhesion coefficients (μ) and for a given slip angle difference $\alpha_R - \alpha_F = 0.04\, rad$.

The relation (3.3) yields the drawings shown in Figure 3.4. It is easy to conclude also intuitively that increasing axle base leads to increase of the critical car speed.

3. Pneumatic Tire Properties

Recall that the tire was invented by Thompson in 1845, its production began in 1888 by Dunlop, and its scientific investigation started in 1915. Mainly rolling resistance was investigated. It turned out that for a fixed steering gear and for a fixed steer angle of steered wheels the curvature radius of a car trajectory increases owing to increase of driving speed. A rubber pneumatic tire is deformed by an action of lateral forces, and therefore a translational wheel movement takes place in the direction of the longitudinal wheel plane axle deviating by the amount of the so-called *slip angle* (see Figure 3.5).

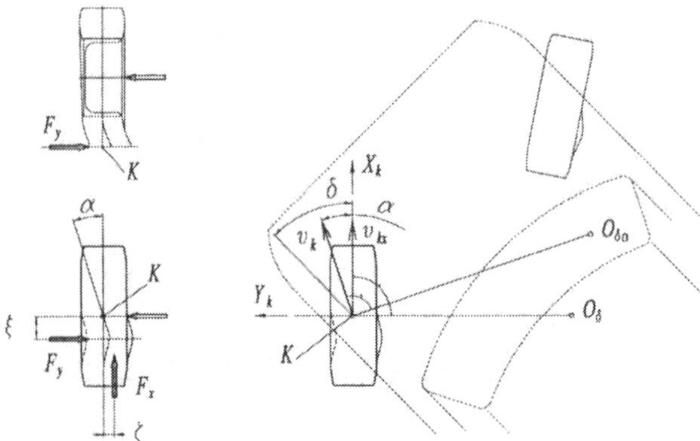

Figure 3.5. A tire deformation and a tire slip angle α.

The tire deformation and the slip tire angle α is shown in Figure 3.5. The action of the so-called *lateral force* causes a tire deformation in the neighbourhood of its contact with a road, and the centre wheel speed vector deviates by the X_k direction (i.e. direction of a longitudinal wheel axle) by the amount of the slip angle α. In what follows the car movement takes place not with respect to instantaneous rotation centre O_δ associated with a wheel steer angle δ, but the instantaneous rotation centre $O_{\delta\alpha}$ associated with two mentioned earlier angles δ and α.

In Figure 3.5 the area of contact between a road and a tire is depicted by K. The following velocity vectors of the artificial point K are distinguished:

(i) v_{kx} is the velocity vector with direction defined by the wheel longitudinal axle. In words, the point K moves in the direction of v_{kx} for a relatively small car speed when a tire slip does not occur;

(ii) v_k is the velocity vector with direction deviating from the longitudinal wheel axle by the amount of the slip angle α. Now the point K moves in the direction of v_k when a car moves along a curve trajectory with a relatively large speed, and when the tires are elastically deformed and the tire slip occurs.

Imagine now that both longitudinal F_x and lateral F_y friction forces act on a tire (see Figure 3.5). The curves associated with the forces F_y and F_x are shifted by the amount of ξ and ζ, respectively. In order to carry out wheel force measurements the action curves of F_y and F_x are reduced virtually to the point K (Fig. 3.5), and to keep the equilibrium condition the so-called *tire stabilization moment* M_k is introduced in the following way:

$$M_k = F_y \cdot \xi + F_x \cdot \zeta \, .$$

In what follows we briefly consider kinematics and dynamics of a tired wheel, when a tire is loaded or unloaded by a transversal force during driving and braking processes.

The velocity vectors and the forces coming from a road surface for a driven and braked tired wheel for $F_y = 0$ is shown in Figure 3.6.

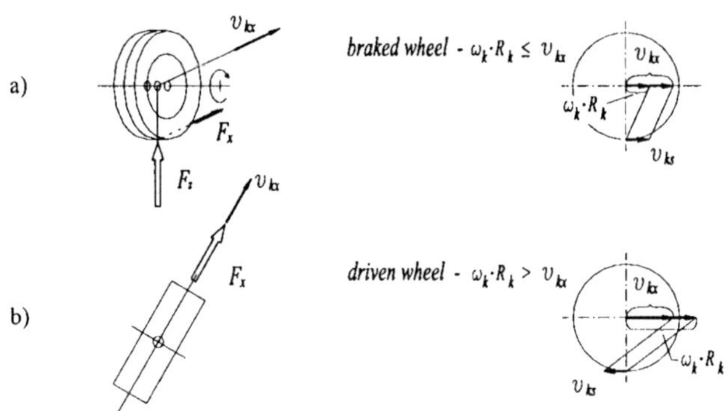

Figure 3.6. Kinematics and dynamics of a braked (a) and driven (b) road wheel.

Let us introduce a relative circumferential wheel longitudinal slip s_x defined in the following way:

$$s_x = 1 - \left(\frac{\omega_k \cdot R_k}{v_{kx}}\right)^n, \quad \text{where:} \quad n = \begin{cases} +1 : \omega_k \cdot R_k \leqslant v_{kx} \\ -1 : \omega_k \cdot R_k > v_{kx} \end{cases}.$$

Observe that $s_x \in [0, 1]$. In the case when $s_x = 0$ the wheel is either not driven or braked, and there are no circumferential forces. In the case of $s_x = 1$ the wheel is braked and it is fixed while the car moves with the velocity $v \neq 0$ and the wheel does not rotate ($\omega_k = 0$). However, the wheel slip $s_x = 1$ may occur when a wheel is driven and it rotates with $\omega_k \neq 0$ while the car stops ($v = 0$). In both cases, when the linear center wheel velocity v_{kx} differs from a linear velocity yielded by the rotational movement $\omega_k \cdot R_k$, the circumferential wheel slip appears and the slip speed is defined by the relation

$$v_{ks} = \begin{cases} v_{kx} - \omega_k \cdot R_k : n = +1 \\ \omega_k \cdot R_k - v_{kx} : n = -1 \end{cases}.$$

3.1 Tire Characteristics

The function $F_x(s_x)$ describing the dependence of the circumferential (longitudinal) friction force versus the relative circumferential slip is called the tire characteristic and it is important for the driving and braking processes. The typical character of this function is shown in Figure 3.7.

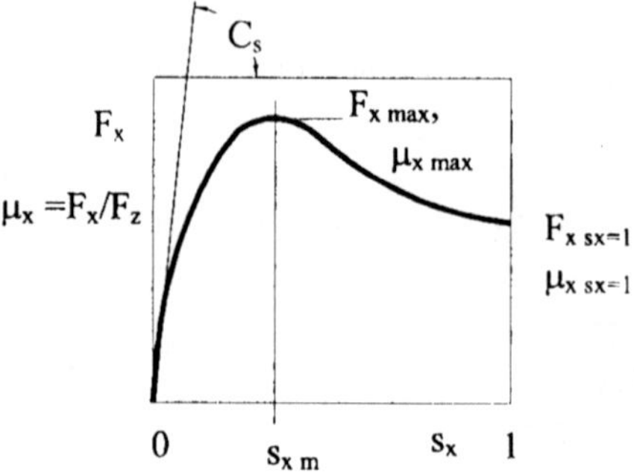

Figure 3.7. The characteristics $F_x(s_x)$ and $\mu_x(s_x)$.

Table 3.1. Essential elements of the characteristics $F_x(s_x)$.

$F_{x\ \max}$	Maximal circumferential friction force generated for a given driving condition
$s_{x\ m}$	Relative slip for $F_{x\ \max}$
C_s	Coefficient of circumferential force F_x: $C_s = \left.\dfrac{dF_x}{ds_x}\right\|_{s_x=0}$

The characteristics reported in Table 3.1 depend on tire properties, road surface, vertical force F_z, car velocity, etc.

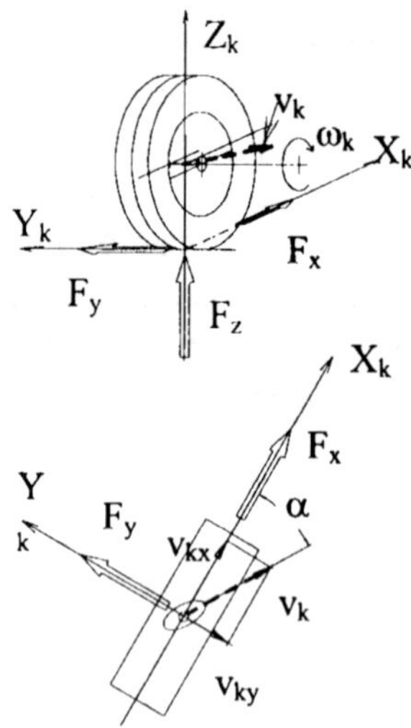

Figure 3.8. Kinematics and dynamics of a tired wheel subjected to transversal force.

Table 3.2. Essential elements of the characteristics $F_y(\alpha)$.

$F_{y\ \max}$	Maximal value of transversal friction force F_y which can be generated for a given driving condition
α_m	Slip angle corresponding to $F_{y\ \max}$
C_α	Coefficient of lateral force: $C_\alpha = \left.\frac{dF_y}{d\alpha}\right\rvert_{\alpha=0}$

The second characteristics $F_y(\alpha)$ are functions of lateral friction forces versus the slip angle and plays an important role during car steering. In Figure 3.8 the speed and force vectors yielded by the road surface and acting on braked and driven wheels are shown. The linear velocity vector of the wheel v_k is deviates by the amount of the slip angle α from the longitudinal wheel axle X_k. The dependence $F_y(\alpha)$ is shown in Figure 3.9.

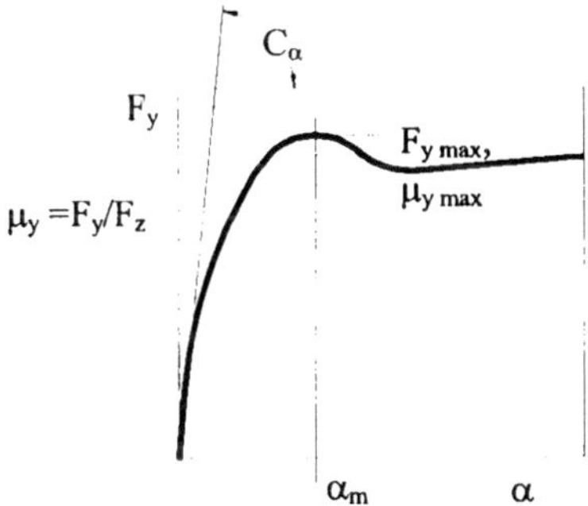

Figure 3.9. The characteristics $F_y(\alpha)$ and $\mu_y(\alpha)$.

Again, the prescribed characteristic values depend essentially on the tire properties, road surface, the vertical force F_z, driving velocity, etc.

Now we briefly describe the *multicomponent tire characteristics*. In Figures 3.10–3.15 typical tire characteristics are presented. One may conclude how a car wheel behaviour depends on: the state and type of the road surface, the circumferential relative slip and tire slip angles, a vertical wheel load, tire pressure, a slope of wheel plane (camber angle), height of the tire tread, and frequency of wheel angle turns.

Observe that the sign of the slope of the wheel plane γ is defined via a relation between the wheel slope and the lateral force vector acting on the wheel. However, in diagnostics this sign is defined in different ways. Namely, it depends on geometrical relations between the car wheel plane and the road surface plane. Owing to Figure 3.16 a negative angle $\gamma(-|\gamma|)$ means that a car wheel is inclined contrary to the force vector F_y (Figure 3.16a). Positive γ means the opposite (Figure 3.16b).

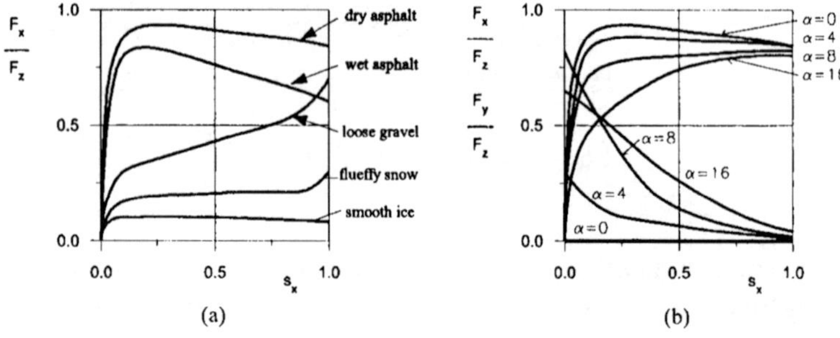

Figure 3.10. Dependence of circumferential friction force on the slip s_x for different types and various states of a road surface (a) and dependence of circumferential and transversal friction forces on the slip s_x for various slip angles α (b).

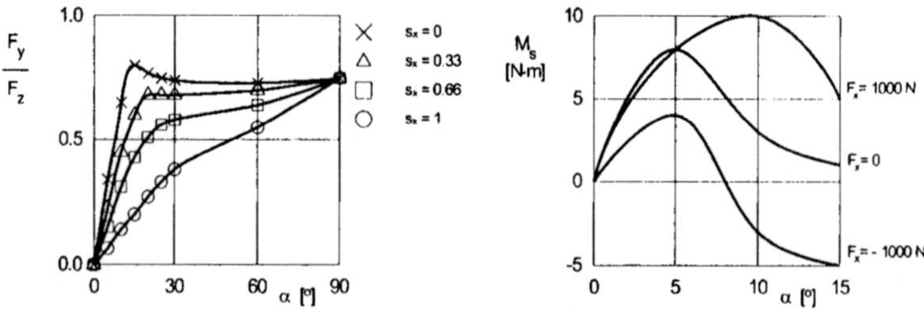

Figure 3.11. Dependence of lateral friction force on slip angles for various circumferential relative slips (a) and dependence of the stabilizing moment on the slip angles for various circumferential friction forces (b).

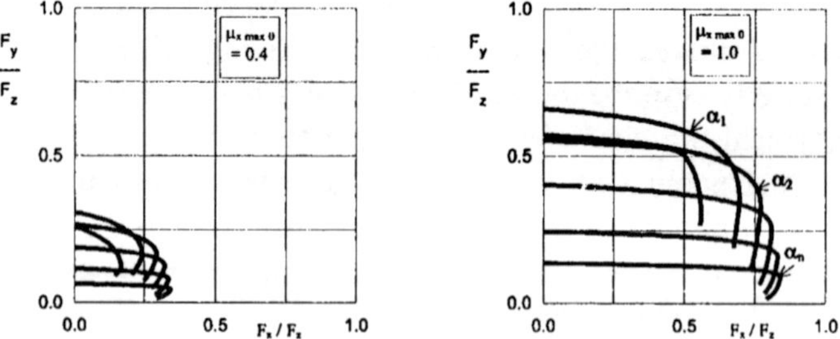

Figure 3.12. Transversal versus circumferential friction force for various slip angles α_i for road surfaces with small and large adhesion coefficients $\mu_{x\ \max}$.

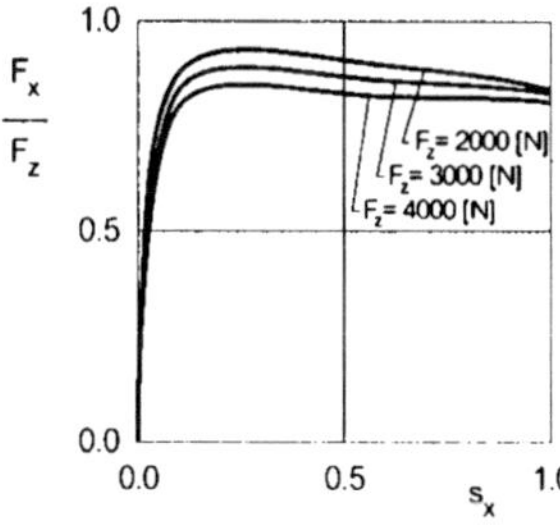

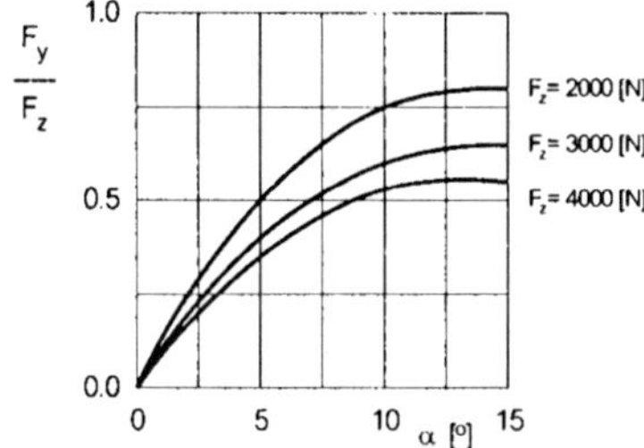

Figure 3.13. Dependence of circumferential friction force on relative slip for various vertical loads (a) and dependence of lateral friction force on the slip angle for various vertical loads F_z (b).

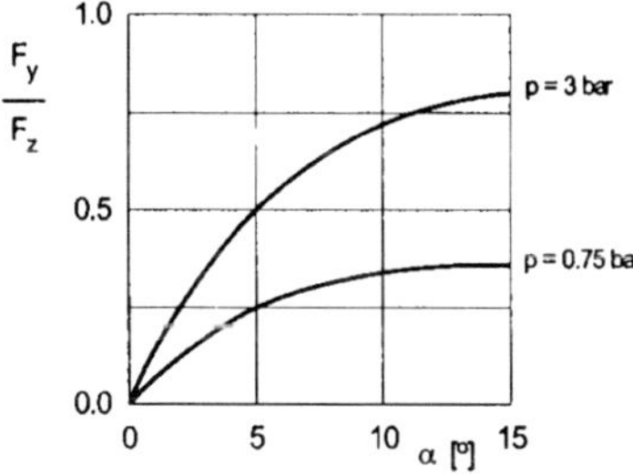

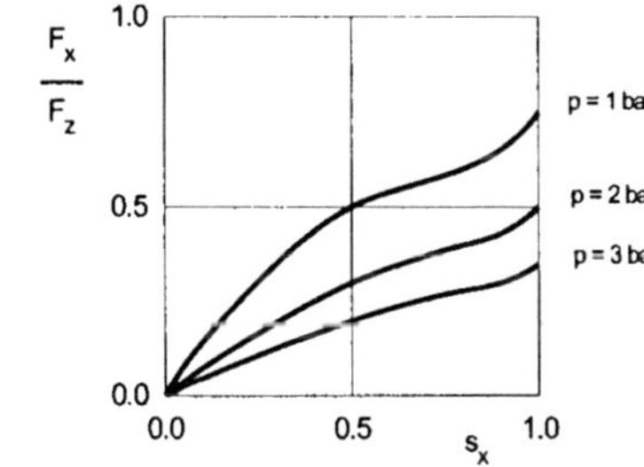

Figure 3.14. Dependence of lateral friction force on the slip angle for various tire pressure (a) and dependence on circumferential friction force on the relative slip for different tire pressures (b).

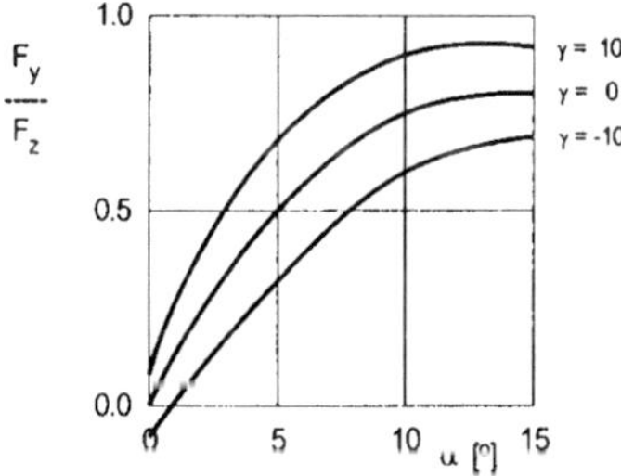

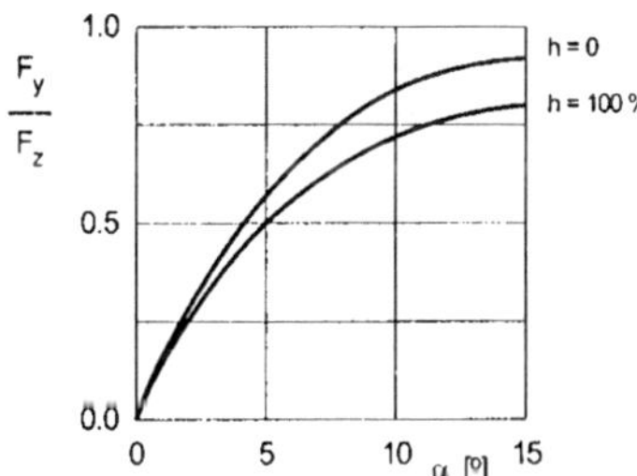

Figure 3.15. Dependence of lateral friction force on the slip angles for various slopes γ (a) and for various tire tread states (b).

Taking into account the characteristic $F_y(\alpha) = \Phi(\gamma)$ shown in Figure 3.15 on the relation reported in Figure 3.16, one can analyse the influence of a car suspension on the car steerability and car resistance to lateral wheel slip of the considered car axle. In Figure 3.17 the *suspension of motor-car body – leading*

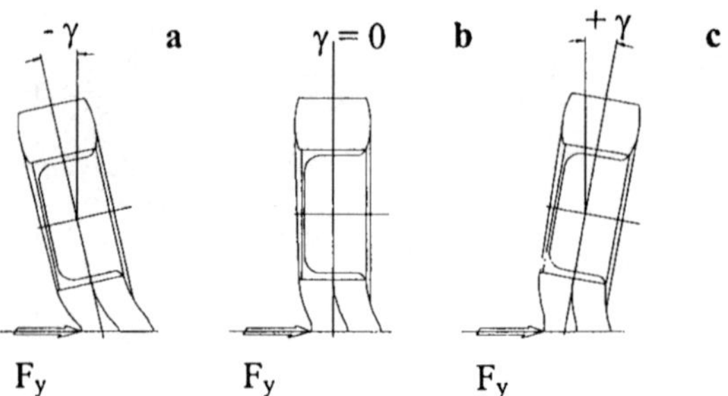

Figure 3.16. Relation between signs "+" and "-" of the γ slope and the vector F_y.

elements of wheel suspension during a transversal force action (for instance, while moving on a turn) for three different suspension designs are shown.

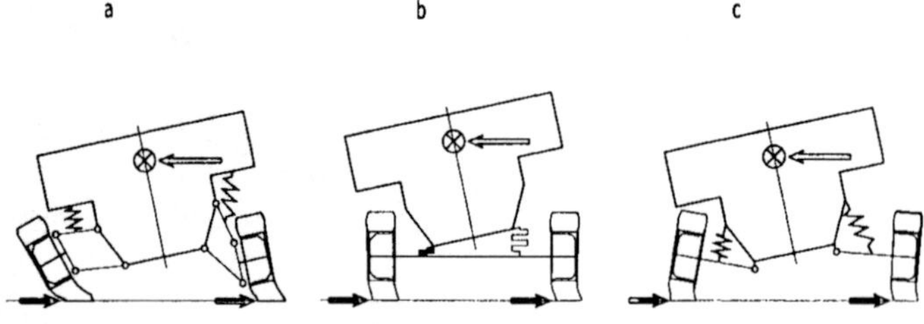

Figure 3.17. Influence of various suspensions on transversal car dynamics during motion on the turns (a) independent suspension on two transversal arms with non-equal length (leads to decrease of resistance against transversal draft; (b) dependent suspension (has no influence on resistance against transversal draft; (c) independent suspension on a single transversal arm (leads to increase of resistance against transversal draft).

An influence of the driving speeds on the dynamical tire characteristics is shown in Figure 3.18. In particularly this influence is important for low driving speeds. An increase of f causes an increase of the relative amplitude changes of the transversal force A_{F_y}, and $A_{F_y} < 1$. For higher speed values, say $80km/h$, influence of frequency f amplitude A_{F_y} is negligible. However, for higher speeds, say $160km/h$, the characteristic is qualitatively changed, since $A_{F_y} > 1$.

3.2 Tires modelling [31, 42, 45, 49, 54, 55, 56, 58, 78, 80, 81, 83, 84]

During numerical simulations of wheeled car dynamics the tire characteristics are mainly described by equations (or rarely by tables). Since there exist large number of industrial type models, here only two of them are addressed.

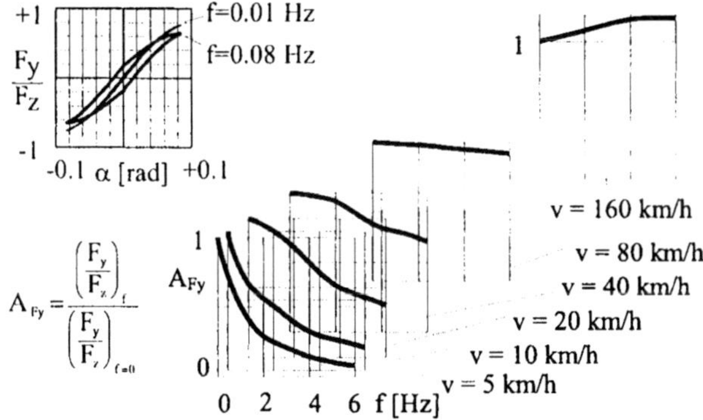

Figure 3.18. Driving speeds influence on the dynamical tire characteristics.

Namely, the driven car wheel model referred as the *Magnum model*, and the braking (driven) and steerable wheel car model called the *Wagner model*.

The Magnum model [54] is related to a driven tired wheel. The key element in this model is represented by the friction between a tire and a road surface, and it includes adhesion, Coulomb and viscous friction parts, and it is governed by the *Stribeck curve* (Figure 3.19).

The friction forces acting on the driven car wheel can be described in the following way:

$$F_{xH\pm} = \mu_{xH\pm} \cdot F_z; \qquad F_{xC\pm} = \mu_{xC\pm} \cdot F_z,$$

where: F_z is the vertical force of interaction between the car wheel and the road surface; μ_{xH}, μ_{xC} are the adhesion and Coulomb friction coefficients, respectively, and F_{xH}, F_{xC} are the adhesion and Coulomb forces, respectively.

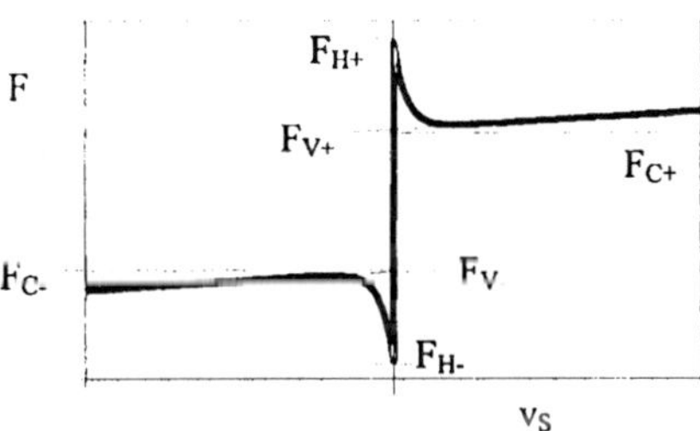

Figure 3.19. The Stribeck curve (F_H – adhesion friction force, F_v – viscous friction force, F_C – Coulomb force, v_s – slip velocity).

In Figure 3.20 a car wheel model, is shown that takes into account pneumatic elasticity behaviour via a separation of a tired tread belt from the wheel disk by adding elasto-viscous elements. The wheel is driven by the moment M_n acting on the axle and wheel disk. The moment

$$M_{og} = M_n - J_1 \cdot \dot{\omega}_k - V_e \cdot \omega_k,$$

where: ω_k - rotation velocity of the wheel disk, J_1 - moment of inertia of the disk and the rotating elements associated with the disk, V_e - coefficient characterizing energy loss in the tire.

In the *Magnum model* F_x is generated in two steps. In the first step, for $v_s = 0$, in the beginning of car wheel movement a slip between tire tread and the road does not exist. The adhesion force F_x increases proportionally to M_{og}, and it can achieve its maximum F_{xH+} (for $v_s = 0$). Further increase of the driven moment can generate a slip $v_s \neq 0$. Now the second step begins ($v_s \neq 0$) and the adhesion force F_x depends on v_s (see Figure 3.19).

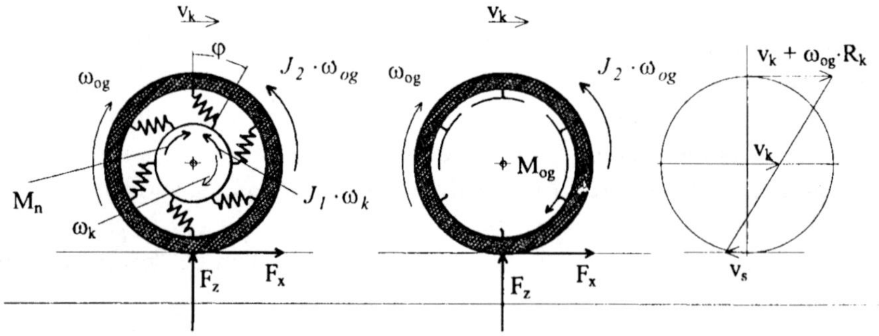

Figure 3.20. Magnum model.

In Figure 3.20 the Magnum model with kinematic and dynamic schemes is shown. In Table 3.3 the block diagram of the Magnum model is reported.

In Figure 3.21 time histories obtained numerically for some kinematic and dynamic quantities with the Magnum tire model are reported.

For some types of road surfaces an increase of the slip velocity is accompanied by a decrease of friction (adhesion). Therefore, in this case the analysed Magnum model does not seem to be the appropriate one, since it uses the Stribeck curve. In this case the so-called "WAVE" friction curve is taken, which is described by the equation

$$F_x = \frac{D \cdot \text{sign}(v_s)}{A \cdot |v_s^3| + B \cdot v_s^2 + C \cdot |v_s| + 1}$$

and is shown in Figure 3.22.

The second "Wagner" Figs. 3.21 and 3.22 model [109] takes into account force connections in lateral and circumferential directions, and it consists of three different zones: adhesion zone, transitional zone and full slip zone.

Table 3.3. Block diagram of the Magnum model.

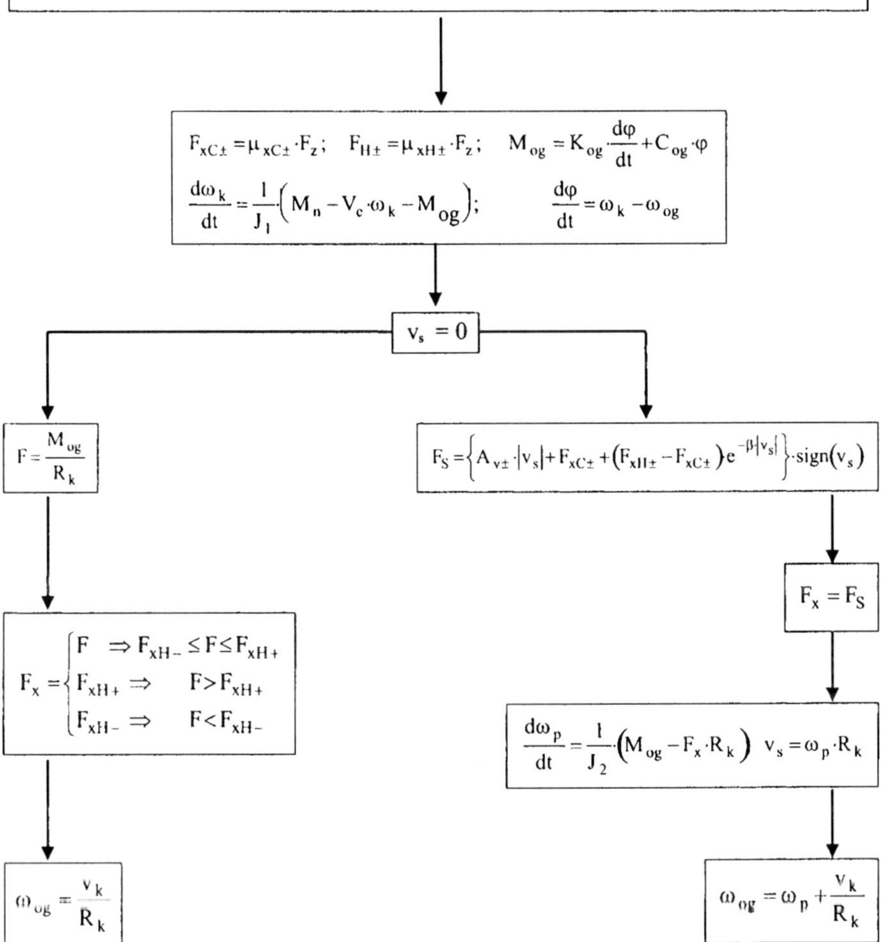

The following assumptions and hypotheses are included in the model:

- the slope of a driven wheel plane is equal to zero,
- road surface is homogeneous and horizontal,

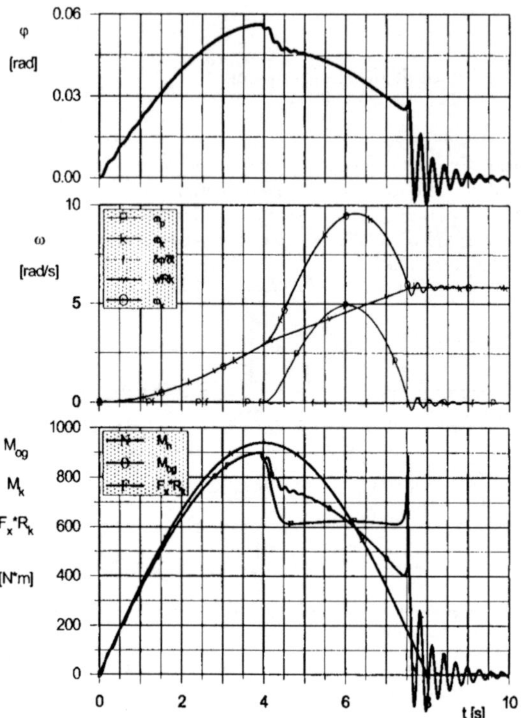

Figure 3.21. Time histories of the driving process for the Magnum tire model.

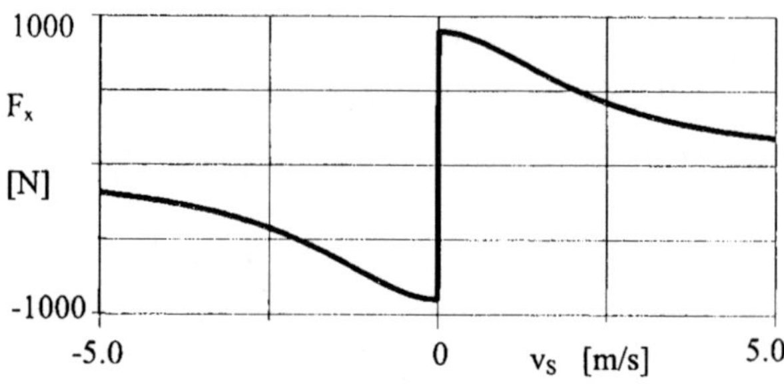

Figure 3.22. The "WAVE" curve for $A = -0.01$, $B = 0.2$, $C = 0.01$, $D = 900$ ($D = F_{xH+}$).

- tire segments are modelled by elastic rectangular blocks coupled radially with the carcass,
- stiff tire carcass moves owing to tire elasticity characterized by longitudinal (k_x) and lateral (k_y) stiffness,

- three zones are analysed, i.e. adhesion, slip and transition zones,
- vertical force F_z is uniquely distributed on the stick surface,
- adhesion coefficient μ depends on a slip velocity linearly.

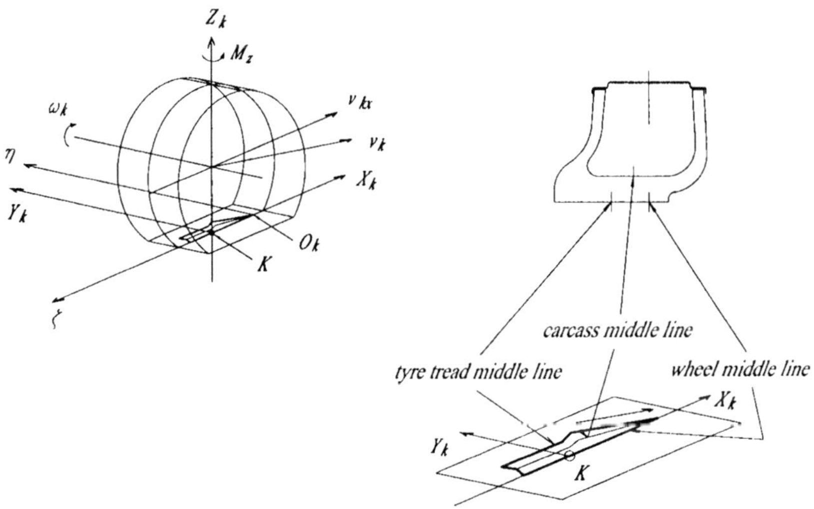

Figure 3.23. "Wagner" tire model and the associated coordinates (see the text).

During further analysis the following notation is used: F_z $[N]$ – vertical load of a wheel; k_x $[N/m]$ – longitudinal tire stiffness coefficient; k_y $[N/m]$ – transversal tire stiffness coefficient; C_α $[N/rad]$ – longitudinal force coefficient; C_s $[N]$ – circumferential load coefficient; $s_y = \tan(\alpha)$ – lateral slip; A_{vs} $[s/m]$ - slip velocity coefficient; μ_{x0} - adhesion coefficient achieving maximum for $v_k \to 0$.

In Figure 3.23 the tire model with the coordinate systems $(X_k Y_k Z_k)$ and (ζ, η) is shown. The coordinates $(X_k Y_k Z_k)$ are associated with the point K, being the idealized contact centre between the tire and the road surface. The coordinate system (ζ, η) is associated with the point O_k characterizing the beginning of the contact between the tire and the road surface.

In spite of the acting forces F_x, F_y, F_z also the stabilizing (aligning) moment M_z appears. The resulting deformation of the tire is represented by the middle curves of both the tire karkas and the tire tread.

In the next Figure 3.24 three friction zones (adhesion, slip and transition) are reported.

The mathematical Wagner model consists of the following equations and coefficients:

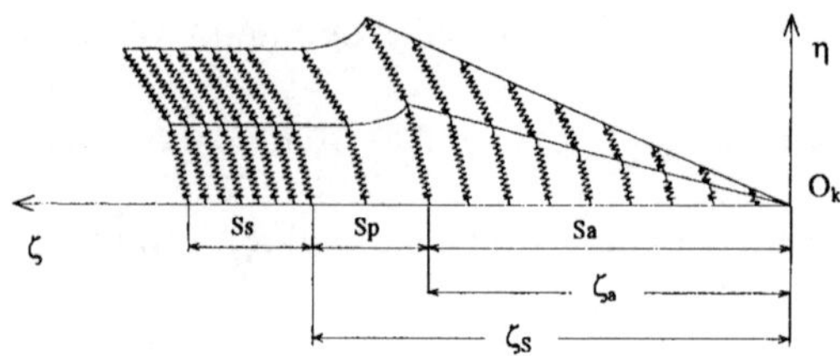

Figure 3.24. Model of deformation in the contact friction pair: tire -road surface (Ss - slip zone, Sp - transitional zone, Sa - adhesion zone).

- circumferential slip velocity

$$v_{ks} = v_k \cdot \cos(\alpha)\sqrt{s_x^2 + s_y^2},$$

- circumferential friction coefficient

$$\mu_x = \mu_{x0} \cdot (1 - A_v \cdot v_{ks}),$$

- constant coefficient A_v,
- relative adhesion zone length

$$Z_{La} = 0.5\,\mu_{x0} \cdot F_z \cdot (1 - s_x) \cdot \left(\sqrt{(s_x \cdot C_s)^2 + (s_y \cdot C_a)^2}\right)^{-1},$$

- relative adhesion zone length in transitional zone

$$Z_{Ls} = 0.5\,\mu_{x0} \cdot F_z \cdot (1 - s_x) \cdot \left(\frac{1}{C_s} + \frac{1}{C_a}\right) \cdot \left(\sqrt{(s_x)^2 + (s_y)^2}\right)^{-1},$$

- relative circumferential force generated in adhesion zone

$$F_{xa} = -C_s \cdot s_x \cdot (Z_{La})^2 \cdot (F_z \cdot (1 - s_x))^{-1},$$

- relative lateral force generated in adhesion zone

$$F_{ya} = -C_a \cdot s_y \cdot (Z_{La})^2 \cdot (F_z \cdot (1 - s_x))^{-1},$$

- relative circumferential force generated in transitional zone

$$F_{xt} = -\frac{C_s \cdot s_x \cdot Z_{La} \cdot (Z_{Ls} - Z_{La})}{F_z \cdot (1 - s_x)}$$

$$-0.5\,\mu_x \cdot s_x \cdot (Z_{Ls} - Z_{La}) \cdot \left(\sqrt{(s_x)^2 + (s_y)^2}\right)^{-1},$$

- relative lateral force generated in transitional zone

$$F_{yt} = -\frac{C_s \cdot s_y \cdot Z_{La} \cdot (Z_{Ls} - Z_{La})}{F_z \cdot (1 - s_x)}$$

$$-0.5\,\mu_x \cdot s_y \cdot (Z_{Ls} - Z_{La}) \cdot \left(\sqrt{(s_x)^2 + (s_y)^2}\right)^{-1},$$

- relative circumferential force generated in full circumferential slip zone

$$F_{xs} = -\mu_x \cdot s_x \cdot (1 - Z_{Ls}) \cdot \left(\sqrt{(s_x)^2 + (s_y)^2}\right)^{-1},$$

- relative lateral force generated in full circumferential slip zone

$$F_{ys} = -\mu_x \cdot s_y \cdot (1 - Z_{Ls}) \cdot \left(\sqrt{(s_x)^2 + (s_y)^2}\right)^{-1},$$

- total relative circumferential force

$$F_x = F_{xa} + F_{xt} + F_{xs},$$

- total relative lateral force

$$F_y = F_{ya} + F_{yt} + F_{ys},$$

- total stability tire moment, aligning torque

$$M_z = \frac{M_{za} + M_{zt} + M_{zs}}{F_z} + \frac{F_x \cdot F_y \cdot (K_y - K_x)}{F_z \cdot K_x \cdot K_y},$$

- relative rotational stabilizing moment generated in adhesion zone

$$M_{za} = -\frac{2L}{3} \cdot \frac{s_r \cdot s_y}{F_z \cdot (1 - s_x)^2} \cdot (C_s - C_a) \cdot (Z_{La})^3$$

$$+\frac{1}{6} \frac{C_a \cdot s_y}{F_z \cdot (1 - s_x)} \cdot (4Z_{La} - 3) \cdot (Z_{La})^2$$

- relative rotational stabilizing moment generated in full slip zone

$$M_{zs} = -\frac{L}{2} \cdot \mu_x \cdot F_z \cdot \left(\left(\frac{1}{C_a} - \frac{1}{C_s}\right) \cdot \mu_x \cdot F_z \cdot s_x - Z_{Ls}\right)$$

Table 3.4. Road surface parameters.

	C_s N/rad	A_{vs} s/m	μ_{x0}
dry asphalt	129060	0.023353	1.016
wet asphalt 1	95620	0.039623	0.8579
wet asphalt 2	80680	0.05605	0.8122
beat down snow	86080	0.04305	0.299
smooth ice	144640	0.03603	0.1238

$$\times s_y \cdot \frac{(1 - Z_{Ls})}{F_z} \cdot \left(\sqrt{s_x^2 + s_y^2}\right)^{-1},$$

- relative rotational stabilizing moment generated in transitional zone

$$M_{zt} = -\frac{\frac{4}{6}L \cdot (Z_{Ls} - Z_{La}) \cdot s_x \cdot s_y \cdot (C_s - C_a) \cdot (Z_{La})^2}{F_z \cdot (1 - s_x)^2} -$$

$$\frac{\frac{1}{6}L \cdot (Z_{Ls} - Z_{La}) \cdot \left(\frac{1}{C_s} + \frac{1}{C_a}\right) \cdot \mu_x \cdot s_x \cdot s_y \cdot (4Z_{La} - 3) \cdot (Z_{La}) \cdot (C_s - C_a)}{F_z \cdot (1 - s_x) \cdot \sqrt{s_x^2 + s_y^2}}$$

$$-\frac{\frac{1}{6}L \cdot (Z_{Ls} - Z_{La}) \cdot \mu_x^2 \cdot F_z \cdot s_x \cdot s_y \cdot (4Z_{La} - 3) \cdot (Z_{La}) \cdot (C_s - C_a)}{C_s \cdot C_a \cdot (s_x^2 + s_y^2)}$$

$$+\frac{\frac{1}{6}L \cdot (Z_{Ls} - Z_{La}) \cdot C_a \cdot s_y \cdot Z_{La} \cdot (4Z_{La} + 2Z_{Ls} - 3)}{F_z \cdot L \cdot (1 - s_x)}$$

$$+\frac{\frac{1}{12}L \cdot (Z_{Ls} - Z_{La}) \cdot \mu_x \cdot s_y \cdot (2Z_{La} + 4Z_{Ls} - 3) \cdot (Z_{La}) \cdot (C_s - C_a)}{\sqrt{s_x^2 + s_y^2}}.$$

In Figure 3.25 the function $F_x(s_x)$ is shown for various road surfaces and the following fixed parameters: $v_k = 7.62 m/s$, $F_z = 4540N$, $k_x = 178578N/m$, $k_y = 89289N/m$, $L = 0.1905m$, $C_\alpha = 45359N/rad$ (wheel parameters), and the road surface parameters are reported in Table 3.4.

In Figure 3.26 the function $F_x(s_x)$ is shown for various slip angles.

4. Travelling System Model

Although the tire characteristics is a very important component of road safety, it is not the only one. The constraints *road surface - car wheel - car body* play also a significant role. In the "flat perspective" they can be shown in a way reported in Figure 3.27.

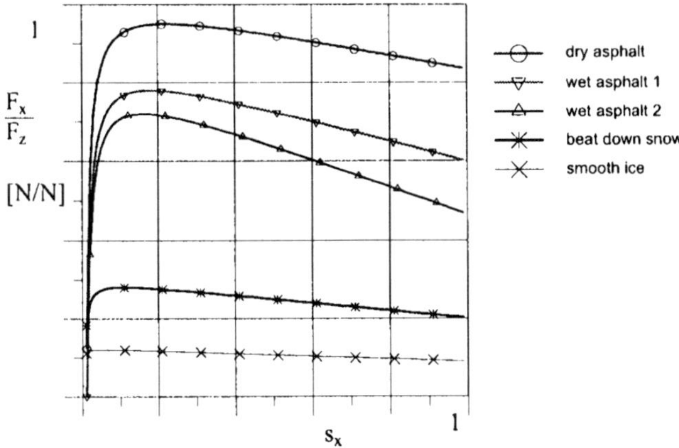

Figure 3.25. The dependence $F_x(s_x)$ for various road surfaces.

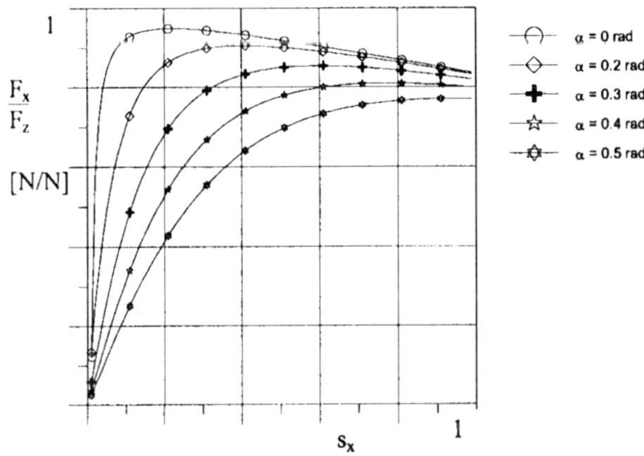

Figure 3.26. The dependence $F_x(s_x)$ for various slip angles α.

First, *kinematics of a stiff wheel-car body system* is considered. Let us begin with the particular case, when the constraints between a wheel and a car body are stiff (see Figure 3.28).

In this figure the following notation is applied: x_K, y_K - coordinates of the unrealistic constant point-between the wheel and the road surface, v_S - linear velocity of the point S being origin of the coordinate system X_SY_S, α - tire slip angle, β - side-slip angle of vehicle, δ - steer angle of front wheels, Ψ - rotation car angle around a vertical axle; yaw angle.

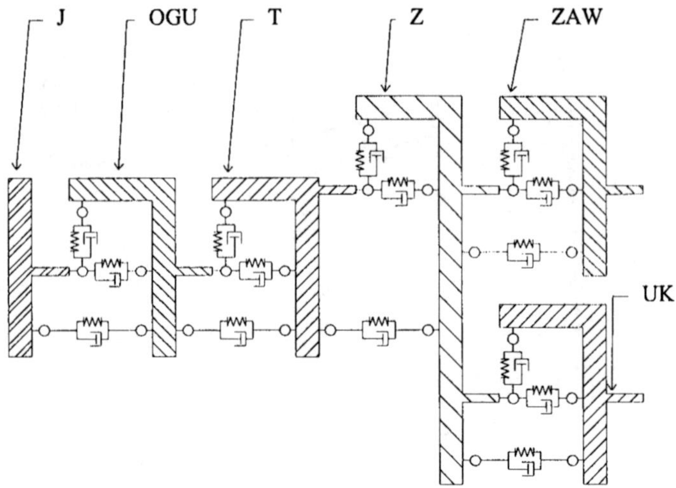

Figure 3.27. Constraints of a flat system representation: J - road surface, OGU - tired wheel, T - dish of the wheel, Z - steering knuckle, ZAW - wheel suspension, UK - steering system.

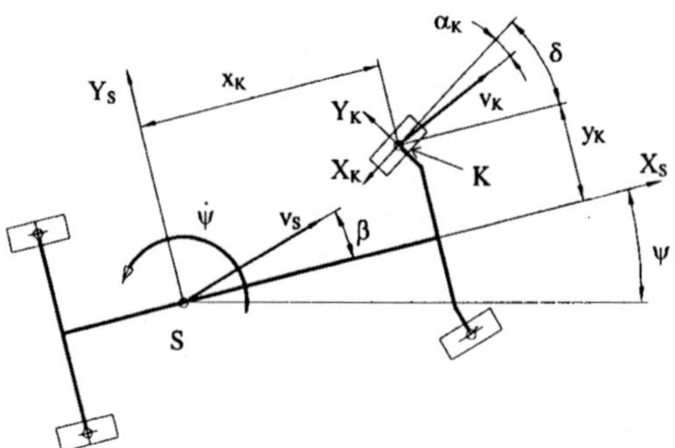

Figure 3.28. Kinematic relations in flat stiff car body-wheel system.

The tire slip angle of the point K is governed by the relation

$$\alpha_K = \delta - \arctan \frac{v_S \cdot \sin \beta + \dot{\Psi} \cdot x_K}{v_S \cdot \cos \beta - \dot{\Psi} \cdot y_K}. \tag{3.4}$$

The above equation is often used during general dynamical car analysis. However, during investigation of lateral car dynamics, one requires a more complex model of the *wheel-car body* system.

Second, kinematics of an elastic flat system wheel-car body is now analysed. The flexible constraints are taken into account owing to elastic properties of wheel bearings, as well as elasticity of a steering system and elasticity of a suspension.

It is assumed, that owing to elasticity of the wheel bearing and on it forces (see projection onto the road surface) the corresponding displacements are shown in Figure 3.29.

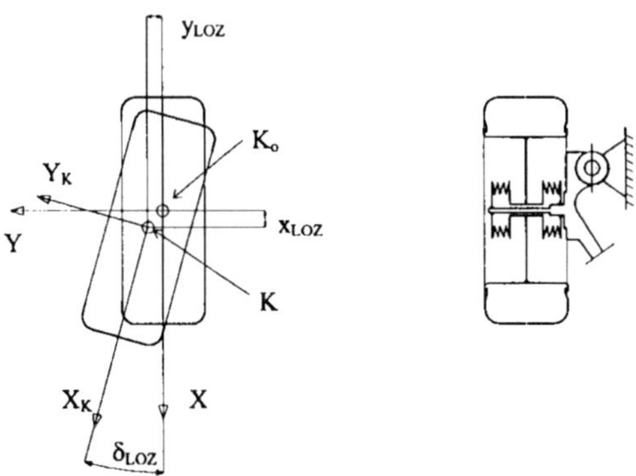

Figure 3.29. Displacement of a wheel owing to bearing elasticity.

Let us consider now wheel displacements yielded by a steering system elasticity (see Figure 3.30).

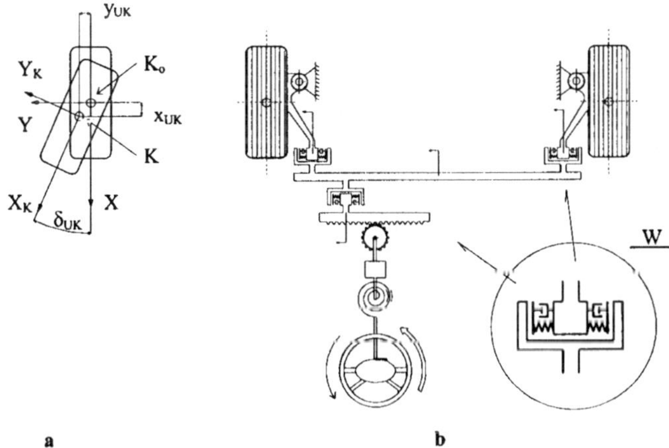

Figure 3.30. Displacements of a wheel yielded by steering system flexibility: (a) scheme of displacements, (b) scheme of the steering system (*W* represents flexible suspension element including stiffness-viscosity-clearance).

In the next step the wheel displacements owing to elasticity properties and kinematics of the wheel suspension are studied (see Figure 3.31).

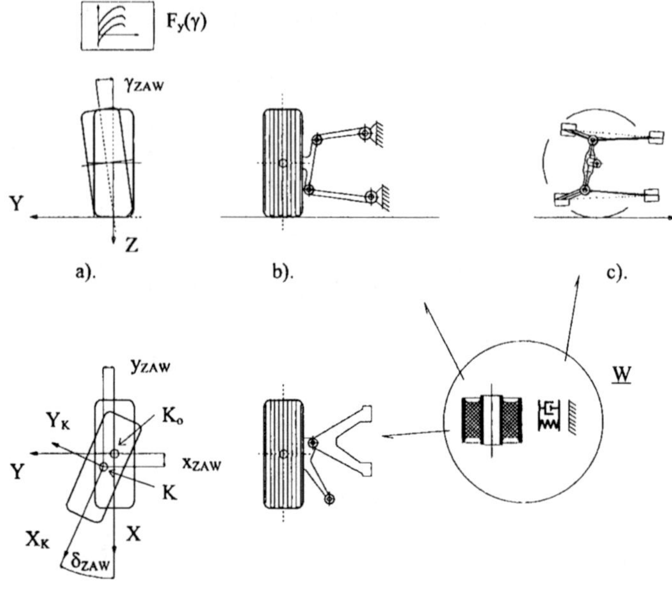

Figure 3.31. The driving wheel displacements yielded by flexibility and kinematics of suspension: (a) camber angle γ; (b), (c), (e) - various examples of suspension; (d) scheme of displacements projected onto the road surface (W - flexible suspension element: stiffness-viscosity-clearance).

Finally, let us consider the displacements of the idealized stick point between the driving wheel and the road surface yielded by the tire elasticity (see Figure 3.32).

The tangential forces F_x and F_y (Fig. 3.5) deform the wheeled tire. It is assumed that the idealized point K (Fig. 3.32) is stiffly attached to the driving wheel disc and lies on the intersection of the driving wheel plane with the surface road (XY plane). The point K moves along x_{OGU} and y_{OGU} in directions of X_K and Y_K, respectively, whereas the stick trace axle rotates (α_{OGU}). In addition, the dynamical process of tire deformation continues during the so-called *uprush* behaviour (increase of α is not simultaneously accompanied by a sudden increase of the transversal force drift, since the latter possesses certain inertia). In what follows a *lateral tire stiffness* C_y is defined by

$$C_y = \frac{dF_y}{dy_{OGU}}.$$

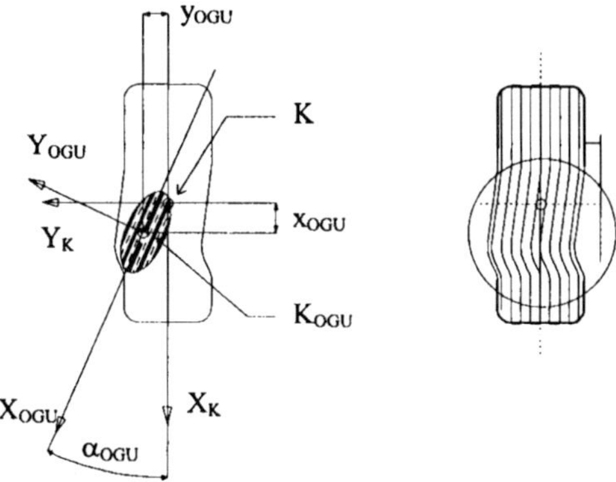

Figure 3.32. Displacements of the unrealistic stick point between the driving wheel and the surface road owing to the tire flexibility.

Owing to the *Schlippe–Dietriech hypothesis* the deformation velocity in the Y_K direction reads

$$v_{SCH-D} = \frac{L_L}{2} \cdot \dot{\alpha}_{OGU} + \frac{\dot{F}_y}{C_{y0}}, \tag{3.5}$$

where: L_L - unrealistic length of stick trace between the tire and the road surface; C_{y0} - transversal tire stiffness for the point $F_y(\alpha)_{\alpha=0}$: $C_{y0} = \left(\frac{dF_y}{dy_{OGU}}\right)_{\alpha=0}$.

Since the analysed system consists of a road-driving wheel-car body with flexible constraints (Figs. 3.29–3.32), its kinematics is composed of many components.

We are going to build a velocity vectors scheme associated with the idealized wheel point K, taking into account bearing elasticities of the steering system elasticities as well as deformation and kinematics of the wheel suspension. In addition, a similar analysis is carried out for the tire point K_{OGU} (see Figure 3.32). Both schemes are shown in Figure 3.33. $\sum\delta$ denotes the sum of all driving wheel rotations including kinematical turn of the steering gear, via deformations of the wheel bearings, and deformations in the steering system, as well as deformations and displacements in the suspension: $\sum\delta = \delta - \delta_{LOZ} - \delta_{UK} - \delta_{ZAW}$.

The slip angle α_K of the velocity vector v_K of the driving wheel point K (see Figure 3.33) is governed by the equation

$$\alpha_K = \delta - \delta_{UK} - \delta_{LOZ} - \delta_{ZAW}$$
$$- \arctan\left(\frac{v_S \cdot \sin\beta + \dot{\Psi} \cdot x_K + \dot{y}_{LOZ} + \dot{y}_{UK} + \dot{y}_{ZAW}}{v_S \cdot \cos\beta - \dot{\Psi} \cdot y_K - \dot{x}_{LOZ} - \dot{x}_{UK} - \dot{x}_{ZAW}}\right). \tag{3.6}$$

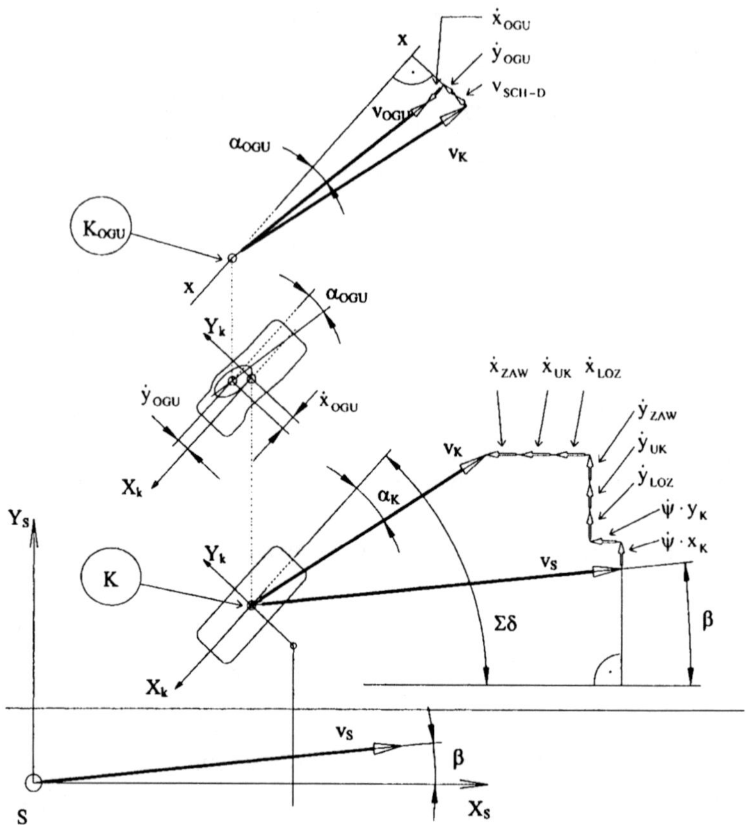

Figure 3.33. The velocity vectors of the points attached to the tire K_{OGU} and the driving wheel K.

Taking into account the *uprush* behaviour and in accordance with the Schlippe–Dietrich hypothesis one gets

$$F_y = C_{F_y/\alpha_{OGU}} \cdot \left(\alpha_{OGU} - \frac{L_L}{2} \cdot \frac{\dot{\alpha}_{OGU}}{v_K} - \frac{\dot{F}_y}{C_{y0} \cdot v_K} \right), \qquad (3.7)$$

where: $C_{F_y/\alpha_{OGU}}$ is the variable coefficient of the transversal force: $C_{F_y/\alpha_{OGU}} = \frac{dF_y}{d\alpha_{OGU}}$.

The scheme in Figure 3.33 yields

$$\alpha_{OGU} = \delta - \delta_{UK} - \delta_{LOZ} - \delta_{ZAW}$$

$$- \arctan\left(\frac{v_S \cdot \sin\beta + \dot{\Psi} \cdot x_K + \sum \dot{y}_{L,U,Z} + (v_{SCH-D} + \dot{y}_{OGU}) \cdot \cos(\sum\delta) - \dot{x}_{OGU} \cdot \sin(\sum\delta)}{v_S \cdot \cos\beta - \dot{\Psi} \cdot y_K - \sum \dot{x}_{L,U,Z} - (v_{SCH-D} + \dot{y}_{OGU}) \cdot \sin(\sum\delta) - \dot{x}_{OGU} \cdot \cos(\sum\delta)} \right),$$

where the velocity v_{SCH-D} is defined by (3.5), and

$$\sum \dot{y}_{L,U,Z} = \dot{y}_{LUZ} + \dot{y}_{UK} + \dot{y}_{ZAW}; \quad \sum \dot{x}_{L,U,Z} = \dot{x}_{LUZ} + \dot{x}_{UK} + \dot{x}_{ZAW}.$$

Observe that two definitions of the slip angle are introduced via the formulas (3.4) and (3.6). The first one is called *the static slip angle*, whereas the second one is called the *dynamic slip angle*. The latter model will be further widely used in our numerical simulations.

5. Introduction to Stability of a Moving Car

A car dynamics is separated into subdynamics, and then these subdynamics are matched together using various approaches. In general the matching is obvious for linear dynamical subsystems, since a superposition rule is applied. However, even if subsystems are non-linear one may apply the so-called asymptotical matching.

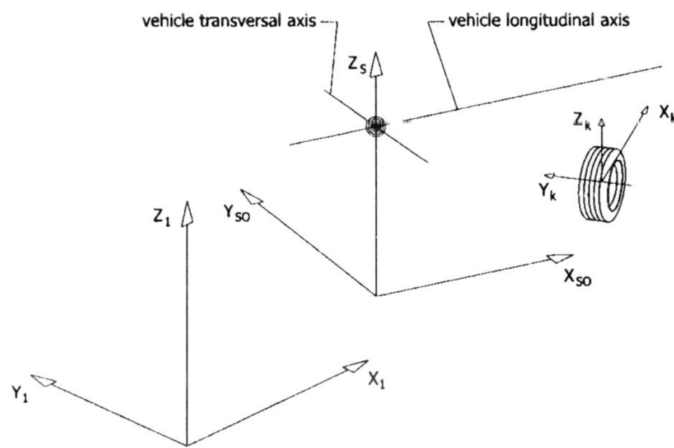

Figure 3.34. Three coordinate systems: inertial $X_1Y_1Z_1$, fixed with a car system $X_{S0}Y_{S0}Z_{S0}$, and fixed with the road wheel system $Y_kZ_kZ_k$.

In order to carry out further analysis the following coordinate systems are introduced (see Figure 3.34):

(i) the inertial system $X_1Y_1Z_1$,

(ii) the fixed with a car $X_{S0}Y_{S0}Z_{S0}$ system (Z_S is perpendicular to a road surface and it includes the car mass centre; X_{S0} lies on the road plane and is to (idealized) longitudinal car axle; Y_{S0} lies in the road plane and is to (idealized) transversal car axle),

(iii) the fixed (with a road wheel) system $X_kZ_kZ_k$ (X_k lies in (idealized) wheel plane and is parallel to the road plane; Z_k intersects the rotation wheel axle, lies in (idealized) wheel plane and is perpendicular to X_k).

Table 3.5. Various vehicle subdynamics.

1	***Longitudinal dynamics***	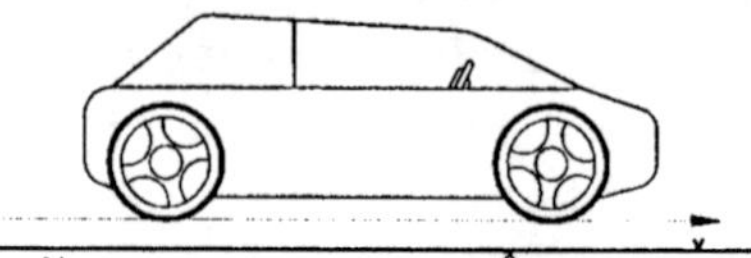	*Driving and braking process*
2	***Lateral dynamics "Yaw"***	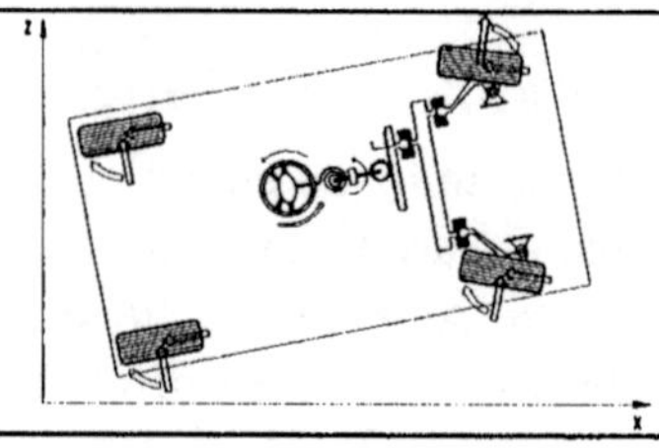	*Driving on a turn, non-symmetric drive, non-symmetric braking*
3	*Vertical dynamics*	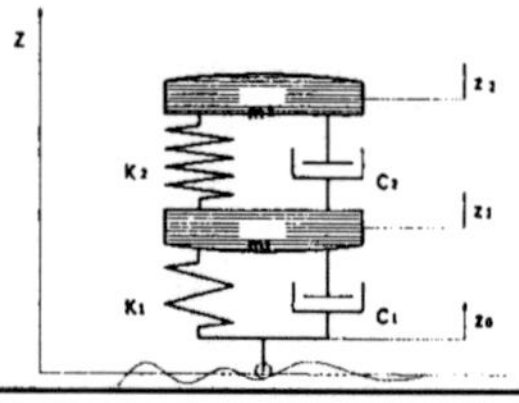	*Rough road surface*
4	*Vertical tilt dynamics "Roll"*	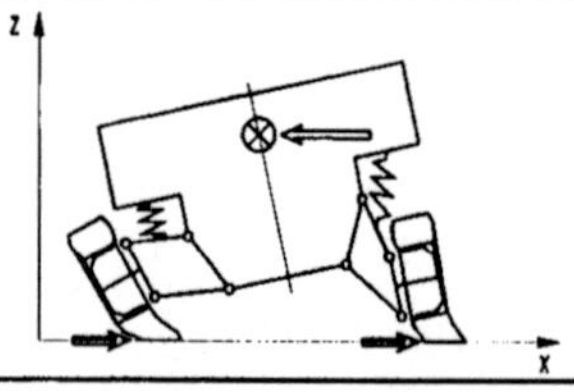	*Driving on a curve and non-smooth surface*
5	*Longitudinal tilt dynamics "Pitch"*	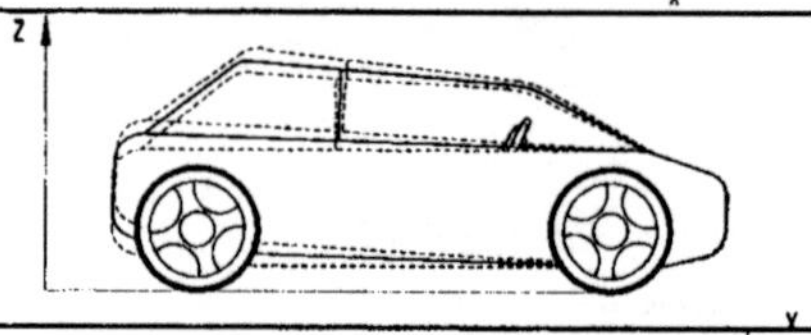	*Drive, braking, negotiate gradients*
6	*Dynamics of road wheel rotations*		*Drive, braking, steering*

In Table 3.5 various subdynamics are included which serve as a basis for drawing general conclusions on dynamics of a wheeled car, and particularly its stability.

Dynamics 1–6 depend on constraints between road surface and tired wheel, and on constraints between tired wheel and car body. Further analysis is devoted to the particular dynamics included in Table 3.5.

Chapter 4

LONGITUDINAL DYNAMICS

Two cases of longitudinal dynamics are studied. The first case represents a general approach devoted to a two axle car, which is braked with and without ABS. The second case is focused on analysis of a truck-tractor with semi-trailer (tank) dynamics filled by a liquid.

1. Longitudinal Dynamics of Two Axle Car

The longitudinal dynamics of a two axle car depends significantly on circumferential horizontal forces on the stick of a tire and a road surface. Assuming that a suspension is stiff the associated forces and moments are shown in Figure 4.1.

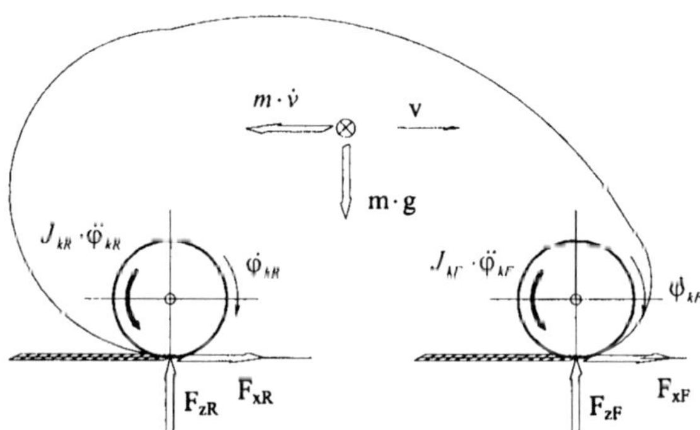

Figure 4.1. External forces associated with longitudinal dynamics of two axle car ($J_{kR,kF}$ - inertial moment of rotating bodies connected with a road wheel, $\varphi_{kF,R}$ - angle of wheel road rotation).

The governing differential equations read

$$\begin{array}{ll} m \cdot \dot{v} - F_{xR} - F_{xF} = 0; & m \cdot g - F_{zR} - F_{zF} = 0 \\ m \cdot \dot{v} \cdot h - m \cdot g \cdot b - F_{zF} \cdot L + J_{kR} \cdot \ddot{\varphi}_R + J_{kF} \cdot \ddot{\varphi}_F = 0 \end{array}. \tag{4.1}$$

The frictional forces F_{xF} and F_{xR} depend on the driven (braking) moments subjected to rotating inertial bodies associated with road wheels and on a developed friction between the road wheel and a road.

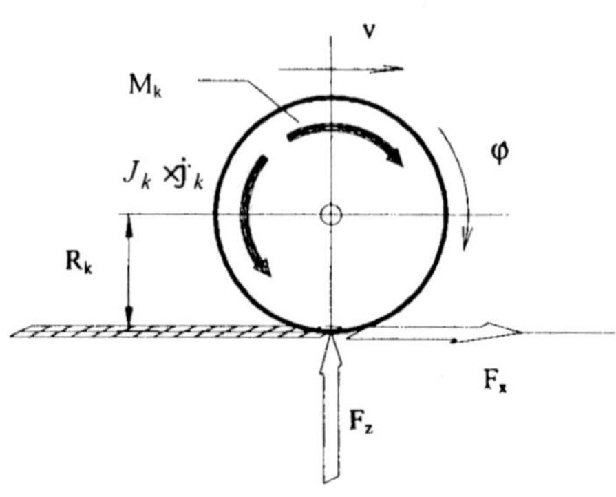

Figure 4.2. External forces acting on a road wheel.

In Figure 4.2 the external forces acting on a road wheel are shown. If a very small deviation of the vertical force F_z can be neglected, then the equations of motion of road wheels of front and rear axles have the form

$$F_{xF} \cdot R_{dF} + J_{kF} \cdot \ddot{\varphi}_{kF} - M_{kF} = 0; \quad F_{xR} \cdot R_{dR} + J_{kR} \cdot \ddot{\varphi}_{kR} - M_{kR} = 0. \tag{4.2}$$

During a drive (brake) of road tired wheels a slip occurs. The coefficients of a relative circumferential slip are defined in the following way (see Section 3):

$$\begin{array}{ll} s_{xF} = 1 - \left(\dfrac{\dot{\varphi}_{kF} \cdot R_{kF}}{v} \right)^{n_F}; & \text{where} \quad n_F = \begin{cases} +1 : \dot{\varphi}_{kF} \cdot R_{kF} \leqslant v \\ -1 : \dot{\varphi}_{kF} \cdot R_{kF} > v \end{cases} \\ s_{xR} = 1 - \left(\dfrac{\dot{\varphi}_{kR} \cdot R_{kR}}{v} \right)^{n_R}; & \text{where} \quad n_R = \begin{cases} +1 : \dot{\varphi}_{kR} \cdot R_{kR} \leqslant v \\ -1 : \dot{\varphi}_{kR} \cdot R_{kR} > v \end{cases} \end{array}. \tag{4.3}$$

The adhesion frictional force F_x depends on the force F_z pressing a wheel to a road surface and on the so-called *circumferential adhesion coefficients*

$$F_{xF} = \mu_{xF} \cdot F_{zF}; \qquad F_{xR} = \mu_{xR} \cdot F_{zR}. \tag{4.4}$$

The mentioned coefficient μ_x characterizing properties of a frictional road wheel - road pair depends on the circumferential slip s_x and a typical shape of this dependence is shown in Figure 4.3.

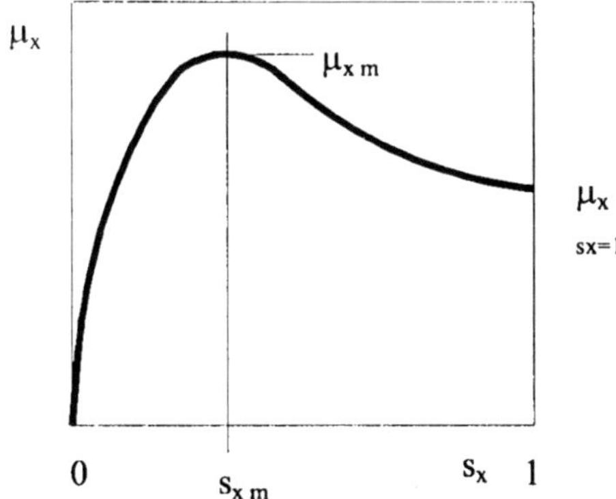

Figure 4.3. Tire characteristics for $v = const$, $F_z = const$, and so on (μ_{xm} - maximal value of μ_x; s_{xm} - relative slip for μ_{xm}; μ_{x1} - slip coefficient for $s_x = 1$, i.e. for a blocked road wheel).

EXAMPLE 4.1 *Consider dynamics of a braked car (Figure 4.1). The rotational moments (see Fig. 4.2 and equation 4.2) have the form*

$$M_{kF} = -M_{HF}; \quad M_{kR} = -M_{HR}, \tag{4.5}$$

where $M_{HF,HR}$ are the rotational moments of interactions of braking devices of the road wheels. From a physical point of view, they are friction moments, and hence (for a blocked road wheel) their values do not depend on a wheel-cylinder brake pressure, but on reaction of force road surface and wheel interaction. The latter is characterized in the following symbolic way

$$M_{HF} = \begin{cases} C_{HF} \cdot p_F \Leftrightarrow \ddot{\varphi}_{kF} \neq 0 \\ F_{xF} \cdot R_{kF} \Leftrightarrow \ddot{\varphi}_{kF} = 0 \end{cases} \quad M_{HR} = \begin{cases} C_{HR} \cdot p_R \Leftrightarrow \ddot{\varphi}_{kR} \neq 0 \\ F_{xR} \cdot R_{kR} \Leftrightarrow \ddot{\varphi}_{kR} = 0 \end{cases}, \tag{4.6}$$

where: $C_{HF,HR}$ are constant coefficients of braking mechanisms of the front and rear wheels, respectively; $p_{F,R}$ is the pressure of a brake working medium.

In this example dynamics of the braking process controlled via an anti-blocked system is studied.

Action algorithm of ABS 6610

In order to introduce a reader to the ABS action, a device constructed at the Institute of Vehicles of Technical University of Lodz is briefly described [104] (see Figure 4.4).

One may conclude from the scheme presented in Figure 4.4, that ABS 6610 applies the generated reference velocity V_{ref}, which can be interpreted as the recommended time history of the braked wheel car velocity. Keeping on this

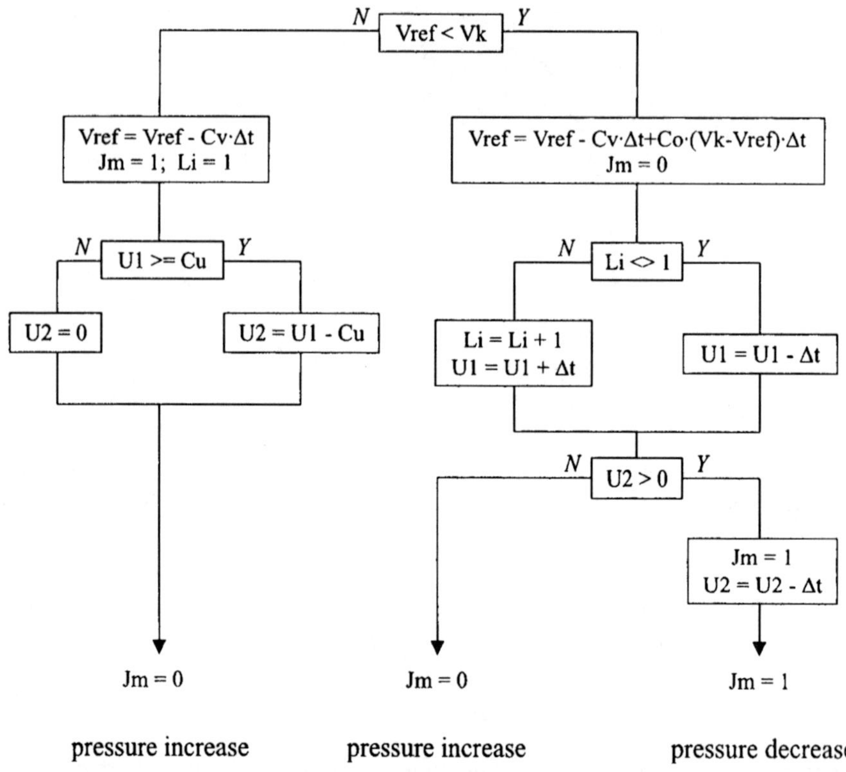

Figure 4.4. Scheme of the algorithm for a control of pressure modulator 6610 (Vk - road wheel velocity, $Vref$ - reference programming velocity, Co, Cv, Cu - constants, Δt - discrete time, Jm - current of modulator control, $U1, U2$ - operational quantities, Li - output counter values ($Li = 1, 2, \ldots$), p - pressure in wheel brake cylinder).

interpretation one may also conclude that if the wheel car velocity is smaller than the referenced velocity, then the controller should decrease the friction moment of the braking mechanism (for instance, by a pressure decrease in the brake cylinder of a given time instant), than the controller should increase frictional moment of the braking mechanism (for instance through a pressure increase in the brake cylinder of a given road wheel). In other words, the following simple rules hold:

- $V_k < V_{ref} \Rightarrow Jm = 1$ (pressure decrease),
- $V_k > V_{ref} \Rightarrow Jm = 0$ (pressure increase).

The reference velocity is programmed during the braking process and its time history depends on the road wheel braking process. It is adjusted to the current road conditions, i.e. road adhesion.

The considered ABS system consists of two subsystems located on:

(i) front axle wheels (two-stage system realizing both pressure increase and decrease);

(ii) rear axle wheels (three-stage system realizing either pressure increase or decrease or a steady state).

In the second case (ii) the pressure achieves the steady state for $V_k < V_{ref}$ and $dV_k/dt > 0$.

Modelling the frictional pair: road wheel - road surface is very important during investigation of dynamics with ABS. Two cases will be further studied, i.e. braking on either homogeneous or non-homogeneous road surfaces.

The following adhesion coefficients are taken

$$\mu_{xF} = \frac{CC1 \cdot s_{xF}}{(CC2 + s_{xF})^2}; \quad \mu_{xR} = \frac{CC1 \cdot s_{xR}}{(CC2 + s_{xR})^2}, \tag{4.7}$$

where: $CC1$ and $CC2$ are constant parameters characterizing the road surface.

It is assumed that the non-homogeneous surface road model includes two variable parameters: lengths of successive road intervals with different properties $\Delta L(\zeta)$, and μ_{max} (see Figure 4.3).

It is also assumed that the two parameters are random values, and that it is the normal random distribution typical for a real behaviour.

The adhesion coefficients are governed also by (4.7), where $CC1$ is taken as the random one and the $CC2$ is the deterministic one. The functions $\mu_x(s_x)$ read

$$\mu_{xF}(\xi) = \frac{CC1_F(\xi) \cdot s_{xF}}{(CC2 + s_{xF})^2}; \quad \mu_{xR}(\xi) = \frac{CC1_R(\xi) \cdot s_{xR}}{(CC2 + s_{xR})^2}, \tag{4.8}$$

where: $CC1_F(\xi)$ and $CC2_R(\xi)$ are random variables.

Owing to the values of the characteristic parameters of the dependence $\mu_x(s_x)$ (Figure 4.3) and analysing (4.8) one gets: $s_{xm} = CC2$, $\mu_{xm} = 0.25CC1/CC2$. Consequently, $\mu_{xm}(\xi) = 0.25CC1(\xi)/CC2$. In the relation (4.8) two parameters $CC1$ are distinguished: $CC1_F(\xi)$ associated with the front wheels, and $CC2_R(\xi)$ associated with the rear wheels. It is assumed further that these two parameters are dependent, which is motivated by the car drive. In other words, the end axial wheels are on the same road interval where the front wheels are for a given time instant after crossing L distance, i.e.

$$CC1_R(\xi) = \Phi\left(CC1_F(\xi), v, L\right),$$

where v is the car velocity.

During computer simulation the pseudorandom function generating uniquely distributed random variables is used. Any normal random distribution can be obtained using the equation

$$zm(\varsigma) = (zm_1 + zm_2 + zm_3 + ... + zm_n - 0.5 \cdot n) \cdot \sqrt{\frac{12}{n}}, \tag{4.9}$$

where: $zm_{1,2,...,n}$ are successively generated random functions. The following program is used to generate a random variable with the defined expected value and the variance:

```
losdelL

DANE
delL0 = 0.5;          //expected value
warianc = 1.1;        //variance
Lc = 20;              //road length
k:=int(Lc/delL0);     //number of road intervals

for i=1 to k
    wspL=0
        for ii=1 to 118;
            wspL:=wspL+rnd;
        next ii
    wspL:=(wspL-59)/3;
    delL(i):=delL0*(1-wspL*mm)
    if delL(i)=<1.2 and delL(i)>1.2 then L1211:=L1211+1
    if delL(i)=<1.1 and delL(i)>1   then L1110:=L1110+1
    if delL(i)=<1   and delL(i)>0.9 then L109:=L109+1
    if delL(i)=<0.9 and delL(i)>0.8 then L89:=L89+1
    if delL(i)=<0.8 and delL(i)>0.7 then L78:=L78+1
    if delL(i)=<0.7 and delL(i)>0.6 then L76:=L76+1
    if delL(i)=<0.6 and delL(i)>0.5 then L65:=L65+1
    if delL(i)=<0.5 and delL(i)>0.4 then L54:=L54+1
    if delL(i)=<0.4 and delL(i)>0.3 then L43:=L43+1
    if delL(i)=<0.3 and delL(i)>0.2 then L32:=L32+1
    if delL(i)=<0.2 and delL(i)>0.1 then L21:=L21+1
    if delL(i)=<0.1 and delL(i)>0 then L10:=L10+1
    L(12):=L1211;L(11):=L1110;L(10):=L109;L(9):=L89;
    L(8):=L78;L(7):=L76;L(6):=L65;L(5):=L54;L(4):=L43;
    L(3):=L32;L(2):=L21:L(1):=L10
    sdelL:=sdelL+delL(i)
next i

OUTPUT VALUES i, delL(i), SdelL, L(i)
```

The above procedure generates an arbitrary number (k) of road intervals with the arbitrary length (Lc). A length of each road interval is obtained from (4.9) consisting of 118 components of the variable "random" ($n = 118$). The random value of the road interval length $delL(i)$ is yielded by 118 values of "random", the expected value "delL0", and the variance "warianc". The procedure yields the step function of the density of the random variable via computation of samples size $L(1), L(2), \ldots$ in the intervals of the random variable values ("length" of the intervals is equal to $0.1m$). The procedure gives also average value of the random variable.

In Figure 4.5 computer simulations of the random value ΔL_i are shown. For the given random process parameters and for the algorithm generating ΔL_i, results close to expected ones have been obtained. The density distribution function is similar to the normal distribution. The average value differs from the expected value by an amount of 0.05, i.e. 10%.

A similar procedure is applied to generate μ_{xm}, i.e. maximal value of the adhesion frictional coefficient (see Figure 4.6). For the given random process parameters and the algorithm generating μ_{xm}, results close to the expected ones are obtained. The density distribution function is close to the normal distribution. The averaged value for 30–40 events differs from the expected value of 0.05, i.e. of 10%.

Now, gathering the successive events ΔL_i and the successive events $\mu_{xm\,i}$ the obtained pairs describe the road surface along its length L_c. In other words, the data taken from Figures 4.5 and 4.6 yields the data shown in Figure 4.7. In Figure 4.7 the random distribution $(\Delta L_i, \mu_{xm})$ characterizes the road model with the parameters: $L_c = 22m$, the expected value $E\,\mu_{xm\,i} = 0.85$, the variance $D^2\mu_{xm\,i} = 0.7$, the expected value $E\,\Delta L_i = 0.55$, and the variance $D^2\Delta L_i = 1.1$.

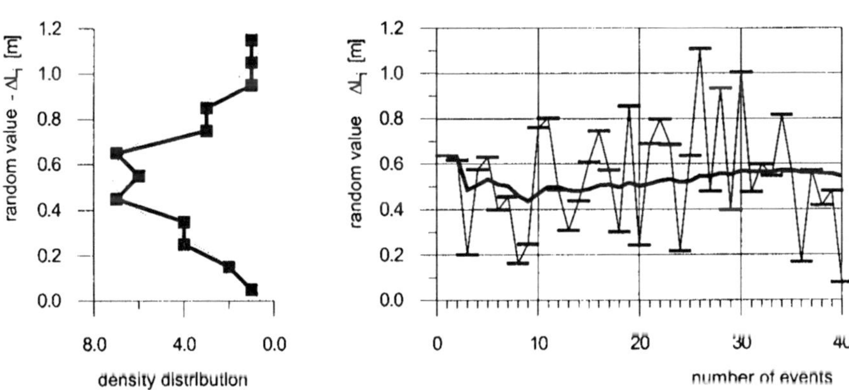

Figure 4.5. Normal distribution of road intervals length ($L_c = 22m$, $E\Delta L_i = 0.55$, $D^2\Delta L_i = 1.1$).

In what follows we consider simulations of a car braking process with ABS. The mathematical car model is governed by the equations (4.1)–(4.8).

In Figures 4.8 and 4.9 some quantities obtained via numerical simulations and characterizing braking processes of the front and rear wheels on the homo-

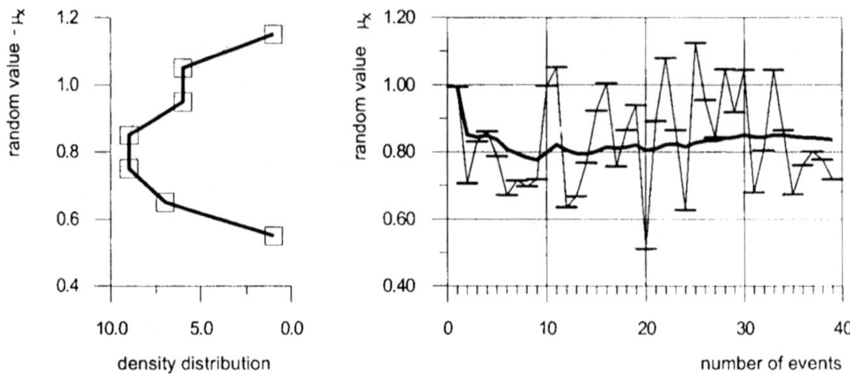

Figure 4.6. Normal distribution of maximal adhesion coefficient ($L_c = 22m$, $E\Delta L_i = 0.85$, $D^2\mu_{xm\,i} = 0.7$).

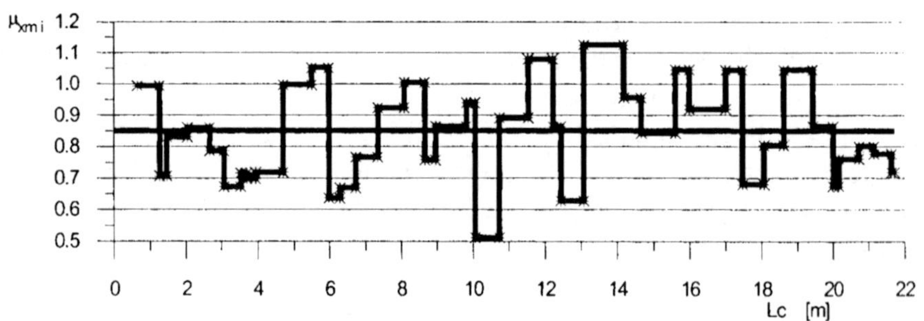

Figure 4.7. Normal distribution of the pairs (ΔL_i, $\mu_{xm\,i}$).

geneous road surface are shown. The obtained results are used next for comparison and evaluation of the distributed processes yielded by a non-homogeneous surface (see Figures 4.10 and 4.11).

Note that comparing the perturbated and unperturbated responses one may draw conclusions about car stability properties.

As a stability measure of the road wheel braking process a value of the relative circumferential slip, which occurs during control via ABS, can be used. In order to get a quantitive definition, the relative slip values are registered beginning from the ABS action and continuing for 2 seconds.

Owing to the simulation results (see Figures 4.8 and 4.11), of the controlled braking process of the road wheel moving on the non-homogeneous road surface ($E\,\mu_{xmi} = 0.85$, $D^2\mu_{xmi} = 0.7$, $E\,\Delta L_i = 0.55$, $D^2\Delta L_i = 1.1$), the following conclusions are formulated:

(i) the process is unstable for two-steps control owing to the relative slip s_x,

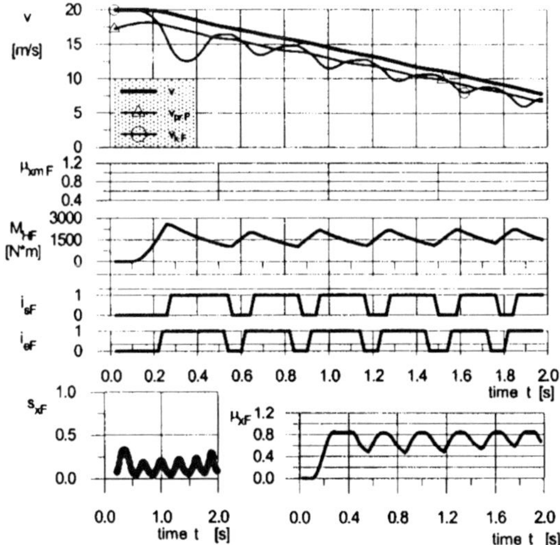

Figure 4.8. Braking process of the front road wheel on the homogeneous road surface (three-steps control via ABS 6610). The used notations: v - car velocity, v_{prF} - front wheel reference speed, v_{kF} - front wheel velocity, M_{HF} - rotary moment of braking mechanism, s_{uF}, s_{sF}, s_{eF} - controlling signals of the pressure modulator.

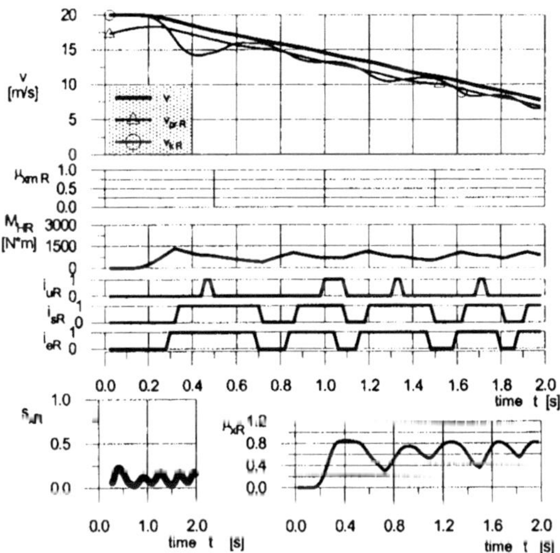

Figure 4.9. Braking process of the end road wheel on the homogeneous road surface (two-steps control via ABS 6610). The used notations: v - car velocity, v_{prF} - front wheel velocity, v_{kF} - front wheel velocity, M_{HF} - rotary moment of braking mechanism, s_{uF}, s_{sF}, s_{eF} - controlling signals of the pressure modulator.

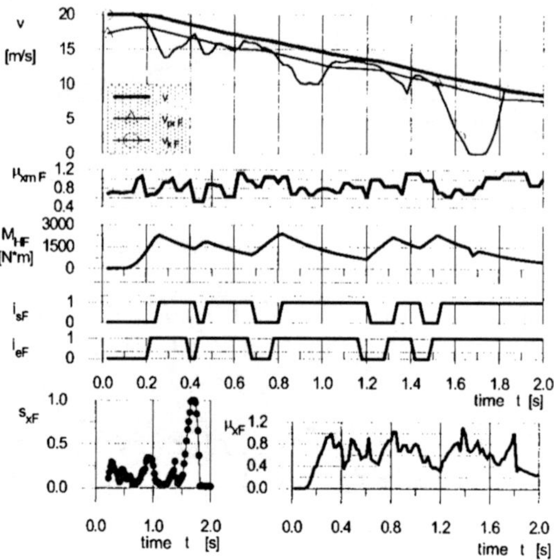

Figure 4.10. Braking process of the front road wheel on the non-homogeneous road surface (three-steps control via ABS 6610). The used notations: v - car velocity, v_{prF} - front wheel velocity, v_{kF} - front wheel velocity, M_{HF} - rotary moment of braking mechanism, s_{uF}, s_{sF}, s_{eF} - controlling signals of the pressure modulator.

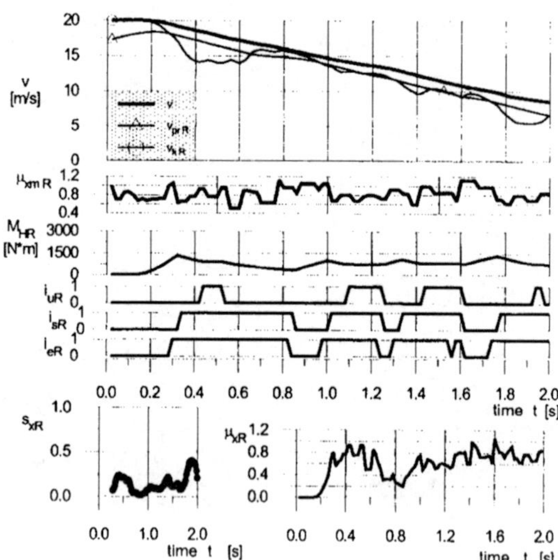

Figure 4.11. Braking process of the end road wheel on the non-homogeneous road surface (two-steps control via ABS 6610). The used notations: v - car velocity, v_{prF} - front wheel velocity, v_{kF} - front wheel velocity, M_{HF} - rotary moment of braking mechanism, s_{uF}, s_{sF}, s_{eF} - controlling signals of the pressure modulator.

(ii) the process is stable for three-steps control owing to the relative slip s_x.

2. Longitudinal Tank Vehicle Dynamics

An elastic model of fluid in a parallelepiped tank is applied (see Figure 4.12). This model consists of two parts [91]:

- first, the fluid part does not move with respect to tank walls,
- second, the fluid part is modeled as a mass-spring system.

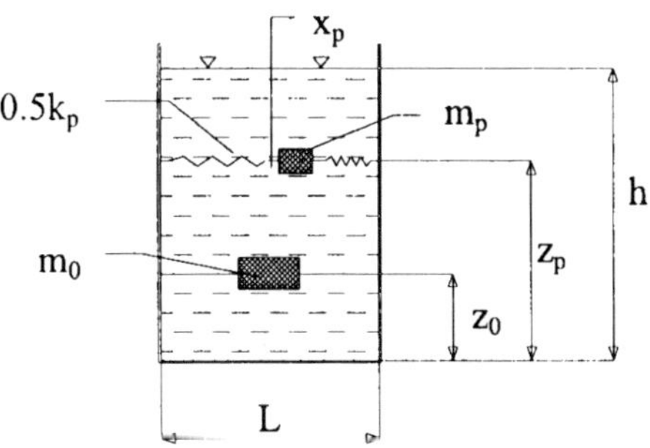

Figure 4.12. Elastic model of a liquid in a rectangular tank.

The following parameters describe the model: m_p - mass of moving liquid part, m_0 - mass of non-movable liquid part, k_p - stiffness coefficient characterizing coupling of movable and non-movable liquid parts, z_p - position of movable part, z_0 - position of non-movable liquid part. The mentioned parameters are defined via the following relations:

$$m_f = \rho \cdot h \cdot L \cdot d, \tag{4.10}$$

$$m_p = 0.258 \cdot m_f \cdot \frac{L}{h} \cdot \tanh\left(\frac{\pi \cdot h}{L}\right), \tag{4.11}$$

$$m_0 = m_f - m_p, \tag{4.12}$$

$$k_p = 0.813 \cdot m_f \cdot \frac{g}{h} \cdot \tanh\left(\frac{\pi \cdot h}{L}\right), \tag{4.13}$$

$$z_p = h - \frac{0.637}{L} \cdot \tanh\left(\frac{0.5\pi \cdot h}{L}\right), \tag{4.14}$$

$$z_0 = 0.5h + (0.5h - z_p) \cdot \frac{m_p}{m_0}, \tag{4.15}$$

where: ρ is the liquid density, h is the liquid level height, L is the tank length, b - is the tank width and m_f is the liquid mass.

In Figures 4.13 and 4.14 for the data $b = 2m$, $h = 1.8m$, $L_c = 10m$, $\rho = 850kg/m^3$ the values of stiffness coefficient k_p and movable part position h are shown.

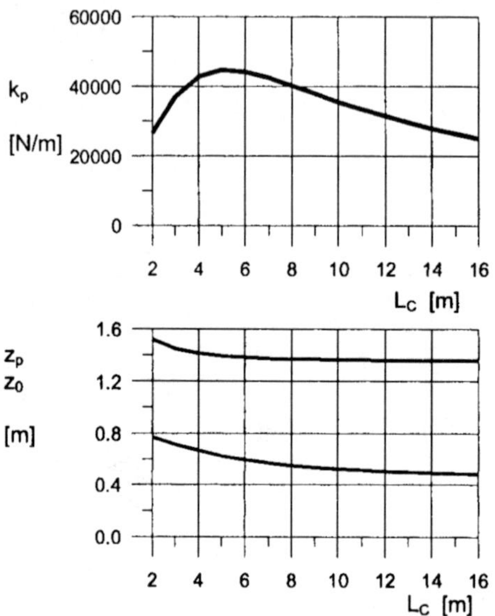

Figure 4.13. Stiffness coefficients (k_p) and position of movable part (z_p) versus L_c.

In the next step an elastic model of free liquid in a cylindrical tank is analysed [91].

In accordance with a Figure 4.15 the liquid is divided into n elements having equal width b and equal length L, and having various heights $h(k)$, which are defined by the following relations:

$$y(k) = (k-1) \cdot b; \quad z(k) = R\sqrt{1 - \left(\frac{k-1}{n}\right)^2}; \quad h(k) = z(k) + (h - R), \tag{4.16}$$

where: R denotes the tank radius, $b = R/n$, $k = 1, 2, \ldots, n$.

Exemplary parameters of liquid partition of a cylindrical tank are shown in Figures 4.16 and 4.17.

Since a liquid contained in the cylindrical tank is divided into k elements, then the cylindrical tank model can be constructed (Figure 4.18).

The cylindrical tank with length L and radius R is filled with liquid up to height h. Non-movable liquid parts are situated at a distance L_0 from the truck. The movable parts are situated at a distance $(L_0 - x_p(k))$ from the truck and in height $z_p(k)$ from the tank bottom.

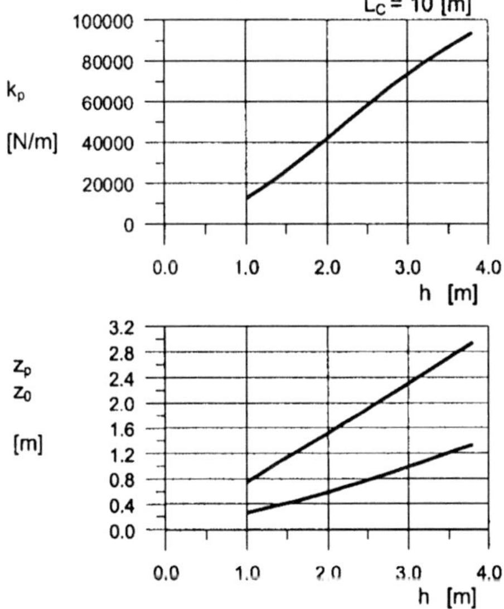

Figure 4.14. Stiffness coefficients (k_p) and position of movable part (z_p) versus h.

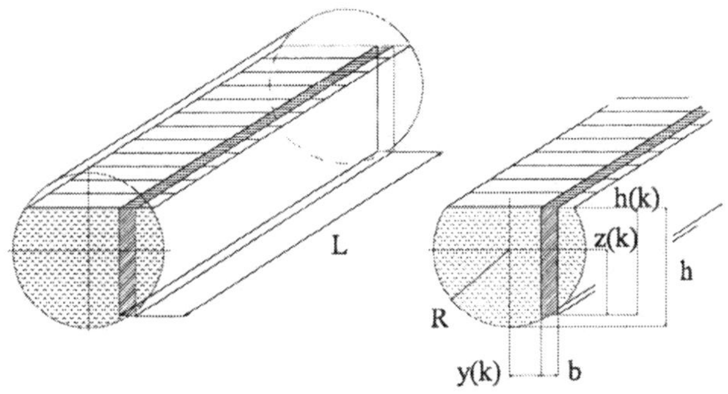

Figure 4.15. Liquid modelling.

Observe that the moving liquid part having the mass $m_p(k)$ undergoes an action of two constraints associated with elasticity $k_p(k)$ and damping $c_p(k)$ (see Figure 4.19).

The equations of motion of the k-th element read

$$m_p(k) \cdot (\ddot{x}_p(k) - a_x) + c_p(k) \cdot \dot{x}_p(k) + k_p(k) \cdot x_p(k) = 0, \qquad (4.17)$$

where: $x_p(k)$ describes displacement (see Figure 4.18).

Taking into account the masses of both the tank construction and of the load-carrying ability of the semi-trailer m_N, and its position h_N and a_N, then the

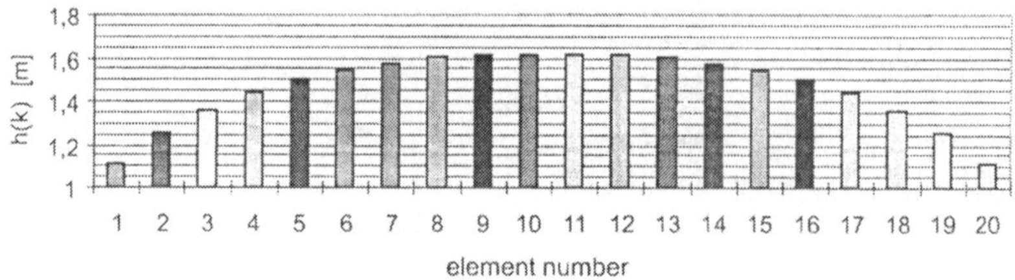

Figure 4.16. Heights of liquid elements obtained from (4.16).

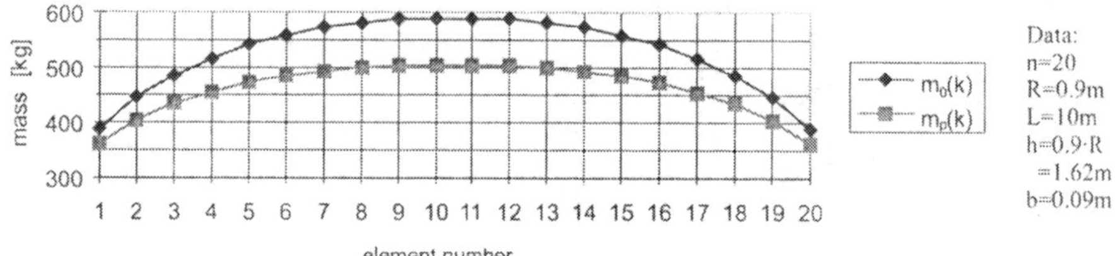

Figure 4.17. Masses of liquid elements obtained from (4.11) and (4.12).

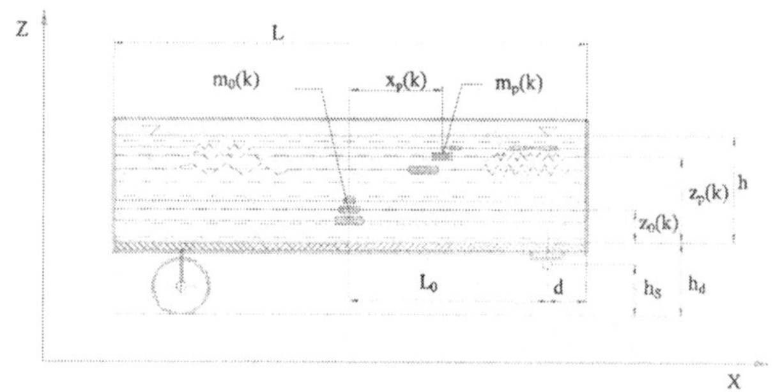

Figure 4.18. Model of free liquid contained in the cylindrical tank.

mechanical model consists of the vectors shown in Figure 4.20. The equations governing dynamics of the braked semi-trailer with the tank filled by liquid read

$$F_{xN} + F_{xS} - m_N \cdot a_x - \sum_{k=1}^{n} [m_0(k)] \cdot a_x - \sum_{k=1}^{n} [m_p(k) \cdot (a_x - \ddot{x}_p(k))] = 0, \tag{4.18}$$

$$F_{zN} + F_{zS} - m_N \cdot g - \sum_{k=1}^{n} [m_0(k)] \cdot g - \sum_{k=1}^{n} [m_p(k)] \cdot g = 0, \tag{4.19}$$

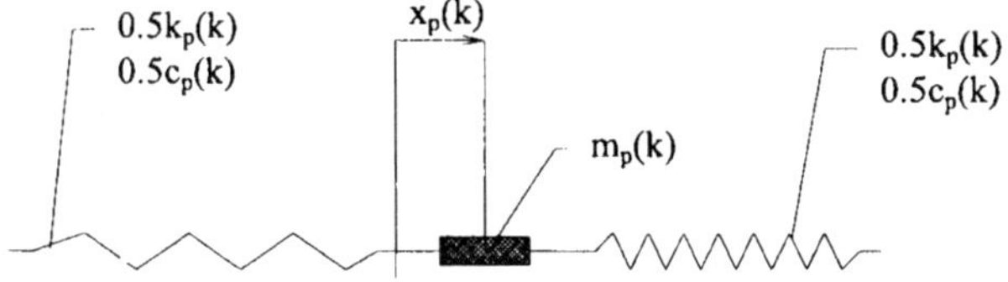

Figure 4.19. Constraints of the movable liquid part.

$$F_{zN} \cdot L_N + F_{xN} \cdot h_S + m_N \cdot a_x \cdot (h_N - h_S) - m_N \cdot g \cdot a_N - \sum_{k=1}^{n} [m_0(k)] \cdot g \cdot L_0$$

$$+ \sum_{k=1}^{n} [(m_0(k) \cdot (h_d - h_S + z_0(k)))] \cdot a_x - \sum_{k=1}^{n} [m_p(k) \cdot (L_0 - x_p(k))] \cdot g$$

$$+ \sum_{k=1}^{n} [(m_p(k) \cdot (h_d - h_S + z_p(k))) \cdot (a_x - \ddot{x}_p(k))] = 0 \,. \tag{4.20}$$

The model of the truck-tractor is shown in Figure 4.21. Its equations of motion have the following form:

$$F_{x1} + F_{x2} - F_{xS} - m \cdot a_x = 0, \tag{4.21}$$

$$F_{z1} + F_{z2} - F_{zS} - m \cdot g = 0, \tag{4.22}$$

$$F_{z1} \cdot L_S - F_{z2} \cdot (L_C - L_S) - F_{x1} \cdot h_S - F_{x2} \cdot h_S$$

$$+ m \cdot a_x \cdot (h_S - h_C) - m \cdot g \cdot (L_S - a_C) = 0 \,. \tag{4.23}$$

EXAMPLE 4.2 *In this example the longitudinal dynamics of the truck tractor with a tank is analysed. The steps of ABS application discussed in section 1 are followed. Taking into account the semi-trailer and tank models, ABS model [4], as well as the models of braking mechanisms of the tractor and the tractor road wheels, the computational model is obtained. The tank is filled by the liquid up to certain height. The tired wheel-road surface friction pair is approximated by Wagner's model (see Section 3).*

The relative friction coefficient F_x/F_z depends on the relative slip of the wheel, on the driving velocity, etc. It is assumed that the car moves on the road surface characterized by small adhesion (see Figure 4.22).

In Figure 4.23 numerical results are shown for the data from Table 4.1.

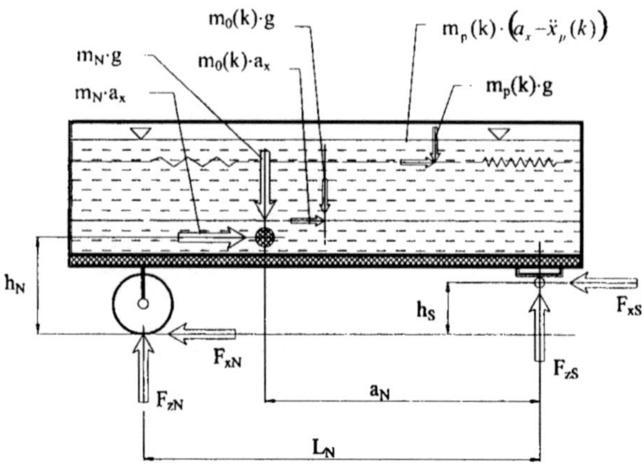

Figure 4.20. Longitudinal dynamics of a tank with the force acting on the braked tractor.

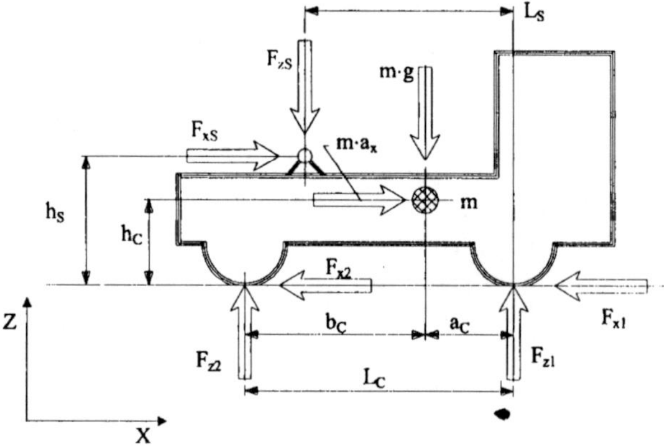

Figure 4.21. Longitudinal dynamics of the braked semi-trailer.

The following conclusions are formulated owing to the carried out analysis of a truck-tractor with a semi-trailer (cistern) filled to 90% by a liquid:

(i) a car deceleration changes widely in the interval: $1\,(1.5) \div 3\,(3.5)\,m/s^2$;

(ii) liquid movement possesses large amplitudes: $2m$ for accelerations $-3 \div +3\,m/s^2$ and frequency $0.29Hz$;

(iii) liquid movement is undamped;

(iv) loads of tractor end wheels and semi-trailer wheels are changed in the intervals of $(120000N \div 140000N)$ and of $(130000 \div 150000N)$, respectively.

Table 4.1. Data for tractor with semi-trailer (cistern).

Quantity, parameter	*Symbol*	*Units*	*Value*
Cistern radius	R	m	0.9
Cistern length	L	m	10
Liquid height $h = 0.9 \cdot 2 \cdot R$	h	m	1.62
Number of partition elements	n	-	20
Element width $b = 2R/n$	b	m	0.09
Height of mass centre position	h_C	m	1
Height position of fifth wheel	h_S	m	1.3
Longitudinal tractor mass centre position distance to front axle	a_C	m	1.6
Longitudinal tractor mass centre position distance to rear axle	b_C	m	2
Tractor wheelbase	L_C	m	3.6
Longitudinal position of non-movable liquid centre	L_S	m	3
Longitudinal position of semi-trailer mass centre	L_0	m	4
Distance fifth wheel to semi-trailer real axle	L_N	m	8
Height of semi-trailer mass centre	h_N	m	1.2
Height position of tank bottom	h_d	m	1.3
Longitudinal position of semi-trailer mass centre	a_N	m	6
Longitudinal position of the fifth wheel	d	m	1
Tractor mass	m	kg	5000
Semi-trailer mass	m_N	kg	7000
Liquid density	ρ	kg/m^3	750

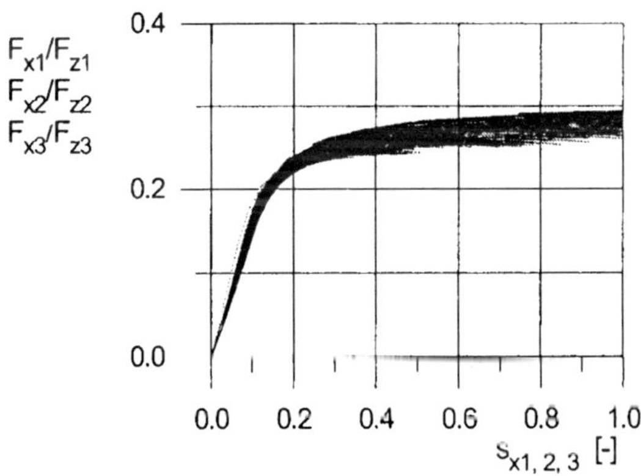

Figure 4.22. Relative friction force F_x/F_z versus relative road wheel slip in the car velocity interval $20 : 0\, m/s$ for Wagner's tire model

Oscillations of vertical load values of road wheels play an important role for stability of its lateral motion. Therefore, the studied longitudinal motion of our car system can be considered as a dangerous one owing to liquid oscillations. To

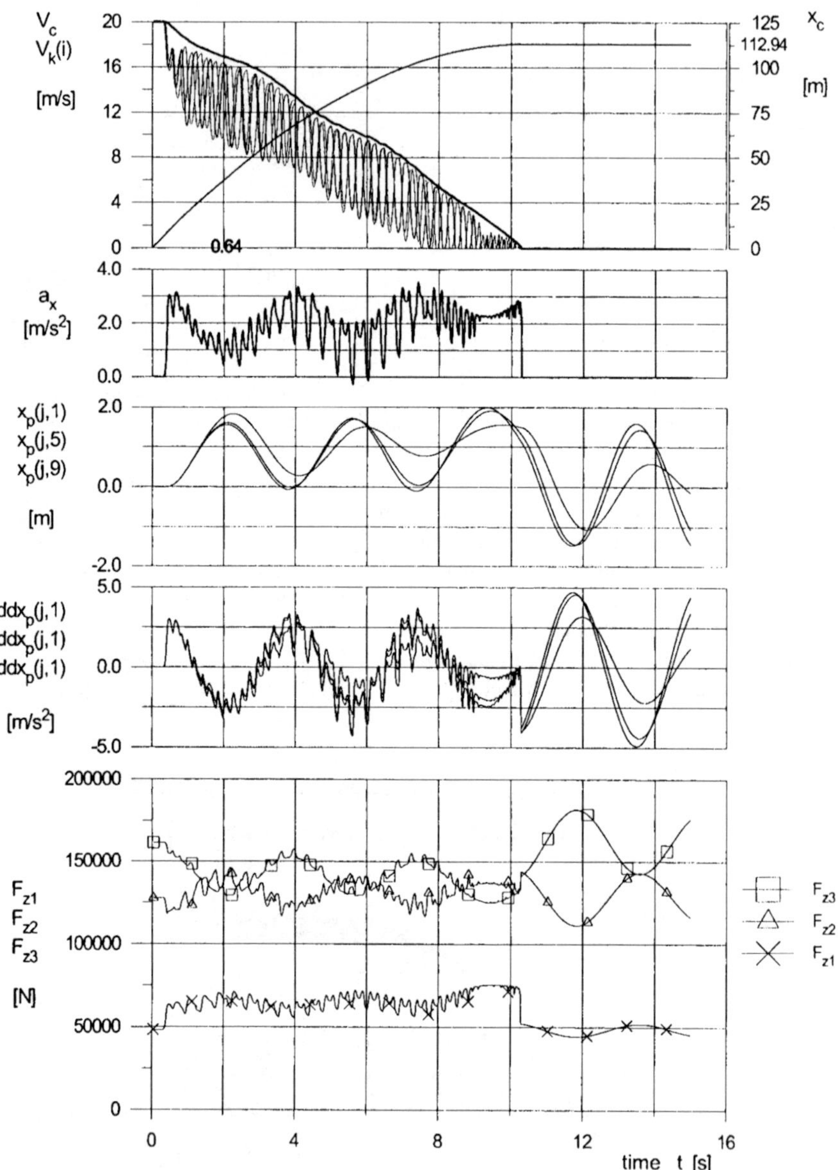

Figure 4.23. Time histories of kinematic and dynamic quantities during braking of the truck-tractor with the semi-trailer-cistern and with ABS (V_c - car velocity, $V_k(i)$ - wheels velocities, a_x - car deceleration, $x_p(j,k)$ - displacements of 1, 3 and 9 liquid slice, $ddx_p(j,k)$ - acceleration of 1, 3 and 9 liquid slice, F_{z1}, F_{z2}, F_{z3} - vertical forces of the road action on the wheels).

avoid these problems long containers are internally separated by the bulkheads. The latter ones significantly change the dynamics. In Figure 4.24 the parameters associated with the system separation into "m" parts are displayed. The non-movable and movable masses of the j-th cistern part and of the k-th slice are denoted by $m_p(j, k)$, respectively. The partition into slices can be easily introduced in a way described earlier for non-bulkhead containers.

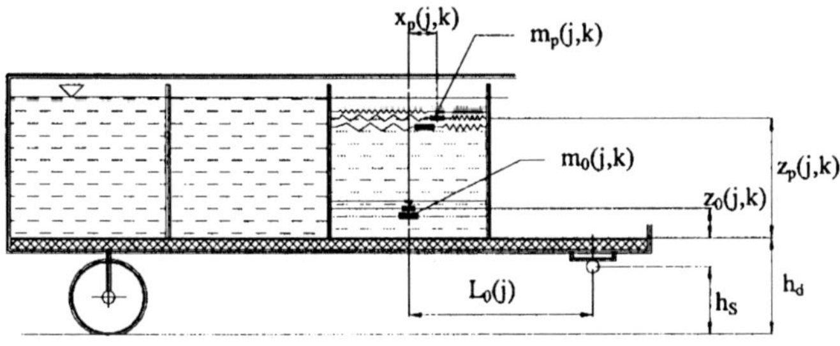

Figure 4.24. Tank divided into "m" parts.

The positions of non-movable liquid parts are governed by the relations (see Figure 4.24)

$$L_0(j) = 0.5L_c(j) + L_c(j+1) + L_c(j+2) + \ldots + L_c(m) - b\,.$$

The equations of motion of the tractor-truck with the semi-trailer-cistern partioned by bulkheads into "m" parts read

$$F_{x1} + F_{x2} - F_{xS} - m \cdot a_x = 0, \tag{4.24}$$

$$F_{z1} + F_{z2} - F_{zS} - m \cdot g = 0, \tag{4.25}$$

$$\begin{aligned} F_{z1} \cdot L_S - F_{z2} \cdot (L_C - L_S) - F_{x1} \cdot h_S - F_{x2} \cdot h_S \\ + m \cdot a_x \cdot (h_S - h_C) - m \cdot g \cdot (L_S - a_C) = 0, \end{aligned} \tag{4.26}$$

$$m_N \cdot a_x \cdot (h_N - h_S) - m_N \cdot g \cdot a_N - \sum_{j=1}^{m} \sum_{k=1}^{n} [m_0(j,k) \cdot L_0(j)] \cdot g +$$

$$F_{z3} \cdot L_N - \sum_{j=1}^{m} \sum_{k=1}^{n} [m_p(j,k) \cdot (L_0(j) - x_p(j,k))] \cdot g$$

$$+ F_{x3} \cdot h_S + \sum_{j=1}^{m} \sum_{k=1}^{n} [(m_0(j,k) \cdot (h_d - h_S + z_0(j,k)))] \cdot a_x$$

$$+ \sum_{j=1}^{m} \sum_{k=1}^{n} [(m_p(j,k) \cdot (h_d - h_S + z_p(j,k))) \cdot (a_x - \ddot{x}_p(j,k))] = 0\,, \tag{4.27}$$

Table 4.2. Data if the cistern is divided into four equal parts.

Quantity, parameter	*Symbol*	*Dimension*	*Value*
Number of partitions	j	-	4
Length of the partitioned container	$L_c(j)$	[m]	
$L_c(1) = L_c(2) = L_c(3) = L_c(4)$			2.5

$$F_{x3} + F_{xS} - m_N \cdot a_x - \sum_{j=1}^{m} \sum_{k=1}^{n} [m_0(j,k)] \cdot a_x$$

$$-\sum_{j=1}^{m} \sum_{k=1}^{n} [m_p(j,k) \cdot (a_x - \ddot{x}_p(j,k))] = 0\,, \tag{4.28}$$

$$F_{z3} + F_{zS} - m_N \cdot g - \sum_{j=1}^{m} \sum_{k=1}^{n} [m_0(j,k)] \cdot g - \sum_{j=1}^{m} \sum_{k=1}^{n} [m_p(j,k)] \cdot g = 0\,, \tag{4.29}$$

$$m_p(j,k) \cdot (\ddot{x}_p(j,k) - a_x) + c_p(j,k) \cdot \dot{x}_p(j,k) + k_p(j,k) \cdot x_p(j,k) = 0. \tag{4.30}$$

The numerical simulation results of the cistern motion for the same parameters as in the last example, but divided into 4 parts (see Table 4.2), are displayed in Figure 4.25.

The numerical simulations (Figure 4.25) are carried out for the tractor-truck with the semi-trailer-cistern separated into 4 equal parts. The cylindrical cistern has diameter $1.8m$ and length $10m$. It is filled with liquid to 90% of its height and it has mass $750\,kg$. The conclusions follow:

(i) car system deceleration is characterized by fewer oscillations in comparison to a non-bulkhead container;

(ii) the tank liquid oscillates with small amplitudes $\sim 0.2m$ with the accelerations $-3 \div +3\,m/s^2$ and the frequency $0.75\,Hz$;

(iii) oscillations of vertical values of the road wheels loads are negligible;

(iv) the liquid movement and the vertical loading of the road wheels do not vanish (they are undamped).

It should be emphasized that the car oscillations caused by the liquid motion separated by three bulkheads moves safely in comparison to our previous example. If the tank partitions were into unequal parts, then the car dynamics would be slightly changed. The latter case will be further analysed for the parameters given in Table 4.3.

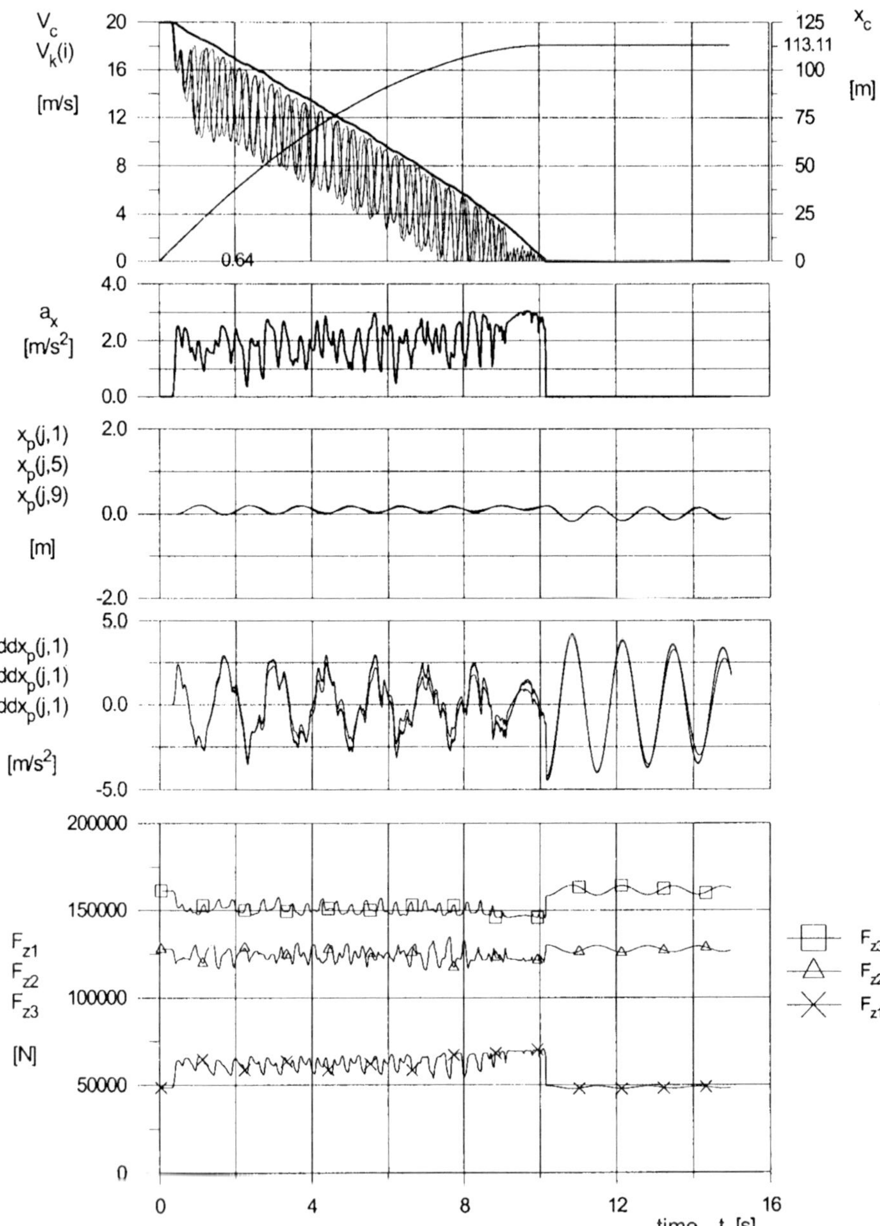

Figure 4.25. Time histories of kinematic and dynamic quantities of the braking process of the tractor-truck with the semi-trailer (cistern) and with ABS. (The used bulkheads divide the tank into four equal parts with a length of $2.5\ m$).

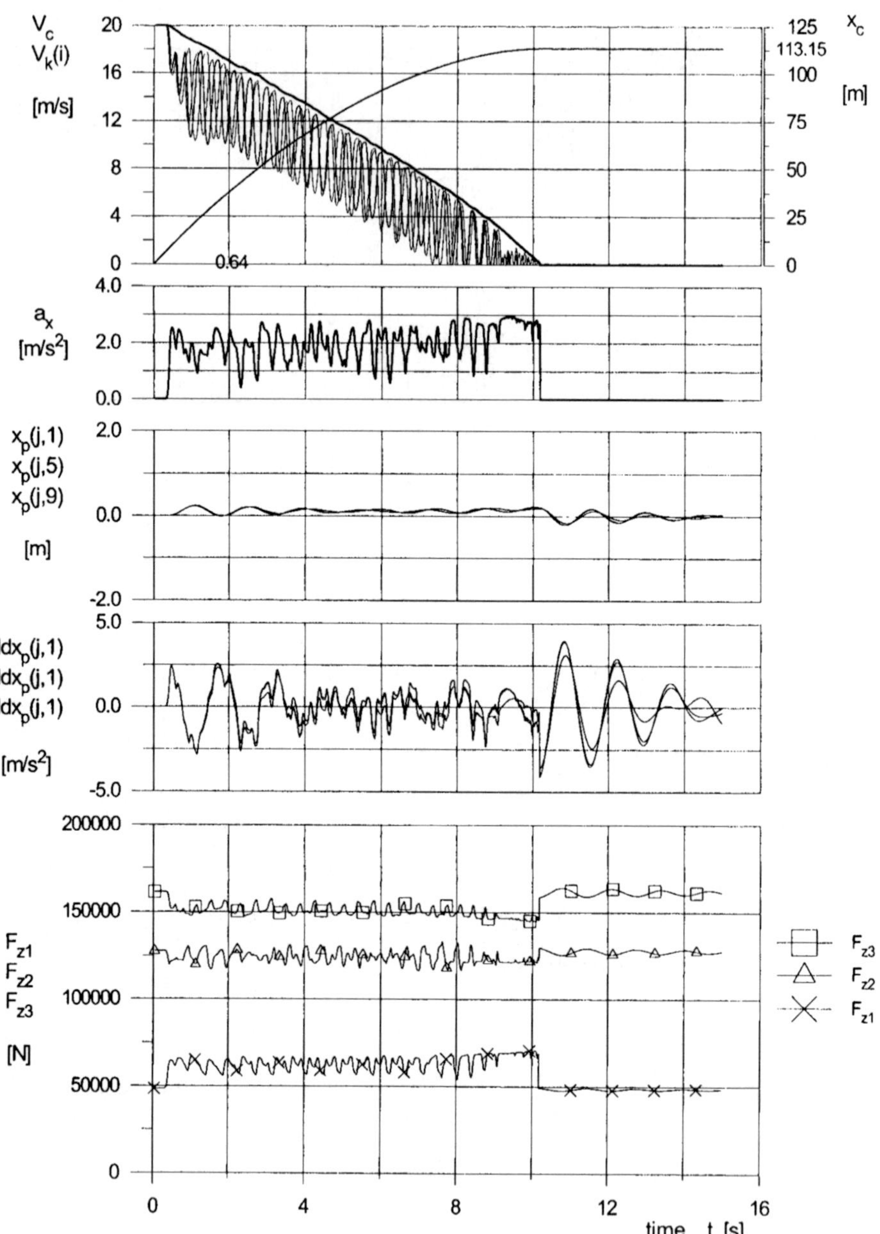

Figure 4.26. Time histories of kinematic and dynamic quantities of the braking process of the tractor-truck with the semi-trailer (cistern) and with ABS. (The used bulkheads divide the tank into four unequal parts with the lengths $2.8\,m$, $2.6\,m$, $2.4\,m$, $1.2\,m$).

Table 4.3. Data of the cistern divided into 4 unequal parts.

Quantity, parameter	*Symbol*	*Dimension*	*Quantity*
Number of container partitions	j	-	4
Length of container parts	$L_c(j)$	[m]	
$L_c(1)$			2.8
$L_c(2)$			2.6
$L_c(3)$			2.4
$L_c(4)$			2.2

The numerical simulation has been carried out for the tractor-truck with the semi-trailer (cistern) partitioned by bulkheads into four unequal parts of the cylindrical tank with the diameter $1.8\,m$, with the length $10\,m$ and filled with liquid to 90% of it's height, and the density $750\,kg/m^3$. The conclusions follow:

(i) car deceleration possesses essentially fewer oscillations in comparison to the full tank (without bulkheads);

(ii) the tank liquid oscillates with small amplitudes $\sim 0.2m$ with the accelerations $-3 \div +3\,m/s^2$ and the frequency $0.75\,Hz$;

(iii) oscillations of vertical values of the road wheels loads are negligible;

(iv) the liquid movement and the vertical loading of the road wheels vanish (they are damped).

In general, the tank partition into unequal parts yields safer motion of the analysed system.

Chapter 5

A TRANSVERSAL DYNAMICS

In this chapter two-axle and multi-elements vehicles are analysed. In particular, the stability concepts in the Lyapunov and Bogusz sense are analysed and stability conditions are derived. Special attention is focused on the dynamical system consisting of a driver and a car, and its stability is addressed. Finally, the shimmy phenomenon of a road wheel is described and stability of the car model is investigated.

1. Stability of a Two Axle Wheeled Vehicle [2, 3, 14, 53, 59, 68, 100, 103]

The so-called *effective* methods are applied to analyse stability of a two axle wheeled vehicle. The name refers to such methods which do not require a knowledge of solutions to the mathematical models, but they yield stability estimation of a *system, and object* or *a process.*

1.1 Two Axle Wheeled Vehicle Model

A stability of two axle wheeled vehicle moving on a flat surface is analysed. The scheme of the introduced model with the marked velocity vectors of the vehicle characteristic points F, S, R is shown in Figure 5.1.

Note that in general the velocity vectors do not have the directions of the longitudinal vehicle axle, since they depend on turn angles of the road wheels and on the tire drift angles.

In Figure 5.2 the kinematics of our model is shown. The front (respectively, rear) road wheel is turned through the angle δ_F (δ_R). The mass centre moves with the velocity v_S, and $v_S = O_{in}S \cdot \dot{\psi}$, where O_{in} is the instantaneous rotation centre, and $\dot{\psi}$ is the vehicle rotational velocity.

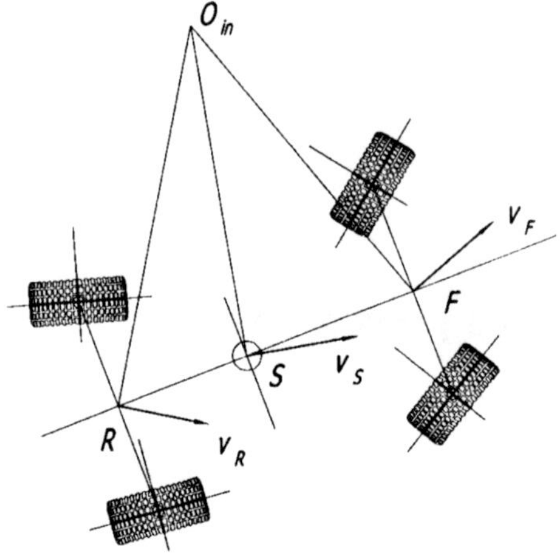

Figure 5.1. Model of two axle vehicle 4WS. (v_F, v_S and v_R correspond to velocity centers of the front, mass centre and rear axles, respectively.)

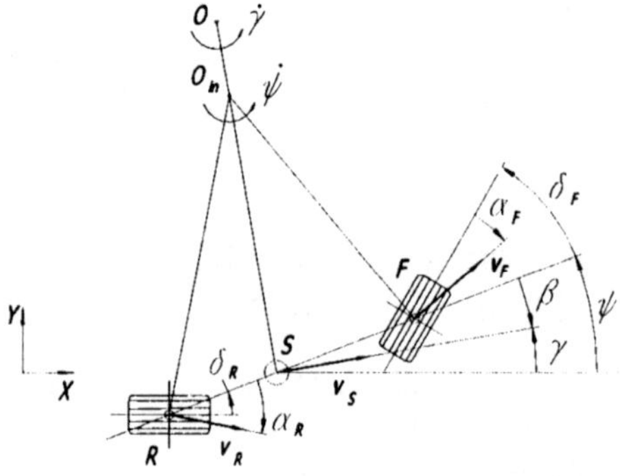

Figure 5.2. Kinematics of the model.

The position of the vector v_S with respect to the longitudinal vehicle axle is defined by the *side slip angle* β. The angle position of the vehicle longitudinal axle RF with respect to an unmovable coordinate system is defined by the rotation angle yaw angle ψ. The angle γ defines a position of the velocity vector v_S, and the following relation holds:

$$\gamma = \psi - \beta \, . \tag{5.1}$$

The path traced by the point S defines the radius $\rho = OS$, and $\rho = v/\dot{\gamma}$, where $\dot{\gamma}$ is the rotational velocity of the vector v revolution around O.

Now we are going to estimate the front road wheel slip δ_F. From the sketches shown in Figure 5.3 one may derive the following relation:

$$y = v_S \cdot \sin(\beta), \quad x = v_S \cdot \cos(\beta), \quad \alpha' = \arctan\left(\frac{\dot{\psi} \cdot a - y}{x}\right).$$

The front road wheel slip angle that we seen reads

$$\alpha_F = \delta_F - \arctan\left(\frac{\dot{\psi} \cdot a - v_S \cdot \sin(\beta)}{v_S \cdot \cos(\beta)}\right).$$

For small $(0.1 - 0.2rad)$ δ, α and β one can approximate $\sin(\cdot) = (\cdot)$, and $\cos(\cdot) = 1$, and we obtain the relation

$$\alpha_F = \delta_F + \beta - \frac{\dot{\psi} \cdot a}{v_S}. \tag{5.2}$$

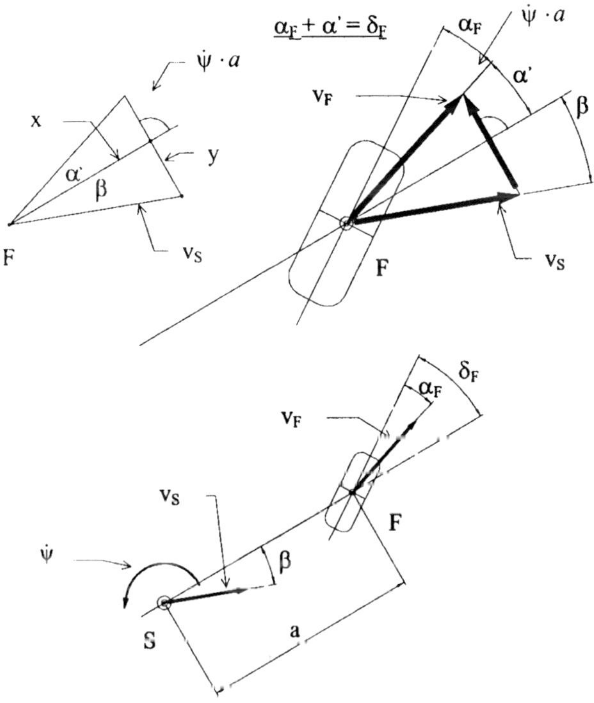

Figure 5.3. Kinematics of the front road wheel.

In a similar way, in the case of the rear road wheel slip angle δ_R, one gets (see Figure 5.4):

$$y = v_S \cdot \sin(\beta), \quad x = v_S \cdot \cos(\beta), \quad \alpha'' = \arctan\left(\frac{\dot{\psi} \cdot b + y}{x}\right),$$

$$\alpha_R = -\delta_R - \arctan\left(\frac{\dot{\psi} \cdot b - v_S \cdot \sin(\beta)}{v_S \cdot \cos(\beta)}\right),$$

and finally

$$\alpha_R = -\delta_R + \beta + \frac{\dot{\psi} \cdot b}{v_S}. \tag{5.3}$$

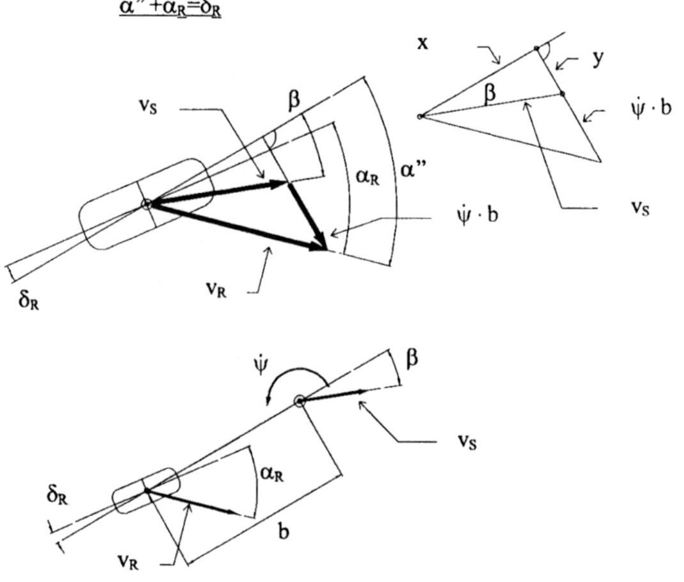

Figure 5.4. Kinematics of the rear road wheel.

In our next step dynamics of the vehicle model is considered. It is assumed, that the vehicle is subjected to the torque M_Z, which is created either by a wind or by a road roughness. Since the vehicle moves on a curve, both centrifugal F_ρ and inertial torque $J_Z \cdot \ddot{\psi}$ act on it. The road wheels are subjected to action of the lateral forces F_{yF} and F_{yR} lying in the plane XY. For small angle (see Figure 5.5) $\sum F = F_\rho - F_{yF} - F_{yR} \equiv 0$, where: $F_\rho = m \cdot v_S \cdot \dot{\gamma}$. Assuming that

$$F_{yF} = k_F \cdot \delta_F \quad \text{and} \quad F_{yR} = k_R \cdot \delta_R,$$

where: k_F and k_R are the resistance coefficients of the road wheel against the lateral slip, one obtains

$$\sum F = m \cdot v \cdot \dot{\gamma} - k_F \cdot \delta_F - k_R \cdot \delta_R \equiv 0 \, .$$

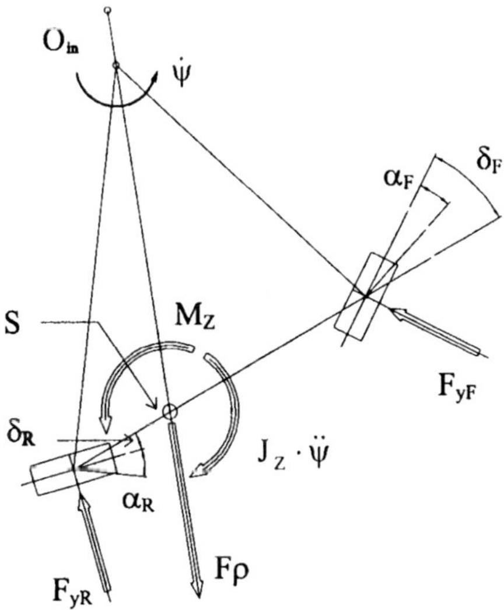

Figure 5.5. Dynamics of the vehicle model.

The relation (5.1) yields $\dot{\gamma} = \dot{\psi} - \dot{\beta}$, and applying (5.2) and (5.3) we obtain the differential equation

$$\left(m \cdot v_S - \frac{k_R \cdot b - k_F \cdot a}{v_S} \right) \cdot \dot{\psi}$$

$$-m \cdot v_S \cdot \beta - (k_F + k_R) \cdot \beta + k_R \cdot \delta_F + k_F \cdot \delta_F = 0. \tag{5.4}$$

On the other hand, from Figure 5.5 one gets (for small angles):

$$\Sigma M_S = J_Z \cdot \ddot{\psi} - M_Z - F_{yF} \cdot a + F_{yR} \cdot b \equiv 0,$$

and taking into account (5.2) and (5.3), we obtain the differential equation

$$J_Z \cdot \ddot{\psi} + \left(k_F \cdot a^2 + k_R \cdot b^2 \right) \cdot \frac{1}{v_S} \cdot \dot{\psi} + (k_R \cdot b - k_F \cdot a) \cdot \beta -$$

$$k_F \cdot a \cdot \delta_F - k_R \cdot b \cdot \delta_R - M_Z(t) \equiv 0 \, . \tag{5.5}$$

Table 5.1. Data associated with three road vehicle variants A, B and C.

Road vehicle $m = 1050\,kg$, $J_Z = 2330\,kg \cdot m$, $v = 10\,m/s$	*Damping coefficient* D_ψ $N \cdot m \cdot s/rad$	*Stiffness coefficient* C_ψ $N \cdot m/rad$
A $a = 1.25\,m, b = 1.25\,m$ $k_F = 50000\,N/rad$, $k_R = 50000\,N/rad$	41565	148809
B $a = 1.5\,m, b = 1.5\,m$ $k_F = 75000\,N/rad$, $k_R = 25000\,N/rad$	41565	24107
C $a = 1.0\,m, b = 1.5\,m$ $k_F = 25000\,N/rad$, $k_R = 75000\,N/rad$	41565	199107

From the above equation one gets

$$\beta = \frac{k_F \cdot a \cdot \delta_F + k_R \cdot b \cdot \delta_R + M_Z(t) - J_Z \cdot \ddot{\psi} - \left(k_F \cdot a^2 + k_R \cdot b^2\right) \cdot \frac{1}{v_S} \cdot \dot{\psi}}{k_R \cdot b - k_F \cdot a}, \tag{5.6}$$

and after differentiation

$$\dot{\beta} = \frac{k_F \cdot a \cdot \dot{\delta}_F + k_R \cdot b \cdot \dot{\delta}_R + \dot{M}_Z(t) - J_Z \cdot \dddot{\psi} - \left(k_F \cdot a^2 + k_R \cdot b^2\right) \cdot \frac{1}{v_S} \cdot \ddot{\psi}}{k_R \cdot b - k_F \cdot a}. \tag{5.7}$$

It has been assumed that during the phenomenon observation the coefficients k_R, k_F and the velocity v_S are fixed. Substituting (5.6) and (5.7) into (5.4) the following third order differential equation is obtained:

$$J_Z \cdot \dddot{\psi} + D_\psi \cdot \ddot{\psi} + C_\psi \cdot \dot{\psi}$$

$$= \dot{M}_Z + H \cdot M_Z + k_F \cdot a \cdot \dot{\delta}_F + E \cdot \delta_F + k_R \cdot b \cdot \dot{\delta}_R + E \cdot \delta_R, \tag{5.8}$$

where: D, C, H and E are fixed and defined by

$$D_\psi = \frac{J_Z \cdot (k_F + k_R)}{m \cdot v_S} + \frac{k_F \cdot a^2 + k_R \cdot b^2}{v_S}, \quad H = \frac{k_F + k_R}{m \cdot v_S},$$

$$C_\psi = \frac{L^2 \cdot k_F \cdot k_R}{m \cdot v_S^2} + k_R \cdot b - k_F \cdot a, \quad E = \frac{L \cdot k_F \cdot k_R}{m \cdot v_S}. \tag{5.9}$$

The equation (5.8) can be integrated in the limits $0 - t$ and the equation of motion of the analysed wheeled vehicle reads

$$J_Z \cdot \ddot{\psi} + D_\psi \cdot \dot{\psi} + C_\psi \cdot \psi = \Phi\left(M_Z,\ \delta_F,\ \delta_R\right). \tag{5.10}$$

The coefficients D_ψ and C_ψ are called damping and stiffness coefficients respectively. The equation (5.10) includes two control units, i.e. the *dynamical* one (with respect to $M_Z(t)$, which can be treated either as control or perturbation quantity), and the *kinematical* one (with respect to steering angles $\delta_{F,R}(t)$). The control function Φ has the form

$$\Phi\left(M_Z, \delta_F, \delta_R\right)$$

$$= M_Z + H\cdot\int_0^t M_Z dt + k_F\cdot a\cdot\delta_F + E\cdot\int_0^t \delta_F dt + k_R\cdot b\cdot\delta_R + E\cdot\int_0^0 \delta_R dt + C_C, \quad (5.11)$$

where C_C is the integration constant defined by the attached initial conditions.

Three families of data associated with the coefficients D_Φ and C_Φ for three vehicle variants A (neutral), B (oversteer) and C (understeer) are included in Table 5.1.

One may conclude while analysing the data in Table 5.1, that the dynamical characteristics of three road vehicle variants are different, and hence they are briefly discussed in the next section.

1.2 The Road Vehicle Properties

First we begin with vibration analysis. The free road vehicle vibrations read

$$J_Z\cdot\frac{d^2\psi}{dt^2} + D_\psi\cdot\frac{d\psi}{dt} + C_\psi\cdot\psi = 0, \quad \psi(t=0) = \psi_0,$$

$$\frac{d\psi}{dt}(t=0) = \left(\frac{d\psi}{dt}\right)_0, \quad (5.12)$$

where: $\Omega_w^2 = C_\psi/J_Z$, $\eta = (0.5D_\psi)/J_Z$. In Figure 5.6 phase plots $\dot\psi(\psi)$ for three different variants A, B, C (see Table 5.1) are reported. In all cases considered $\psi_0 = \dot\psi_0 = 0.02\,rad$, and the vehicle velocities have been taken as $10\,m/s$ and $25\,m/s$. For the smaller velocity $10\,m/s$ (see Figure 5.6a) all three trajectories approach the origin $(0,0)$. However, for the larger velocity $25\,m/s$ (Figure 5.6b) and for the oversteered vehicle case the trajectory leaves the marked zone $(-0.1, 0.2), (-0.04, 0.04)$.

Since the equation (5.10) governs the vehicle dynamics with excitation (control) generated by the road wheels' steering angles δ_F, δ_R and by the rotational moment M_Z (see Figure 5.5), the following sinusoidal control is assumed:

$$M_Z(t) = M_{Zm}\cdot\sin(v_M\cdot t), \;\; \delta_F(t) = \delta_{F,m}\cdot\sin(v_\delta\cdot t), \;\; \delta_R(t) = \delta_{Rm}\cdot\sin(v_\delta\cdot t).$$

First, the control via $M_Z(t)$ is investigated. For $\delta_F = \delta_R = 0$, the equation (5.10) yields

$$J_Z\cdot\frac{d^2\psi}{dt^2} + D_\psi\cdot\frac{d\psi}{dt} + C_\psi\cdot\psi = M_{Zm}\cdot\sin(\nu_M\cdot t) - \frac{M_{Zm}\cdot H}{\nu_M}\cdot cos(\nu_M\cdot t), \quad (5.13)$$

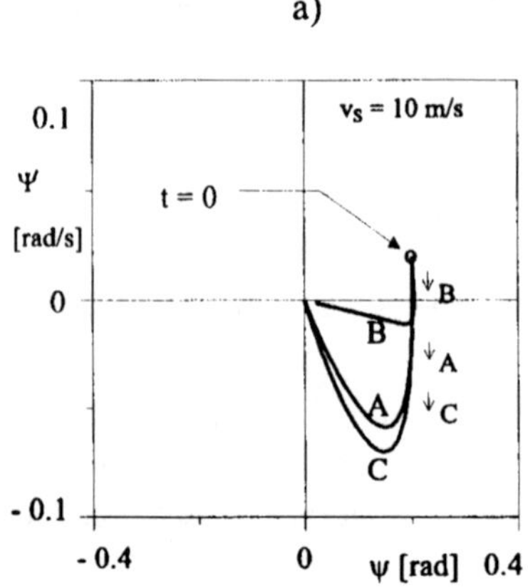

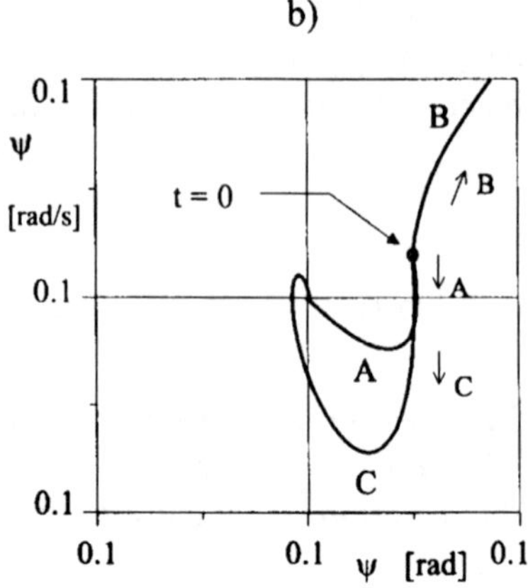

Figure 5.6. Trajectories in the plane $\dot{\psi}(\psi)$ for three vehicle variants A, B, C for $v_S = 10\,m/s$ (a), and $v_S = 25\,m/s$ (b), where $\psi, \dot{\psi}$ denote the rotational angle and speed of the vehicle, respectively.

where: H is the constant defined by relation $H = \dfrac{k_F + k_R}{m \cdot v_S}$.

A particular solution of the non-homogeneous equation (5.13) has the following form:

$$\psi = A_M \cdot \sin(\nu_M \cdot t) + \psi_M) + B_M \cdot \cos(\nu_M \cdot t + \psi_M),$$

where:

$$A_M = M_{Zm} \cdot \frac{(C_\psi - J_Z \cdot \nu_M^2) - H \cdot D_\psi}{(D_\psi \cdot \nu_M)^2 + (C_\psi - J_Z \cdot \nu_M^2)^2},$$

$$B_M = -\,M_{Zm} \cdot \frac{(C_\psi - J_Z \cdot \nu_M^2) \cdot \frac{H}{\nu_M} + \nu_M \cdot D_\psi}{(D_\psi \cdot \nu_M)^2 + (C_\psi - J_Z \cdot \nu_M^2)^2}$$

and the amplitude $\psi_M = \sqrt{A_M^2 + B_M^2}$.

The relation $\psi_M(v_M)$ is reported in Figure 5.7. Since the model (5.8) is strongly damped via D_ψ, the resonance effects for $v_M = \Omega$ is not observed. The largest oscillation amplitudes occur for the oversteered vehicle B. On the other hand, $\psi_M \rightarrow \infty$ for $v_M \rightarrow 0$ for all considered vehicle variants. However, from the practical point of view, the case when $v_M \rightarrow 0$ can be physically understood as an action of impulse within a bounded time interval and possessing infinitesimal value.

In the second case, control of steering angles of the front and rear wheels is applied. Namely, control of the steering angle of front wheels $\delta_F(t)$ is analysed, and control of the end steering angle wheels $\delta_R = e \cdot \delta_F$, where e denotes a

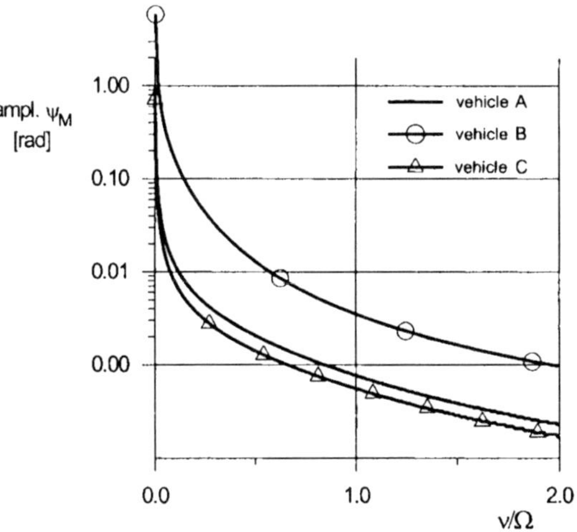

Figure 5.7. Amplitude versus excitation frequency for three vehicle variants A, B and C.

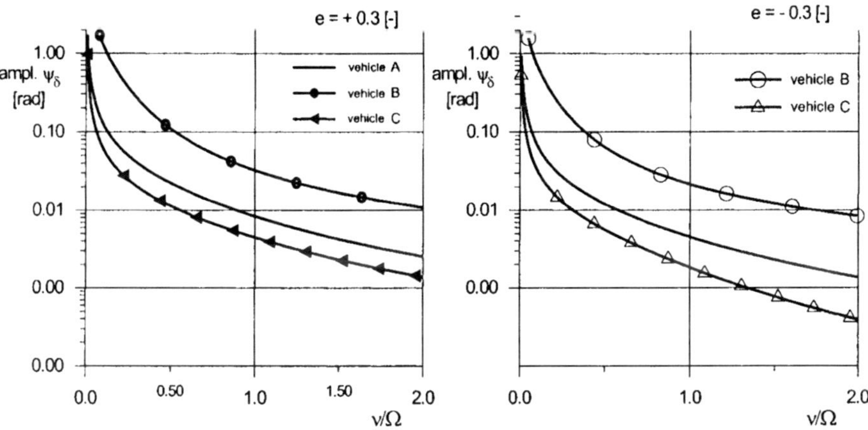

Figure 5.8. Amplitude ψ_δ versus frequency ν_δ for control via δ_F and δ_R for $e > 0$ (a) and $e < 0$ (b).

constant coefficient, is also investigated. The equation (5.10) for $M_Z = 0$ and for the introduced control is transformed to the form

$$J_Z \cdot \frac{d^2\psi}{dt^2} + D_\psi \cdot \frac{d\psi}{dt} + C_\psi \cdot \psi$$

$$= \delta_{Fm} \cdot (k_F \cdot a + k_R \cdot b \cdot e) \cdot \sin(\nu_\delta \cdot t) - \delta_{Fm} \cdot \frac{E}{\nu_\delta} (1 + e) \cdot \cos(\nu_\delta \cdot t). \quad (5.14)$$

A particular solution to the equation (5.14) has the form

$$\psi = A\delta_{FR} \cdot \sin(\nu_\delta \cdot t + \psi_\delta) + B\delta_{FR} \cdot \cos(\nu_\delta \cdot t + \psi_\delta),$$

where:

$$A\delta_{FR} = \delta_{Fm} \cdot \frac{(C_\psi - J_Z \cdot \nu_\delta^2) \cdot (k_F \cdot a + k_R \cdot b \cdot e) - E \cdot (1+e) \cdot D_\psi}{(D_\psi \cdot \nu_\delta)^2 + (C_\psi - J_Z \cdot \nu_\delta^2)^2},$$

$$B\delta_{FR} = -\,\delta_{Fm} \cdot \frac{(C_\psi - J_Z \cdot \nu_\delta^2) \cdot \frac{E}{\nu_\delta} \cdot (1+e) + \nu_\delta \cdot D_\psi \cdot (k_F \cdot a + k_R \cdot b \cdot e)}{(D_\psi \cdot \nu_\delta)^2 + (C_\psi - J_Z \cdot \nu_\delta^2)^2},$$

and $\psi_\delta = \sqrt{A^2_{\delta_{FR}} + B^2_{\delta_{FR}}}$.

In Figure 5.8 the function $\psi_\delta(v_\delta)$ for the discussed vehicle variants A, B and C is shown. Again, resonance effects are not observed, since the system is strongly damped (D_ψ). The largest amplitudes of oscillations appear for the oversteered vehicle (variant B), and the amplitude $\psi_\delta \to \infty$ for $v_\delta \to 0$. In addition, the control function $\delta_R = e \cdot \delta_F$ allows us to achieve large amplitudes of ψ for $e = 0.3$.

In order to complete the two axle vehicle characterization the *steerability characteristics* for three analysed variants are displayed in Figure 5.9.

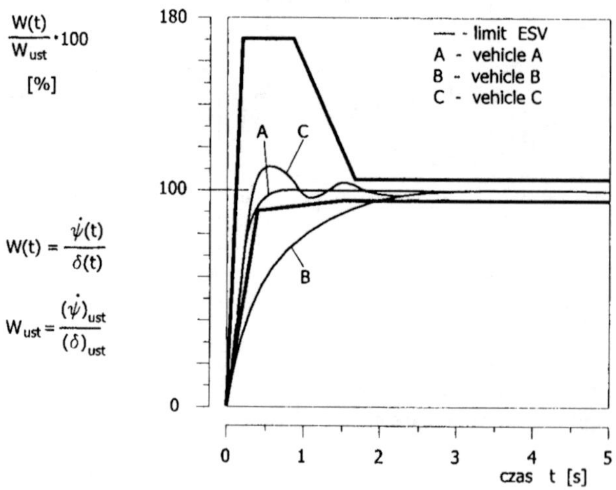

Figure 5.9. Factors of steerability $W(t)$ for three vehicle models.

By a steerability characteristic we mean behaviour of a transitional state from the straight line drive to the drive on the curved trajectory with a constant steering angle value (see Figure 5.9).

A transition from the initial state is realized via rotational movement of a steering gear in order to achieve constant turn velocity of the road wheels of the value $2\,rad/s$. The steerability factor $W(t)/W_{ust} \cdot 100\%$ is shown in Figure 5.9. The area of its proper behaviour is bounded by the limit ESV (experimental safety vehicle). Although both A and C vehicle variants satisfy the ESV requirements, the B vehicle variant does not, because $W(t)$ for $t < 2\,s$ is outside the admissible area.

1.3 Stability Investigation of a Two Axle Wheeled Vehicle

Stability in the Lyapunov Sense

In order to have stable second order linear differential equations with constant coefficients, it is necessary and sufficient (in the Lyapunov sense) that all coefficients of the homogeneous associated equation are positive. This observation is yielded by the Hurwitz criterion, i.e. a condition requiring that all main corner minors must be positive. In the case of the considered equation (5.10) they read:

$$C_\psi > 0, \quad D_\psi > 0 \quad \text{and} \quad \begin{vmatrix} D_\psi & C_\psi \\ 0 & J_Z \end{vmatrix} > 0\,.$$

So

$$C_\psi > 0, \quad D_\psi > 0 \quad and \quad J_Z > 0\,,$$

and hence the Lyapunov stability conditions read

$$D_\psi = \frac{J_Z \cdot (k_F + k_R)}{m \cdot v_S} + \frac{k_F \cdot a^2 + k_R \cdot b^2}{v_S} > 0;$$

$$C_\psi = \frac{L^2 \cdot k_F \cdot k_R}{m \cdot v_S^2} + k_R \cdot b - k_F \cdot a > 0. \tag{5.15}$$

Assuming that $D_\psi > 0$, one may introduce the so-called *Lyapunov stability factor* (or stability margin factor) $C_{LAP} = C_\psi$. For our case

$$C_{LAP} = \frac{L^2 \cdot k_F \cdot k_R}{m \cdot v_S^2} + k_R \cdot b - k_F \cdot a: \quad C_{LAP} = \begin{cases} C_\psi > 0: & \text{stable vehicle} \\ C_\psi = 0: & \text{unstable vehicle}\,, \\ C_\psi < 0: & \text{unstable vehicle} \end{cases}$$

and a wheeled vehicle is stable ($C_\psi > 0$), or unstable ($C_\psi < 0$ or $C_\psi = 0$).

The Lyapunov stability factor C_{LAP} as a function of driving velocity and of axle base is reported in Figure 5.10 for the parameters given in Table 5.1. The values of the stability factor C_{LAP} decrease with an increase of the driving velocity. For the A vehicle (neutral) this decrease is relatively high and $v \to \infty$ for $C_{LAP} \to 0$. For C variant this decrease is a relatively small one and hence this variant is characterized by the largest stability. In the case B (oversteered case) the vehicle is unstable for the velocity $v_S > 12\,m/s$. The values of C_{LAP}

increase for increase of the axle base L. For the variants A and C, the influence of L is similar and both vehicles are stable for the considered L values. The oversteered vehicle variant is unstable for $L < 2\,m$.

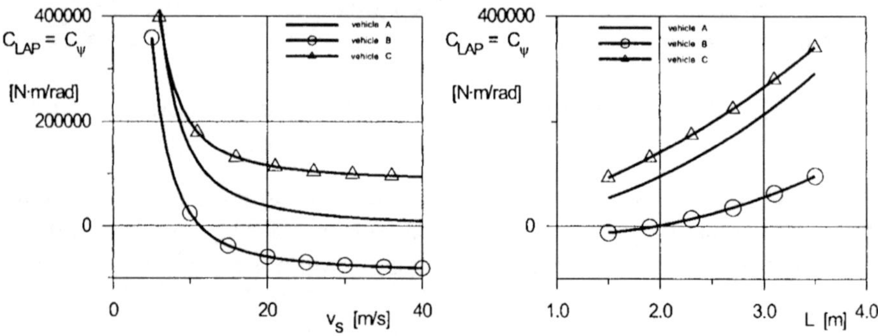

Figure 5.10. Influence of a velocity v_S (a) and a axle base L (b) on vehicle stability in the Lyapunov sense.

Very often tires and constraints between a wheel and a vehicle body are not treated as stiff ones. For example, the tire *wear out coefficient* k_i occurring in the relations (5.11) can represent either only tires elasticity (when it is assumed that the constraints between a road wheel and a vehicle body are stiff) or both tires and constraints elasticity (when an elasticity in a steering system and wheel suspension is taken into account). In the first case $k_i > 0$. The full angle displacement of the velocity vector v_k direction with respect to "k-k" being a direction of a road wheel axle is equal to $\alpha = F_y/k_{tire}$ (see Figure 5.11a). In the second case, the velocity vector v_k is additionally shifted with respect to "k-k" on the angle constraints strain, and hence the full angle displacement reads

$$\alpha' = \frac{F_y}{k_{tire}} + \frac{F_y}{k_{susp}},$$

where: k_{susp} is the resistance coefficient. The angle α' is called the *artificial slip angle*, whereas the coefficient $k_\Sigma = (1/k_{tire} + 1/k_{susp})^{-1}$ is said to be the *artificial slip coefficient*. However, k_Σ can be either positive or negative. In a wheeled vehicle a special suspension can be constructed that $k_\Sigma < 0$ (see Figure 5.11b).

In the vehicle system shown in Figure 5.11b an angle road wheel displacement in the opposite direction from the tire drift occurs. As a result, the velocity vector direction deviates with respect to "k-k" on the "negative" angle α'. This negative artificial drift is yielded by negative value of the coefficient k_{susp} and by the relation:

$$\text{for } k_{tire} > 0 \text{ and } k_{susp} < 0, \quad |k_{susp}| > k_{tire}.$$

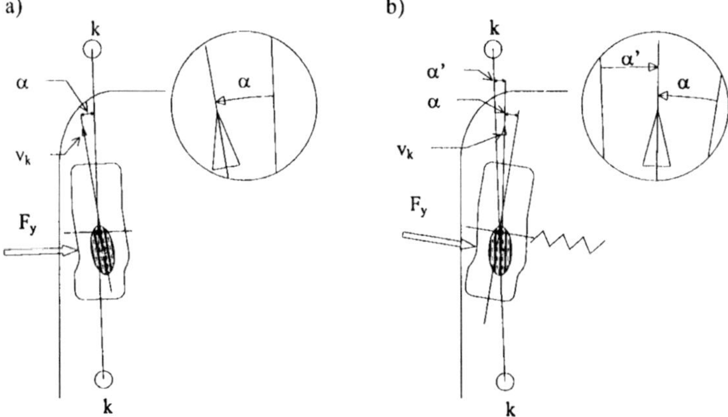

Figure 5.11. Two action effects of a lateral force: a positive (a) and a negative (b) road wheel drift.

The effect can be obtained applying the elasto-kinematic systems of wheel suspension, i.e. Multi-Link Axle.

Stability in the Bogusz Sense [9, 10, 11, 12]

Recall that in the Lyapunov stability concept the stability of a homogeneous and a non-homogeneous differential equation makes no sense. A non-homogeneous equation is stable, when the corresponding homogeneous equation is stable. However, the situation changes drastically when one applies the Bogusz stability approach. It is especially important during vehicle dynamic stability investigations. In other words, one may distinguish and analyse both straight and curved vehicle trajectories.

First, stability of a two-axle wheeled vehicle model driving on a straight line is analysed. Our considerations follow the model described in Section 2 with a requirement of admissible straight line motion. Therefore a technical stability is investigated for the steering wheels angles $\delta_F(t) = \delta_R(t) = 0$. It is assumed that the observation time is sufficiently short, i.e. $t_0 < t < t_0 + T$, where $T \neq \infty$, and the perturbations are continuously acting on the system in the form of the torque $M_Z < \lambda$ (see Figure 5.5) perpendicular to the road surface. It is assumed that the velocity vector value v_S (see Figure 5.1) and the values of lateral force coefficients k_F and k_R do not change in time interval $t_0 < t < t_0 + T$. Applying the introduced assumptions to the system governed by (5.10) the following equation is obtained:

$$J_Z \cdot \ddot{\psi} + D_\psi \cdot \dot{\psi} + C_\psi \cdot \psi = (1 + H \cdot t) \cdot M_Z, \tag{5.16}$$

where: D_ψ, C_ψ and H are defined by (5.9). The equation (5.16) governs dynamics of the perturbed straight line vehicle motion. As an input we get the vehicle angle rotation - yaw angle - ψ. Assuming $M_Z = \lambda$, the equation (5.16) is transformed to two first order differential equations

$$\dot{x} = y,$$

$$\dot{y} = -\frac{D_\psi}{J_Z} \cdot y - \frac{C_\psi}{J_Z} \cdot x + \frac{(1 + H \cdot t)}{J_Z} \cdot \lambda \,. \tag{5.17}$$

Let us first define the initial condition zone ω. The initial system state is defined by $\psi_0 = \dot{\psi}_0 = 0$. Since in the real conditions the deviations $\Delta\psi_0$ and $\Delta\dot{\psi}_0$ appear, then the initial conditions can be presented in the following way:

$$\omega = \left\{ \begin{matrix} \psi & -\Delta\psi_0 < \psi < +\Delta\psi_0 \\ \dot{\psi} & -\Delta\dot{\psi}_0 < \dot{\psi} < +\Delta\dot{\psi}_0 \end{matrix} \right\}.$$

In the next step the admissible solutions zone Ω is defined. This zone is determined using an assumption that a vehicle not controlled by a driver can deviate from a given straight trajectory on a certain angle and hence a certain angular velocity occurs. Let us assume now that the admissible parameter deviations are known for given driving conditions. Therefore, a zone of admissible solutions (after time T) is defined in the following way:

$$\Omega = \left\{ \begin{matrix} \psi & -\Delta\psi_Z < \psi < +\Delta\psi_Z \\ \dot{\psi} & -\Delta\dot{\psi}_Z < \dot{\psi} < +\Delta\dot{\psi}_Z \end{matrix} \right\}.$$

For the system governed by equations (5.17) the following Bogusz function is defined:

$$V_B(x, y) = 0.5\frac{C_\psi}{J_Z} \cdot x^2 + 0.5y^2, \quad \text{where } x = \psi,\ y = \dot{\psi} \,.$$

In accordance with Theorem 1 of Section 2.6:

$$C_\psi > 0. \tag{5.18}$$

The Bogusz function satisfies three following conditions (see Theorem 2.14 of Section 2.6)

1°. $V_B(\psi, \dot{\psi}) \leq C_0$ for $\psi, \dot{\psi} \in \omega$

The number which satisfies the above equality is

$$C_0 = 0.5\frac{C_\psi}{J_Z} \cdot (\Delta\psi_0)^2 + 0.5\left(\Delta\dot{\psi}_0\right)^2 .$$

2°. $V_B(\psi, \dot{\psi}) \geq C_1$ for $\psi, \dot{\psi} \notin \Omega/\omega$
The number which satisfies the above equality is

$$C_1 = 0.5\frac{C_\psi}{J_Z} \cdot (\Delta\psi_Z)^2 + 0.5\left(\Delta\dot{\psi}_Z\right)^2 .$$

3°. $\frac{dV_B}{dt} < \frac{C_1 - C_0}{T}$ for $\psi, \dot{\psi} \notin \Omega/\omega$ and for $t_0 \leq t \leq t_0 + T$
A derivative of the Bogusz function V_B along the solutions of the system (5.16) has the form

$$\frac{dV_B}{dt}(x,\ y) = \frac{C_\psi}{J_Z} \cdot x \cdot \dot{x} + y \cdot \dot{y} .$$

Taking into account (5.17) one gets

$$\frac{dV_B}{dt}(x, y) = \frac{C_\psi}{J_Z} \cdot x \cdot y + y \cdot \left(-\frac{D_\psi}{J_Z} \cdot y - \frac{C_\psi}{J_Z} \cdot x + \frac{(1 + H \cdot t)}{J_Z} \cdot \lambda\right),$$

where: $x = \psi$, $y = \dot{\psi}$. A maximum of this derivative is defined for $\psi, \dot{\psi} \notin \omega$ by

$$\inf_{\psi, \dot{\psi} \notin \omega} \frac{dV_B}{dt}(\psi,\ \dot{\psi}) = -\frac{D_\psi}{J_Z} \cdot \left(\Delta\dot{\psi}_0\right)^2 + \frac{(1 + H \cdot (t_0 + T))}{J_Z} \cdot \left(\Delta\dot{\psi}_Z\right) \cdot \lambda.$$

The condition 3° can be realized in the following manner ($t_0 = 0$)

$$-\frac{D_\psi}{J_Z} \cdot \left(\Delta\dot{\psi}_0\right)^2 + \frac{(1 + H \cdot T)}{J_Z} \cdot \left(\Delta\dot{\psi}_Z\right) \cdot \lambda$$

$$< \frac{1}{T} \cdot \left[0.5\frac{C_\psi}{J_Z} \cdot (\Delta\psi_Z)^2 + 0.5\left(\Delta\dot{\psi}_Z\right)^2 - 0.5\frac{C_\psi}{J_Z} \cdot (\Delta\psi_0)^2 - 0.5\left(\Delta\dot{\psi}_0\right)^2\right]. \tag{5.19}$$

Owing to the condition (5.18) the quantity called a *first technical stability factor* $FT1$ is introduced. For partial reasons, the stability factor defined by (5.18) reads

$$FT1 = \frac{C_\psi}{E}, \tag{5.20}$$

where C_ψ and E are defined by (5.9).

A *second technical stability factor* $FT2$ is defined by (5.19) and it has the form

$$FT2 = \frac{D_\psi \cdot T}{J_Z} \cdot \left(\Delta\dot{\psi}_0\right)^2 - \frac{(1 + H \cdot T) \cdot \lambda \cdot T}{J_Z} \cdot \left(\Delta\dot{\psi}_Z\right)$$

$$+\frac{C_\psi}{2 \cdot J_Z} \cdot \left[(\Delta\psi_Z)^2 - (\Delta\psi_0)^2 \right] + \frac{1}{2} \cdot \left[\left(\Delta\dot{\psi}_Z\right)^2 - \left(\Delta\dot{\psi}_0\right)^2 \right]. \qquad (5.21)$$

On the basis of earlier considerations the wheeled two axle vehicle is stable if $FT1 \wedge FT2 > 0$ ($FT1 \vee FT2 \leq 0$). Technical stability increases, and it depends on the vehicle motion parameters (v_S), statical vehicle properties (a, b, m, J_Z), dynamical vehicle properties (k_F, k_R), continuously acting perturbations ($\lambda; M_Z \leq \lambda$), technical parameters of motion monitoring ($T, \Delta\psi_0, \Delta\dot{\psi}_0$), and technical parameters for a zone of admissible solutions ($\Delta\psi_2, \Delta\dot{\psi}_2$).

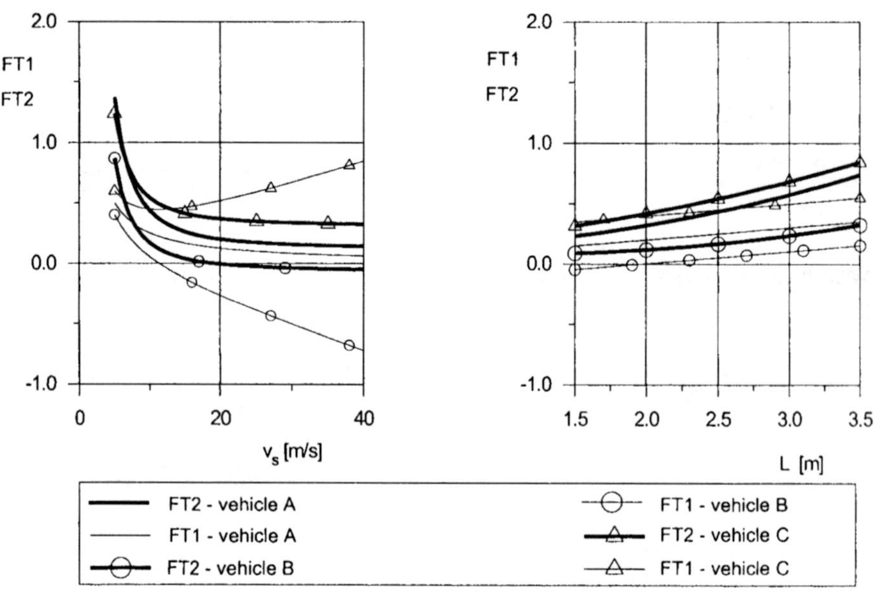

Figure 5.12. Technical stability factors $FT1$ and $FT2$ as functions of a vehicle velocity (a) and of a axle base L (b).

In Figure 5.12 investigation results of technical stability in the Bogusz sense for the vehicle variants A, B and C (see Table 5.1) are reported. The input data are shown in Table 5.2).

The factor values decrease when the vehicle velocity is increased. For the variant A a decrease is relatively large (for $v_S \to \infty$, $FT1, FT2 \to 0$). For the variant C a decrease is relatively small, i.e. the vehicle is characterized by a largest stability. For the oversteered vehicle (case C), the vehicle is unstable for the driving velocity $v_S > 10\, m/s$. Both factors $FT1$ and $FT2$ increase when the axle base L is increased. The vehicle variants A and C are stable for

Table 5.2. Stability conditions data.

Monitoring time	$T = 1\,s$	
Perturbations, steerings	$\lambda\,(M_Z \leq \lambda) = 5\,N \cdot m$	
Zone of initial conditions	ω	$\Delta\dot{\psi}_0 = 0.01\,rad/s$, $\Delta\ddot{\psi}_0 = 0.012\,rad/s$
Zone of admissible conditions	Ω	$\Delta\dot{\psi}_Z = 0.1\,rad/s^2$, $\Delta\ddot{\psi}_Z = 0.5\,rad/s^2$

the given interval of L values, whereas the oversteered vehicle is unstable for $L < 1.9\,m$.

Now our attention is focused on the stability of a two-axle wheeled vehicle moving on a curved trajectory and the vehicle model described in Section 5.1 is analysed. The technical stability is investigated for constant non-zero values of δ_F and δ_R. The monitoring time is assumed to be relatively short $t_0 \leq t \leq t_0 + T, T \neq \infty$, and the perturbations are acting continuously in the form of the rotational moment M_Z (see Figure 5.5), which is perpendicular to a road surface ($M_Z < \lambda$). It is assumed additionally that v_S (see Figure 5.1) and k_F, k_R do not change in time in the duration $t_0 \leq t \leq t_0 + T$. The described assumptions applied to the model governed by equation (5.8) yield

$$J_Z \cdot \dddot{\psi} + D_\psi \cdot \ddot{\psi} + C_\psi \cdot \dot{\psi} = H \cdot \lambda + E \cdot \delta_F + E \cdot \delta_R, \tag{5.22}$$

where: D_ψ, C_ψ, H and E are defined by (5.9).

The equation (5.22) describes the perturbed vehicle motion with the initial circular trajectory. The inputs are: δ_F, δ_R and M_Z, whereas ψ is the output. Introducing $x = \dot{\psi}$, $y = \ddot{\psi}$ we get

$$\dot{x} = y;$$

$$\dot{y} = -\frac{D_\psi}{J_Z} \cdot y - \frac{C_\psi}{J_Z} \cdot x + \frac{H \cdot \lambda}{J_Z} + \frac{E \cdot \delta_F}{J_Z} + \frac{E \cdot \delta_R}{J_Z}. \tag{5.23}$$

Proceeding in a way similar to the previous one observe that the initial system state is defined by t_0, $\dot{\psi}_0 = x_0$ and $\ddot{\psi}_0 = y_0$. In the real world behaviour the following deviations of these values are expected: $\Delta\dot{\psi}_0$ and $\Delta\ddot{\psi}_0$. Therefore the initial zone

$$\omega = \left\{ \begin{array}{ll} \dot{\psi} & \dot{\psi}_0 - \Delta\dot{\psi}_0 < \dot{\psi} < \dot{\psi}_0 + \Delta\dot{\psi}_0 \\ \ddot{\psi} & \ddot{\psi}_0 - \Delta\ddot{\psi}_0 < \ddot{\psi} < \ddot{\psi}_0 + \Delta\ddot{\psi}_0 \end{array} \right\},$$

where: $\dot{\psi} = x$, $\ddot{\psi} = y$.

An admissible solutions zone Ω is defined owing to the assumptions: $\dot{\psi}_0 =$ constant, $\ddot{\psi}_0 = 0$. The deviations of these values read: $\Delta\dot{\psi}_Z$ and $\Delta\ddot{\psi}_Z$, and finally

$$\Omega = \left\{ \begin{matrix} \dot{\psi} & \dot{\psi}_0 - \Delta\dot{\psi}_Z < \dot{\psi} < \dot{\psi}_0 + \dot{\Delta}\psi_Z \\ \ddot{\psi} & \ddot{\psi}_0 - \Delta\ddot{\psi}_Z < \ddot{\psi} < \ddot{\psi}_0 + \Delta\ddot{\psi}_Z \end{matrix} \right\}.$$

Now, according to Theorem 1 (see Section 2.6) the coefficient

$$C_\psi > 0\,. \tag{5.24}$$

For the system governed by equation (5.23) the following Bogusz function is defined:

$$B(x,y) = 0.5\frac{C_\psi}{J_Z}\cdot x^2 + 0.5y^2,$$

where: $x = \dot{\psi}$, $y = \ddot{\psi}$.

The Bogusz function satisfies the following three conditions (see Theorem 2 in Section 2.6):

1°. $V_B(x,y) \le C_0$ for $x = \dot{\psi}$, $y = \ddot{\psi} \in \omega$
The number satisfying the above inequality is

$$C_0 = 0.5\frac{C_\psi}{J_Z}\cdot(\psi_0 + \Delta\psi_0)^2 + 0.5\left(\dot{\psi}_0 + \Delta\dot{\psi}_0\right)^2.$$

2°. $V_B(x,y) \ge C_1$ for $x = \dot{\psi}$, $y = \ddot{\psi} \notin \Omega/\omega$, and

$$C_1 = 0.5\frac{C_\psi}{J_Z}\cdot\left(\psi_0 + \dot{\psi}_0\cdot T - \Delta\psi_Z\right)^2 + 0.5\left(\dot{\psi}_0 - \Delta\dot{\psi}_Z\right)^2.$$

3°. $\dfrac{dV_B}{dt} < \dfrac{C_1 - C_0}{T}$ for $x = \dot{\psi}$, $y = \ddot{\psi} \notin \Omega/\omega$ and for $t_0 \le t \le t_0 + T$
A derivative of the Bogusz function V_B along the solutions of the system (5.23) has the form

$$\frac{dV_B}{dt}(x,\ y) = \frac{C_\psi}{J_Z}\cdot x\cdot\dot{x} + y\cdot\dot{y}\,.$$

Substituting (instead of $\dot{x}$ and $\dot{y}$) the relations defined by (5.17) one gets

$$\frac{dV_B}{dt}(x,y) = \frac{C_\psi}{J_Z}\cdot x\cdot y + y\cdot\left(-\frac{D_\psi}{J_Z}\cdot y - \frac{C_\psi}{J_Z}\cdot x + \frac{H\cdot M_Z}{J_Z} + \frac{E\cdot\delta_F}{J_Z} + \frac{E\cdot\delta_R}{J_Z}\right),$$

where: $x = \dot{\psi}$, $y = \ddot{\psi}$.

The maximal value of the derivative reads

$$\inf_{\notin\,\omega}\frac{dV_B}{dt}(\dot{\psi},\ddot{\psi}) = -\frac{D_\psi}{J_Z}\cdot\left(\Delta\ddot{\psi}_0\right)^2 + \frac{H\cdot\lambda}{J_Z}\left(\Delta\ddot{\psi}_Z\right) + \frac{E\cdot\delta_F}{J_Z}\left(\Delta\ddot{\psi}_Z\right) + \frac{E\cdot\delta_R}{J_Z}\cdot\left(\Delta\ddot{\psi}_Z\right),$$

Table 5.3. Data of stability conditions.

Monitoring time	$T = 1\,s$	
Perturbations, steerings	$\lambda\,(M_Z \leq \lambda) = 5\,N \cdot m,$ $\delta_F = 0.05\,rad,\ \delta_R = 0.1 \cdot \delta_F$	
Zone of initial conditions	ω	$\Delta\dot{\psi}_0 = 0.1\,rad/s,\ \Delta\ddot{\psi}_0 = 0.12\,rad/s$
Zone of admissible conditions	Ω	$\Delta\dot{\psi}_Z = 2.5\,rad/s^2,\ \Delta\ddot{\psi}_Z = 5\,rad/s^2$

and finally the condition 3° is defined in the following way:

$$-\frac{D_\psi}{J_Z}\cdot\left(\Delta\ddot{\psi}_0\right)^2 + \frac{H\cdot\lambda}{J_Z}\cdot\left(\Delta\ddot{\psi}_Z\right) + \frac{E\cdot\delta_F}{J_Z}\cdot\left(\Delta\ddot{\psi}_Z\right) + \frac{E\cdot\delta_R}{J_Z}\cdot\left(\Delta\ddot{\psi}_Z\right)$$

$$< \frac{1}{T}\cdot\left[0.5\frac{C_\psi}{J_Z}\cdot\left(\Delta\dot{\psi}_Z\right)^2 + 0.5\left(\Delta\ddot{\psi}_Z\right)^2 - 0.5\frac{C_\psi}{J_Z}\cdot\left(\Delta\dot{\psi}_0\right)^2 - 0.5\left(\Delta\ddot{\psi}_0\right)^2\right]. \tag{5.25}$$

For the considered case

$$FT1 = C_\psi/E, \tag{5.26}$$

$$FT2 = \frac{D_\psi\cdot T}{J_Z\cdot H\cdot v_S}\cdot\left(\Delta\ddot{\psi}_0\right)^2 - \frac{\lambda\cdot T}{J_Z\cdot v_S}\cdot\left(\Delta\ddot{\psi}_Z\right) - \frac{E\cdot T\cdot(\delta_F+\delta_R)}{J_Z\cdot H\cdot v_S}\cdot\left(\Delta\ddot{\psi}_Z\right)$$

$$+\frac{C_\psi}{2\cdot J_Z\cdot H\cdot v_S}\cdot\left[\left(\Delta\dot{\psi}_Z\right)^2 - \left(\Delta\dot{\psi}_0\right)^2\right] + \frac{1}{2\cdot H\cdot v_S}\cdot\left[\left(\Delta\ddot{\psi}_Z\right)^2 - \left(\Delta\ddot{\psi}_0\right)^2\right]. \tag{5.27}$$

The same observation of the technical stability factors holds now as in the previous cases, and it will not be repeated here. For the data from Table 5.1, investigations of the technical stability results are reported in Figure 5.13 for the same parameters as earlier, i.e. v_S and L. The data are included in Table 5.3.

Analysing the results shown in Figure 5.13 the following conclusions are derived. The vehicle model C is the most stable one. The oversteered vehicle (B) is only stable for relatively small velocities, and for $v_S > 9\,m/s$ the technical instability appears. However, for $v_S = 10\,m/s$ the vehicle is technically stable when its axle base is increased up to the value of $L = 2.7\,m$.

It is worth noticing that the technical stability concepts in the Bogusz sense are very suitable for analysis of a wheeled vehicle, and they include the adequate system parameters while defining the so-called *admissible solutions zone*. The technical stability concepts include: (i) an occurrence of input data motion monitoring scatter; (ii) continuous perturbations; (iii) steering depended on turn angles of steered road wheels.

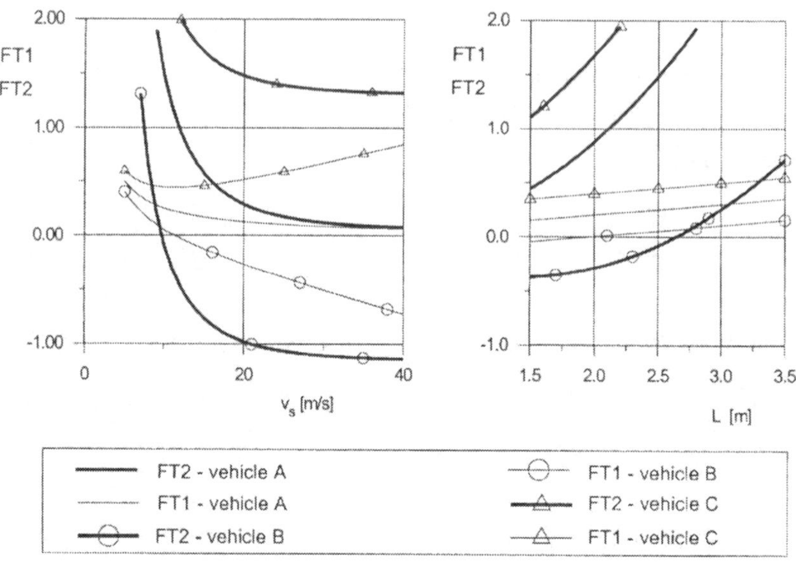

Figure 5.13. Influence of a driving velocity v_S (a) and a axle base L (b) on a vehicle moving on a curved trajectory stability ($FT1$ and $FT2$ are the technical stability factors).

2. Stability of Wheeled Articulated Vehicles

A wheeled articulated vehicle has been known for a long time. In 1894 a car (a tractor with a semi-trailer) took part in a rally and won. However, in the beginning it was difficult to overcome all difficulties. For example, production of a car with a passenger semi-trailer caused the collapse of the Charles Duryea company. The producers did not predict that the applied windowless semi-trailer was wrong. The passengers were attacked by dust and wind.

Figure 5.14. A tractor with a passenger semi-trailer produced by the Charles Duryea company in 1898.

The model of two articulated vehicles coupled by a ball-and-socket joint with are vehicle pulling the other, is universally applied to such examples as the truck-tractor (see Figures 5.15 and 5.17), an *articulated bus* (Figure 5.16) and a passenger car with a sidecar. A linking joint can be mounted either in a vicinity at the end or on a roof (see Figure 5.18). A special character is seen in a vehicle devoted to log transport (Figure 5.19). Its peculiarity is characterized by the load being transported (for instance, pipes, balks, etc.), which becomes a vehicle loading element. However, the loading element does not guarantee a constant stiffness of the system tractor-load-semi-trailer, since now the system stiffness depend not only on the loading properties, but also on the way of its mounting to the tractor and semi-trailer.

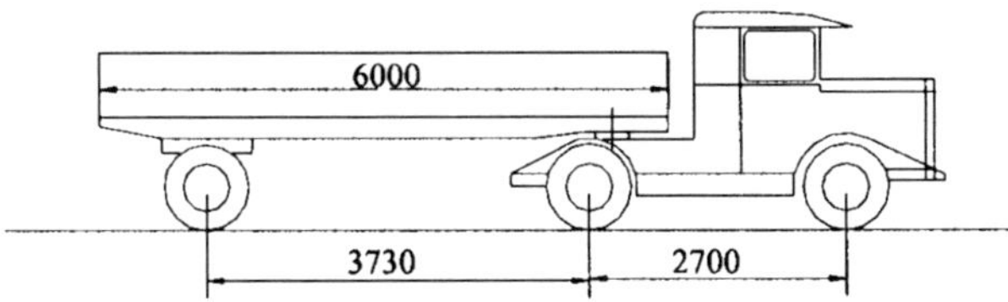

Figure 5.15. A truck tractor-semitrailer from 1932.

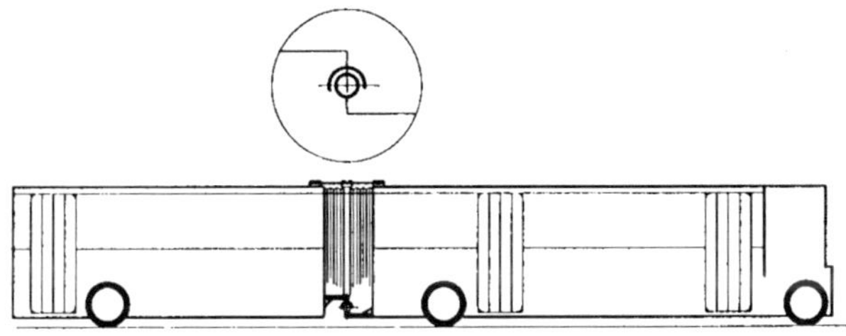

Figure 5.16. An articulated bus with a ball-and-socket joint.

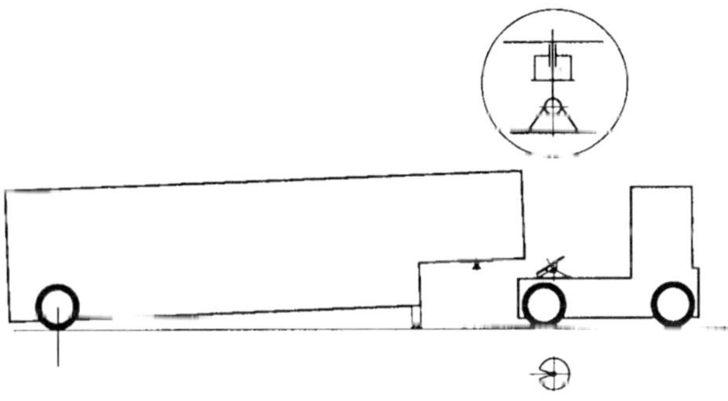

Figure 5.17. A truck tractor with a semi-trailer with a "coupler" – fifth wheel scheme.

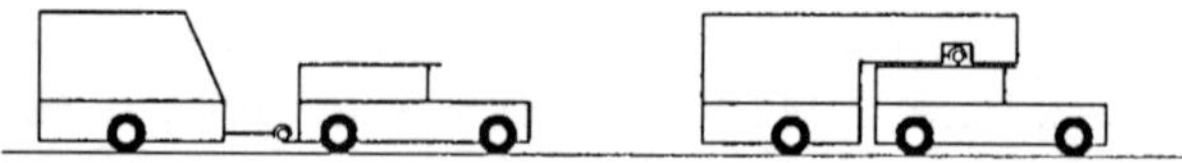

Figure 5.18. A passenger car with a semi-trailer (b) and a trailer (a).

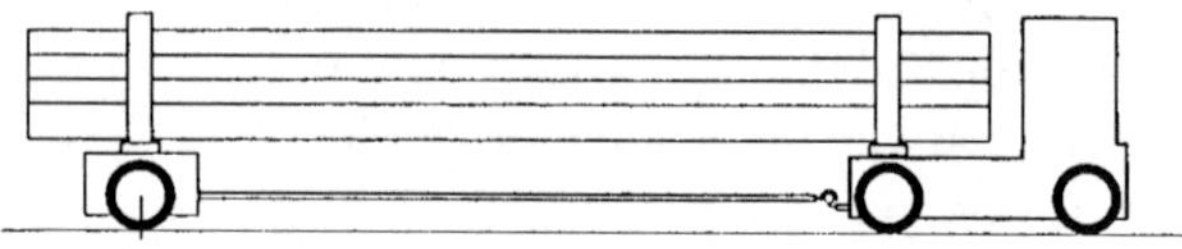

Figure 5.19. An articulated car for a log transport - truck tractor pole trailer.

Unfortunately, stability of the articulated vehicles is not yet satisfactorily solved. In particular, the so-called *jack knifing* phenomenon seems to be very dangerous [13, 15, 69, 93].

2.1 Articulated Vehicle Model (a Truck Tractor-Semitrailer)

A simplified one-truck vehicle model valid for small transversal angle displacements φ (Figures 5.20–5.25) is now considered. In Figure 5.20 representative vehicle bodies and their dimensions are shown.

A perturbed straight line motion owing to an angle displacement of a steered wheel α_F, as well as to torsional moments acting on the truck (M_{ZC}) and on the semi-trailer (M_{ZN}), is analysed. Note that the mentioned quantities are now treated as perturbations, and not as controls, and frictional phenomenona such as joint (fifth wheel) is accounted for. In Figures 5.24 and 5.25 the frictional joint moment M_S is treated as control ($M_{sS} = M_{sS} \cdot \text{sign}(\dot{\varphi})$).

In Figures 5.21 and 5.22 the following kinematical parameters are introduced: α_F - tire slip of a front tractors axle; α_R – tire slip of a rear tractors axle; α_M - tire slip of a semi-trailer axle; ψ - tractor angle rotation (yaw); α_C - slip angle of velocity tractor; φ - semi-trailer rotational angle with respect to a tractor.

Owing to dimensions a, L, L_N, L_s, L_0 included in Figure 5.20, and taking into account Figures 5.21 and 5.22, a mathematical model of a joint vehicle kinematics is built. It includes the following parameters:

$$\alpha_F = \delta_F - arctg\left(\frac{v_C \cdot \sin(\alpha_c) + \dot{\psi} \cdot a}{v_C \cdot \cos(\alpha_C)}\right); \tag{5.28}$$

$$\alpha_R = arctg\left(\frac{-v_C \cdot \sin(\alpha_c) + \dot{\psi} \cdot (L - a)}{v_C \cdot \cos(\alpha_C)}\right); \tag{5.29}$$

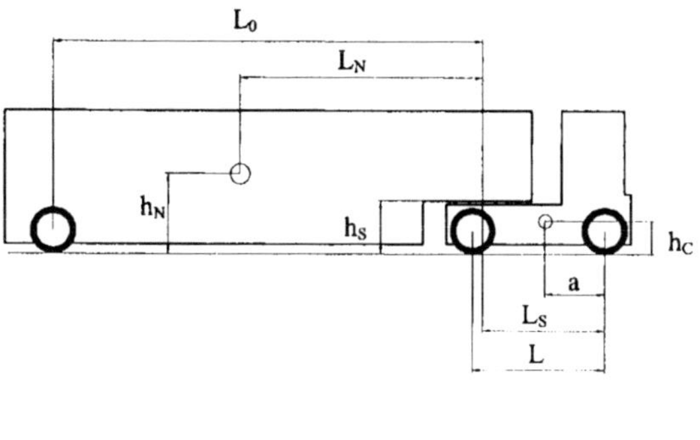

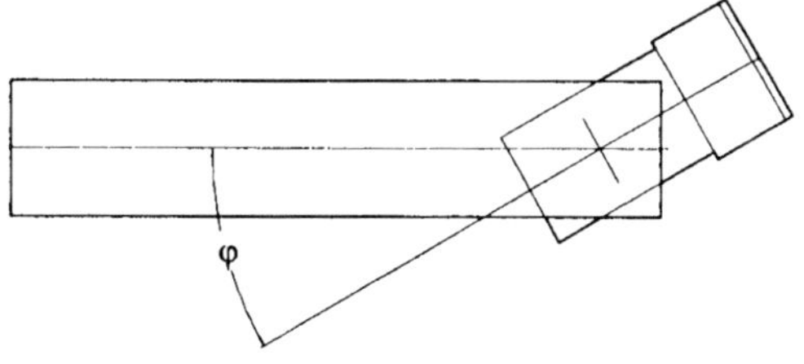

Figure 5.20. Main dimension of a tractor-track with a semi-trailer.

$$\alpha_S = arctg\left(\frac{v_C \cdot \sin(\alpha_c) - \dot{\psi} \cdot (L_S - a)}{v_C \cdot \cos(\alpha_C)}\right); \tag{5.30}$$

$$v_S = \sqrt{(v_C \cdot \cos(\alpha_C))^2 + \left(v_C \cdot \sin(\alpha_C) - \dot{\psi} \cdot (L_S - a)\right)^2}; \tag{5.31}$$

$$\alpha_N = arctg\left(\frac{-v_S \cdot \sin(\alpha_S + \varphi) + \left(\dot{\psi} - \dot{\varphi}\right) \cdot L_N}{v_S \cdot \cos(\alpha_S + \varphi)}\right); \tag{5.32}$$

$$v_N = \sqrt{\left(v_S \cdot \sin(\alpha_S + \varphi) - \left(\dot{\psi} - \dot{\varphi}\right) \cdot L_N\right)^2 + (v_S \cdot \cos(\alpha_S + \varphi))^2}; \tag{5.33}$$

$$\alpha_M = arctg\left(\frac{-v_S \cdot \sin(\alpha_S + \varphi) + \left(\dot{\psi} - \dot{\varphi}\right) \cdot L_0}{v_S \cdot \cos(\alpha_S + \varphi)}\right); \tag{5.34}$$

where: α_S defines the angle position of the fifth wheel velocity v_S; α_N side slip angle - defines the angle position of the semi-trailer mass centre velocity v_N (more details are given in Figures 5.21 and 5.22).

Owing to schemes of dynamics shown in Figures 5.23–5.25 the following relations are obtained:

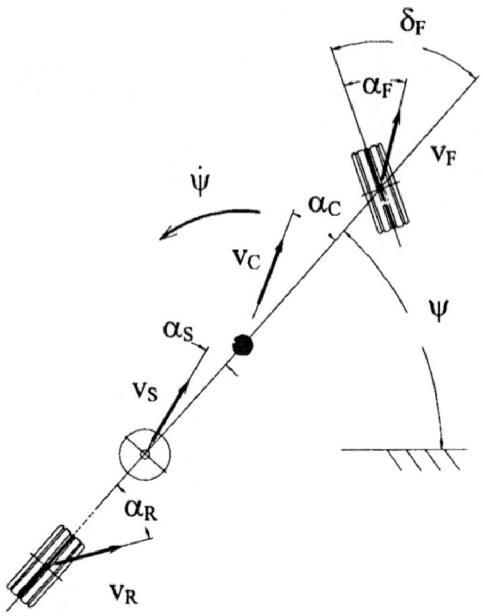

Figure 5.21. A tractor kinematics.

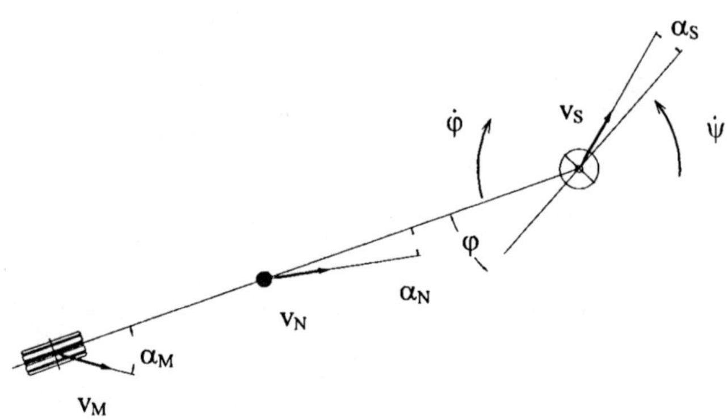

Figure 5.22. A semi-trailer kinematics.

(i) tractor inertial forces

$$P_{BC} = -m_C \cdot \dot{v}_C; \quad M_{BC} = J_{zC} \cdot \ddot{\psi}. \tag{5.35}$$

(ii) semi-trailer inertial forces

$$P_{BN} = -m_N \cdot \dot{v}_N; \quad M_{BN} = J_{zN} \cdot (\ddot{\psi} - \ddot{\varphi}). \tag{5.36}$$

(iii) centripetal force acting on the tractor

$$P_{DC} = m_C \cdot v_C \cdot (\dot{\psi} + \dot{\alpha}_C). \tag{5.37}$$

(iv) centripetal force acting on the semi-trailer

$$P_{DN} = m_N \cdot v_N \cdot (\dot{\psi} - \dot{\varphi} - \dot{\alpha}_N). \tag{5.38}$$

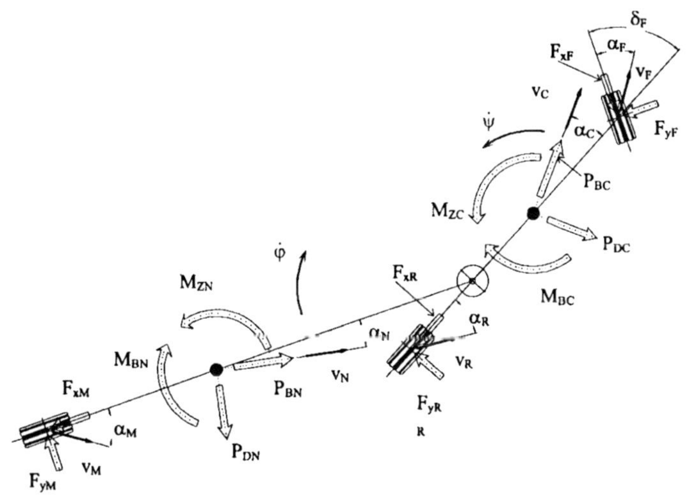

Figure 5.23. Dynamics of the truck tractor with the semi-trailer.

Usually a tire model including the holonomic constraints is defined via the function (for small slip angles)

$$F_{yi} = k_i \cdot \alpha_i, \tag{5.39}$$

where: F_{yi} is the road surface lateral force action on the road wheel, and k_i is the "resistance" coefficient to a transversal drift (tire angle stiffness).

In what follows applying the physical model (Figure 5.20–5.25) and the relations (5.28)–(5.39) one may derive the function $\Phi(\psi, \varphi, \alpha_C, t)$ and the equations of motion. The mathematical model of the truck tractor consists of three second order non-homogeneous differential equations of the form

$$a_{11}{\cdot}\ddot{\psi} + b_{11}{\cdot}\dot{\psi} + c_{11}{\cdot}\psi + a_{12}{\cdot}\ddot{\varphi} + b_{12}{\cdot}\dot{\varphi} + c_{12}{\cdot}\varphi + a_{13}{\cdot}\ddot{\alpha} + b_{13}{\cdot}\dot{\alpha}_C + c_{13}{\cdot}\alpha_C = e_1;$$

$$a_{21}{\cdot}\ddot{\psi} + b_{21}{\cdot}\dot{\psi} + c_{21}{\cdot}\psi + a_{22}{\cdot}\ddot{\varphi} + b_{22}{\cdot}\dot{\varphi} + c_{22}{\cdot}\varphi + a_{23}{\cdot}\ddot{\alpha} + b_{23}{\cdot}\dot{\alpha}_C + c_{23}{\cdot}\alpha_C = e_2;$$

$$a_{31}\ddot{\psi} + b_{31}\dot{\psi} + c_{31}\psi + a_{32}\ddot{\varphi} + b_{32}\dot{\varphi} + c_{32}\varphi + a_{33}\ddot{\alpha} + b_{33}\dot{\alpha}_C + c_{33}\alpha_C = e_3; \tag{5.40}$$

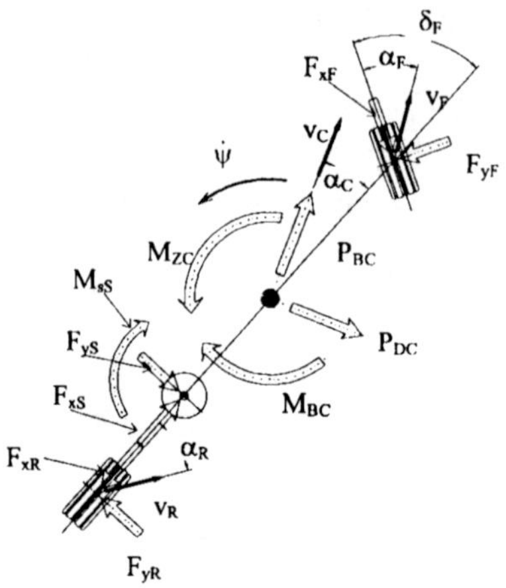

Figure 5.24. Dynamics of the truck tractor.

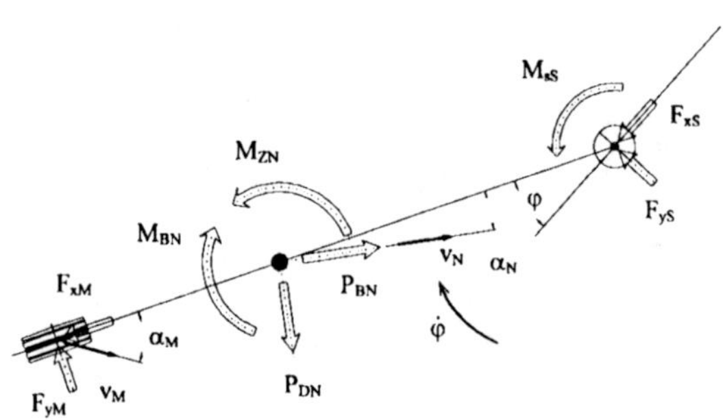

Figure 5.25. The semi-trailer dynamics.

where:

$$a_{11} = -(L_S + L_N - a) \cdot m_N;$$

$$b_{11} = v_C \cdot (m_N + m_C) - \frac{(L_S + L_0 - a) \cdot k_M + (L - a) \cdot k_R - a \cdot k_F}{v_C};$$

$$c_{11} = 0; \quad a_{12} = L_N \cdot m_N; \quad b_{12} = \frac{k_M \cdot L_0}{v_C}; \quad c_{12} = k_M; \quad a_{13} = 0;$$

$$b_{13} = v_C \cdot (m_N + m_C); \quad c_{13} = k_F + k_R + k_M; \quad e_1 = \delta_F \cdot k_F; \quad (5.41)$$

$$a_{21} = J_{zC}; \quad b_{21} = m_C \cdot (L_S - a) \cdot v_C + \frac{a \cdot k_F \cdot L_S + (L - a) \cdot (L - L_S)}{v_C};$$

$$c_{21} = 0; \quad a_{22} = 0; \quad b_{22} = 0; \quad c_{22} = 0; \quad a_{23} = 0;$$

$$b_{23} = m_C \cdot v_C \cdot (L_S - a); \quad c_{23} = k_F \cdot L_S - k_R \cdot (L - L_S);$$
$$e_2 = \delta_F \cdot k_F \cdot L_S + M_{zC} - M_{sS}; \tag{5.42}$$

$$a_{31} = (L_S + L_N - a) \cdot (L_0 - L_N) \cdot m_N - J_{zN};$$
$$b_{31} = \frac{(L-a) \cdot L_0 \cdot k_R - a \cdot L_0 \cdot k_F}{v_C} - v_C \cdot (L_0 - L_N) \cdot m_N - v_C \cdot L_0 \cdot m_C;$$
$$c_{31} = 0; \quad a_{32} = J_{zN} - L_N \cdot (L_0 - L_N) \cdot m_N; \quad b_{32} = 0; \quad c_{32} = 0; \quad a_{33} = 0;$$
$$b_{33} = -v_C \cdot L_0 \cdot m_C - v_C \cdot (L_0 - L_N) \cdot m_N; \quad c_{33} = -k_F \cdot L_0 - k_R \cdot L_0;$$
$$e_3 = -\delta_F \cdot k_F \cdot L_0 - M_{zN} - M_{sS}. \tag{5.43}$$

Since some of the coefficients vanish and since the following notation is introduced

$$x_1 = \dot{\psi}, \; \dot{x}_1 = \ddot{\psi}, \; x_2 = \varphi, \; x_3 = \dot{\varphi}, \; \dot{x}_3 = \ddot{\varphi}, \; x_4 = \alpha_C, \; \dot{x}_4 = \dot{\alpha}_C,$$

the following equations are further analysed

$$\dot{x}_1 = cc_1 \cdot x_1 + cc_2 \cdot x_2 + cc_3 \cdot x_3 + cc_4 \cdot x_4 + cc_5;$$
$$\dot{x}_2 = x_3;$$
$$\dot{x}_3 = bb_1 \cdot x_1 + bb_2 \cdot x_2 + bb_3 \cdot x_3 + bb_4 \cdot x_4 + bb_5;$$
$$\dot{x}_4 = aa_1 \cdot x_1 + aa_2 \cdot x_2 + aa_3 \cdot x_3 + aa_4 \cdot x_4 + aa_5;$$

where:

$$aa_1 = \frac{a_{21} \cdot b_{11} \cdot a_{32} - a_{11} \cdot b_{21} \cdot a_{32} + a_{31} \cdot b_{21} \cdot a_{12} - a_{21} \cdot b_{31} \cdot a_{12}}{a_{11} \cdot b_{23} \cdot a_{32} - a_{21} \cdot b_{13} \cdot a_{32} + a_{21} \cdot b_{33} \cdot a_{12} - a_{31} \cdot b_{23} \cdot a_{12}};$$
$$aa_2 = \frac{c_{12} \cdot a_{21} \cdot a_{32}}{a_{11} \cdot b_{23} \cdot a_{32} - a_{21} \cdot b_{13} \cdot a_{32} + a_{21} \cdot b_{33} \cdot a_{12} - a_{31} \cdot b_{23} \cdot a_{12}};$$
$$aa_3 = \frac{b_{12} \cdot a_{21} \cdot a_{32}}{a_{11} \cdot b_{23} \cdot a_{32} - a_{21} \cdot b_{13} \cdot a_{32} + a_{21} \cdot b_{33} \cdot a_{12} - a_{31} \cdot b_{23} \cdot a_{12}};$$
$$aa_4 = \frac{a_{21} \cdot c_{13} \cdot a_{32} - a_{11} \cdot c_{23} \cdot a_{32} + a_{31} \cdot c_{23} \cdot a_{12} - a_{21} \cdot c_{33} \cdot a_{12}}{a_{11} \cdot b_{23} \cdot a_{32} - a_{21} \cdot b_{13} \cdot a_{32} + a_{21} \cdot b_{33} \cdot a_{12} - a_{31} \cdot b_{23} \cdot a_{12}};$$
$$aa_5 = \frac{a_{21} \cdot e_3 \cdot a_{12} - a_{31} \cdot e_2 \cdot a_{12} + a_{11} \cdot a_{32} \cdot e_2 - e_1 \cdot a_{12} \cdot a_{32}}{a_{11} \cdot b_{23} \cdot a_{32} - a_{21} \cdot b_{13} \cdot a_{32} + a_{21} \cdot b_{33} \cdot a_{12} - a_{31} \cdot b_{23} \cdot a_{12}}; \tag{5.44}$$

$$bb_1 = \frac{a_{31} \cdot b_{21} - a_{21} \cdot b_{31} + a_{31} \cdot b_{23} \cdot aa_1 - a_{21} \cdot b_{33} \cdot aa_1}{a_{32} \cdot a_{21}};$$
$$bb_2 = \frac{a_{31} \cdot b_{23} - a_{21} \cdot b_{33}}{a_{32} \cdot a_{21}} \cdot aa_2; \quad bb_3 = \frac{a_{31} \cdot b_{23} - a_{21} \cdot b_{33}}{a_{32} \cdot a_{21}} \cdot aa_3; \tag{5.45}$$

$$bb_4 = \frac{a_{31} \cdot b_{23} \cdot aa_4 - a_{21} \cdot b_{33} \cdot aa_4 + a_{31} \cdot c_{23} - a_{21} \cdot c_{33}}{a_{32} \cdot a_{21}};$$

$$bb_5 = \frac{a_{31} \cdot b_{23} - a_{21} \cdot b_{33}}{a_{32} \cdot a_{21}} \cdot aa_5 + \frac{a_{21} \cdot e_3 - a_{31} \cdot e_2}{a_{32} \cdot a_{21}}; \tag{5.46}$$

$$cc_1 = -\frac{b_{21} + b_{23} \cdot aa_1}{a_{21}}; \quad cc_2 = -\frac{b_{23} \cdot aa_2}{a_{21}}; \quad cc_3 = -\frac{b_{23} \cdot aa_3}{a_{21}};$$

$$cc_4 = -\frac{c_{23} + b_{23} \cdot aa_4}{a_{21}}; \quad cc_5 = -\frac{b_{23}}{a_{21}} \cdot aa_5 + \frac{e_2}{a_{21}}. \tag{5.47}$$

The frequency characteristic of the system (5.44)–(5.47) with the excitation $\delta_F(t)$ is shown in Figure 5.26.

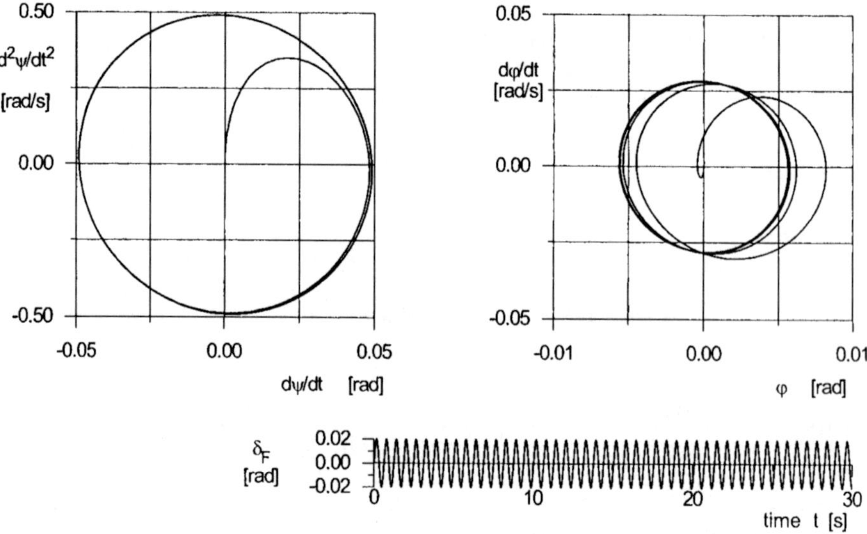

Figure 5.26. The characteristics $\dot{\psi}(\psi)$ and $\dot{\varphi}(\varphi)$ for the harmonic excitation.

The obtained system motion is periodic, and the $\dot{\psi} = \varphi = \dot{\varphi} = 0$ defines its equilibrium state.

2.2 Stability in the Lyapunov Sense

In our case, the problem is reduced to analysis of the following equilibrium state: $\dot{\psi} = 0$, $\varphi = 0$, $\dot{\varphi} = 0$, $\alpha_C = 0$. The equilibrium state of the unit vehicle can be interpreted as the realization of an arbitrary straight line trajectory of the vehicle path on the XY plane (see Figure 5.23). It means that the vehicle leaving the required path does not mean a loss of stability, but can be rather treated as a violation of a certain constraint. An investigation of stability in the sense of Lyapunov of the system (5.47) is reduced to investigation of the homogeneous system defined by the homogeneous part of (5.47), i.e. the non-homogeneous terms cc_5, bb_5 and aa_5 – are omitted.

Since the matrix A is defined by the coefficients of the homogeneous part of (5.47)

$$A = \begin{bmatrix} cc_1 & cc_2 & cc_3 & cc_4 \\ 0 & 0 & 1 & 0 \\ bb_1 & bb_2 & bb_3 & bb_4 \\ aa_1 & aa_2 & aa_3 & aa_4 \end{bmatrix},$$

and because the determinant W reads

$$W = \begin{vmatrix} cc_1 - s & cc_2 & cc_3 & cc_4 \\ 0 & -s & 1 & 0 \\ bb_1 & bb_2 & bb_3 - s & bb_4 \\ aa_1 & aa_2 & aa_3 & aa_4 - s \end{vmatrix},$$

the characteristic equation has the form

$$\begin{gathered} s^4 - (cc_1 + bb_3 + aa_4) \cdot s^3 \\ + (cc_1 \cdot bb_3 + aa_4 \cdot (cc_1 + bb_3) - aa_1 \cdot cc_4 - aa_3 \cdot bb_4 - cc_3 \cdot bb_1 - bb_2) \cdot s^2 \\ + (aa_1 \cdot cc_4 \cdot bb_3 + aa_3 \cdot bb_4 \cdot cc_1 + aa_4 \cdot bb_1 \cdot cc_3 - cc_1 \cdot bb_3 \cdot aa_4) \cdot s \\ + (-cc_3 \cdot bb_4 \cdot aa_1 + cc_4 \cdot bb_1 \cdot aa_3 + bb_2 \cdot aa_4 - aa_2 \cdot bb_4 - bb_1 \cdot cc_2) \cdot s \\ -bb_2 \cdot cc_1 \cdot aa_4 - cc_2 \cdot bb_4 \cdot aa_1 - cc_4 \cdot bb_1 \cdot aa_2 + aa_1 \cdot bb_2 \cdot cc_4 \\ +aa_2 \cdot bb_4 \cdot cc_1 + aa_4 \cdot bb_1 \cdot cc_2 = 0. \end{gathered}$$

Denoting

$$\begin{gathered} a_n = 1; \quad a_{n1} = -(cc_1 + bb_3 + aa_4); \\ a_{n2} = (cc_1 \cdot bb_3 + aa_4 \cdot (cc_1 + bb_3) - aa_1 \cdot cc_4 - aa_3 \cdot bb_4 - cc_3 \cdot bb_1 - bb_2); \\ a_{n3} = aa_1 \cdot cc_4 \cdot bb_3 + aa_3 \cdot bb_4 \cdot cc_1 + aa_4 \cdot bb_1 \cdot cc_3 - cc_1 \cdot bb_3 \cdot aa_4 \\ -cc_3 \cdot bb_4 \cdot aa_1 - cc_4 \cdot bb_1 \cdot aa_3 + bb_2 \cdot cc_1 + bb_2 \cdot aa_4 - aa_2 \cdot bb_4 - bb_1 \cdot cc_2; \\ a_{n4} = -bb_2 \cdot cc_1 \cdot aa_4 - cc_2 \cdot bb_4 \cdot aa_1 - cc_4 \cdot bb_1 \cdot aa_2 \\ +aa_1 \cdot bb_2 \cdot cc_4 + aa_2 \cdot bb_4 \cdot cc_1 + aa_4 \cdot bb_1 \cdot cc_2 \end{gathered} \tag{5.48}$$

and applying the Routh–Hurwitz criterion, the following stability conditions are obtained (for $a_n > 0$ and $n \geq 1$)

$$a_{n1} > 0; \quad \begin{vmatrix} a_{n1} & 1 \\ a_{n3} & a_{n2} \end{vmatrix} > 0; \quad \begin{vmatrix} a_{n1} & 1 & 0 \\ a_{n3} & a_{n2} & a_{n1} \\ 0 & a_{n4} & a_{n3} \end{vmatrix} > 0;$$

$$\begin{vmatrix} a_{n1} & 1 & 0 & 0 \\ a_{n3} & a_{n2} & a_{n1} & 1 \\ 0 & a_{n4} & a_{n3} & a_{n2} \\ 0 & 0 & 0 & a_{n4} \end{vmatrix} > 0; \tag{5.49}$$

which yield

$$a_{n1} > 0; \quad a_{n1} \cdot a_{n2} - a_{n3} > 0;$$

$$a_{n1} \cdot a_{n2} \cdot a_{n3} - a_{n4} \cdot a_{n1} \cdot a_{n1} - a_{n3} \cdot a_{n3} > 0; \quad a_{n4} > 0.$$

Introducing the notation

$$WS1 = a_{n1}; \quad WS2 = a_{n1} \cdot a_{n2} - a_{n3};$$

$$WS3 = a_{n1} \cdot a_{n2} \cdot a_{n3} - a_{n4} \cdot a_{n1} \cdot a_{n1} - a_{n3} \cdot a_{n3}; \quad WS4 = a_{n4}; \tag{5.50}$$

the stability conditions in the Lyapunov sense are given by the inequalities

$$WS1 > 0 \text{ and } WS2 > 0 \text{ and } WS3 > 0 \text{ and } WS4 > 0. \tag{5.51}$$

EXAMPLE 5.1 *Analyse stability in the sense of Lyapunov for the vehicle model governed by the equations* (5.44)–(5.47).

The following data are taken: $m_C = 6250\,kg$, $m_N = 6160\,kg$, $J_{zC} = 10000\,kg \cdot m^2$, $J_{zN} = 25000\,kg \cdot m^2$, $a = 1.27\,m$, $L_S = 2.87\,m$, $L = 3.6\,m$, $L_N = 4\,m$, $L_0 = 7.29\,m$, $v_C = 16\,m/s$, $k_F = 300000\,N/rad$, $k_R = 800000\,N/rad$, $k_M = 1000000\,N/rad$, $M_{zC} = 10\,N \cdot m$, $M_{zN} = 10\,N \cdot m$, $M_{sS} = 10\,N \cdot m$.

The obtained results are shown in Figure 5.27, where the relative coefficients $WSi/WSi_{\max}$ *(*$WSi_{\max}$ *denoting the maximal value of* WSi*) are reported.*

The obtained results will be discussed further together with the results of Example 5.2 (see Section 2.3).

2.3 Stability in the Sense of Bogusz

The equilibrium state of the equations (5.47) governing dynamics of the unit vehicle (see Figures 5.20–5.25) is analysed. The initial conditions zone ω and the admissible solutions zone Ω are defined in the following way:

$$\omega \equiv \left\{x_1^2 < r1^2, x_2^2 < r2^2, x_3^2 < r3^2, x_4^2 < r4^2\right\}; \tag{5.52}$$

$$\Omega \equiv \left\{x_1^2 \leqslant R1^2, x_2^2 \leqslant R2^2, x_3^2 \leqslant R3^2, x_4^2 \leqslant R4^2\right\}. \tag{5.53}$$

The scalar Bogusz function $V_B(x_1, x_2, x_3, x_4)$ has the form

$$V_B(x_1, x_2, x_3, x_4) = 0.5\left(AA \cdot x_1^2 + BB \cdot x_2^2 + CC \cdot x_3^2 + DD \cdot x_4^2\right), \tag{5.54}$$

where: $AA > 0$, $BB > 0$, $CC > 0$ and $DD > 0$.

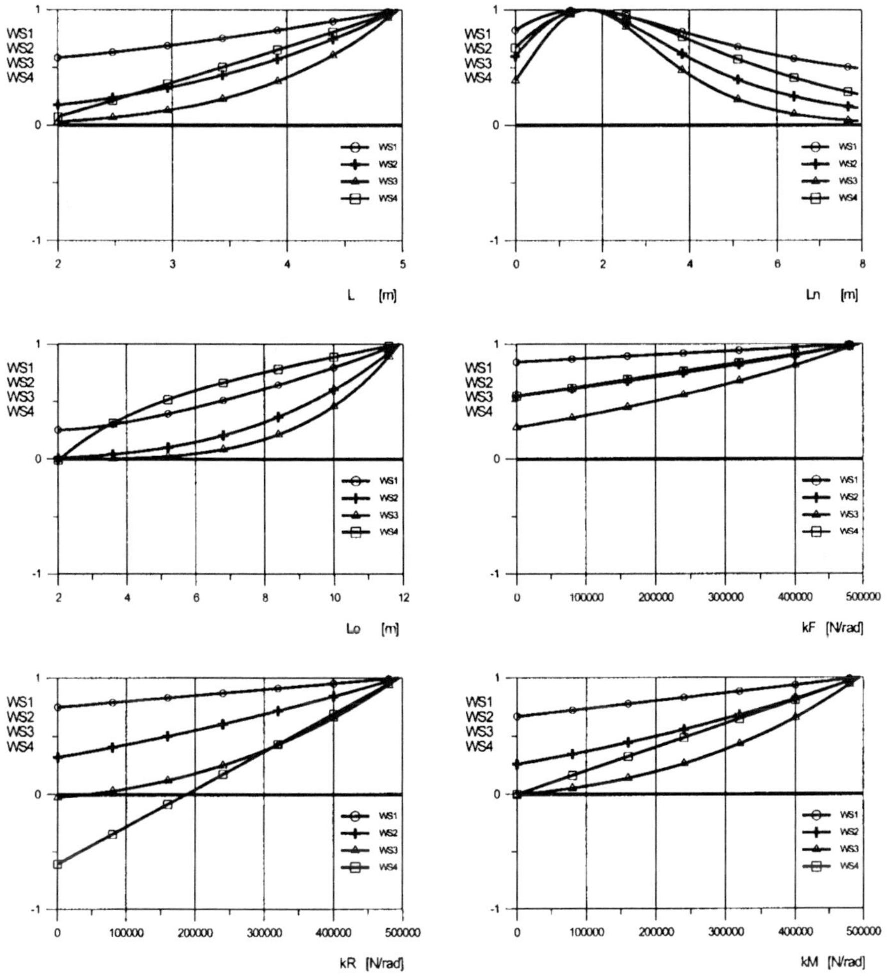

Figure 5.27. a See 5.27b.

The stability conditions for the introduced zones ω and Ω are defined below.

If there exists a number C_0 satisfying the inequality $C_0 \geq V_B(x_1, x_2, x_3, x_4)$ for (x_1, x_2, x_3, x_4) belonging to zone ω, $C_0 > 0$, i.e.

$$C_0 = \sup_{x_i \in \omega} V_B(x_1, x_2, x_3, x_4) =$$

$$0.5\left(AA \cdot r1^2 + BB \cdot r2^2 + CC \cdot r3^2 + DD \cdot r4^2\right) \qquad (5.55)$$

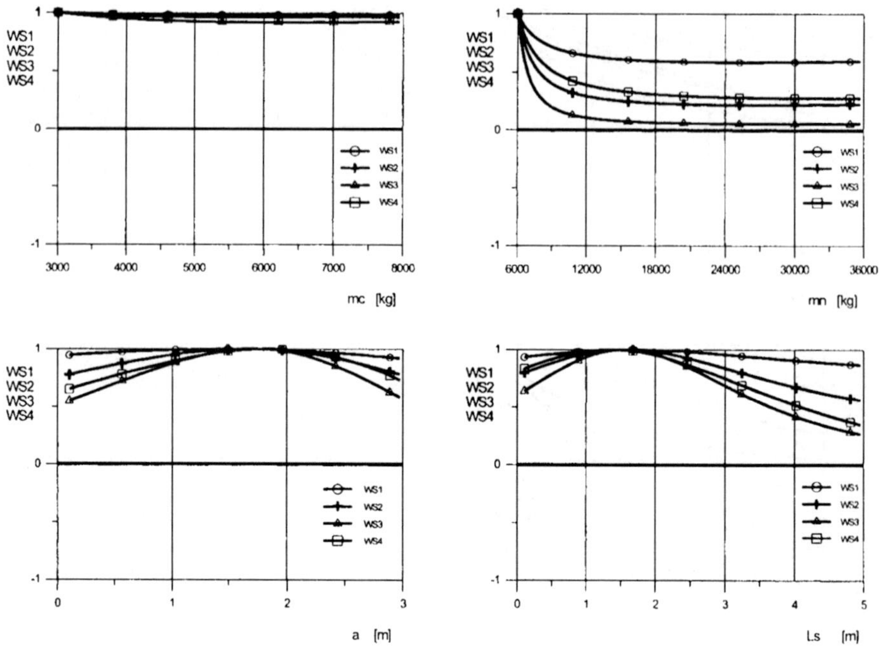

Figure 5.27. *b* Relative stability coefficients versus the parameters m_C, m_N, a, L_S, L, L_N, L_0, k_F, k_R and k_M.

and if there exists a number $C_1 \leq V_B(x_1, x_2, x_3, x_4)$ for (x_1, x_2, x_3, x_4) belonging to zone Ω, $C_1 > 0$, i.e.

$$C_1 = \inf_{x_i \notin \Omega} V_B(x_1, x_2, x_3, x_4)$$

$$= 0.5 \left(AA \cdot R_1^2 + BB \cdot R_2^2 + CC \cdot R_3^2 + DD \cdot R_4^2 \right), \tag{5.56}$$

then for dV_B/dt along the solutions of (5.47), where

$$\frac{dV_B}{dt} = AA{\cdot}cc_1{\cdot}x_1^2 + CC{\cdot}bb_3{\cdot}x_3^2 + DD{\cdot}aa_4{\cdot}x_4^2 + AA{\cdot}cc_5{\cdot}x_1 + CC{\cdot}bb_5{\cdot}x_3$$
$$+DD{\cdot}aa_5{\cdot}x_4 + AA{\cdot}cc_2{\cdot}x_1{\cdot}x_2 + (AA{\cdot}cc_3 + CC{\cdot}bb_1){\cdot}x_1{\cdot}x_3 + DD{\cdot}aa_2{\cdot}x_2{\cdot}x_4$$
$$+ (AA{\cdot}cc + DD{\cdot}aa_1)x_1x_4 + (BB + CC{\cdot}bb_2)x_2x_3 + (CC{\cdot}bb_4 + DD{\cdot}aa_3)x_3x_4, \tag{5.57}$$

a stability condition in the sense of Bogusz reads

$$\sup_{x_i \in \Omega/\omega, t_1 \leqslant t \leqslant t_1 + T} \left\{ \frac{dV_B}{dt} \right\} < \frac{C1 - C0}{T}. \tag{5.58}$$

Table 5.4.

	ω			Ω	
$r_1(\dot{\psi})$	0.10	rad/s	$R_1(\dot{\psi})$	0.50	rad/s
$r_2(\varphi)$	0.01	rad	$R_2(\varphi)$	0.05	rad
$r_3(\dot{\varphi})$	0.02	rad/s	$R_3(\dot{\varphi})$	0.25	rad/s
$r_4(\alpha_c)$	0.01	rad	$R_4(\alpha_c)$	0.05	rad

In practice, a check of the validity of (5.58) is carried out by the numerical simulation. In fact, a numerical check is reduced to analysis of the function dV_B/dt in the space $\{x_1, x_2, x_3, x_4\} \in \Omega/\omega$, i.e. in the space

$$\frac{\Omega}{\omega} \equiv \left\{r1^2 < x_1^2 \leqslant R1^2, r2^2 < x_2^2 \leqslant R2^2, r3^2 < x_3^2 \leqslant R3^2, r4^2 < x_4^2 \leqslant R4^2\right\}.$$

As the technical stability factor the following number is taken: $TS = (C_1 - C_0)/T - dV/dt$.

Accounting for (5.56), (5.57) and (5.58) one gets

$$TS = \frac{C_1 - C_0}{T} - AA \cdot cc_1 \cdot x_1^2 - CC \cdot bb_3 \cdot x_3^2 - DD \cdot aa_4 \cdot x_4^2 - AA \cdot cc_5 \cdot x_1$$
$$-CC \cdot bb_5 \cdot x_3 - DD \cdot aa_5 \cdot x_4 - AA \cdot cc_2 \cdot x_1 \cdot x_2 - (AA \cdot cc_3 + CC \cdot bb_1) \cdot x_1 \cdot x_3$$
$$-DD \cdot aa_2 \cdot x_2 \cdot x_4 - (AA \cdot cc_4 + DD \cdot aa_1) \cdot x_1 \cdot x_4$$
$$-(BB + CC \cdot bb_2) \cdot x_2 \cdot x_3 - (CC \cdot bb + DD \cdot aa_3) \cdot x_3 \cdot x_4 \qquad (5.59)$$

and a positive value of TS denotes that stability is conserved.

EXAMPLE 5.2 *Derive stability conditions in the Bogusz sense.*

The following data are applied: $m_C = 6250\,kg$, $m_N = 6160\,kg$, $J_{zC} = 10000\,kg \cdot m^2$, $J_{zN} = 25000\,kg \cdot m^2$, $a = 1.27\,m$, $L_S = 2.87\,m$, $L = 3.6\,m$, $L_N = 4\,m$, $L_0 = 7.29\,m$, $v_C = 16\,m/s$, $k_F = 300000\,N/rad$, $k_R = 800000\,N/rad$, $k_M = 1000000\,N/rad$, $M_{zC} = 10\,N \cdot m$, $M_{zN} = 10\,N \cdot m$, $M_{sS} = 10\,N \cdot m$ *and monitoring time* $T = 2\,s$.

Zone of initial and admissible solutions are taken intuitively, and they are given in Table 5.4.

An influence of the chosen vehicle parameters governed by the system (5.47) *is investigated. The analysis is carried out on the basis of the introduced theory and in accordance with the procedures defined below.*

Subroutine of initial data

1. *Vehicle data*
   $m_C, m_N, J_{zC}, J_{zN}, a, L_S, L, L_N, L_0, v_C, k_F, k_R, k_M, M_{zC}, M_{zN}, M_{sS}$

2. Data defining technical stability

2.1. Zones of initial ω and admissible Ω conditions

Procedure for finding maximum of $\frac{dV_B}{dt}$, $x_i \in \omega$

1. *Calculation of dV_B/dt for the starting values $x_i = -r_i$*

$$\left(\frac{dV_B}{dt}\right)_{start} = AA\cdot cc_1\cdot x_1^2 + CC\cdot bb_3\cdot x_3^2 + DD\cdot aa_4\cdot x_4^2 - AA\cdot cc_5\cdot x_1$$

$$-CC{\cdot}bb_5{\cdot}x_3 - DD{\cdot}aa_5{\cdot}x_4 + AA{\cdot}cc_2{\cdot}x_1{\cdot}x_2 + (AA\cdot cc_3 + CC\cdot bb_1){\cdot}x_1{\cdot}x_3$$

$$+(AA\cdot cc_4 + DD\cdot aa_1){\cdot}x_1{\cdot}x_4 + (BB + CC\cdot bb_2){\cdot}x_2{\cdot}x_3 + DD{\cdot}aa_2{\cdot}x_2{\cdot}x_4$$

$$+(CC\cdot bb_4 + DD\cdot aa_3)\cdot x_3\cdot x_4.$$

2. *A search for maximum in the interval $\omega\langle -r_i, +r_i\rangle$*
for $x_1 = -r_1$ to r_1 step h_1 for $x_2 = -r_2$ to r_2 step h_2
for $x_3 = -r_3$ to r_3 step h_3 for $x_4 = -r_4$ to r_4 step h_4

$$\frac{dV_B}{dt} = AA\cdot cc_1\cdot x_1^2 + CC\cdot bb_3\cdot x_3^2 + DD\cdot aa_4\cdot x_4^2 + AA\cdot cc_5\cdot x_1$$

$$+CC{\cdot}bb_5{\cdot}x_3 + DD{\cdot}aa_5{\cdot}x_4 + AA{\cdot}cc_2{\cdot}x_1{\cdot}x_2 + (AA\cdot cc_3 + CC\cdot bb_1){\cdot}x_1{\cdot}x_3$$

$$+(AA\cdot cc_4 + DD\cdot aa_1){\cdot}x_1{\cdot}x_4 + (BB + CC\cdot bb_2){\cdot}x_2{\cdot}x_3 + DD{\cdot}aa_2{\cdot}x_2{\cdot}x_4$$

$$+(CC\cdot bb_4 + DD\cdot aa_3)\cdot x_3\cdot x_4$$

if $\frac{dV_B}{dt} > \left[\frac{dV_B}{dt}\right]_{\max}$ then $\left[\frac{dV_B}{dt}\right]_{\max} = \frac{dV_B}{dt}$

next next next next

Procedure of stability investigation and calculation of the FT factor

1. *Calculation of the numbers C_0 and C_1*

$$C_0 = 0.5\left(AA\cdot r_1^2 + BB\cdot r_2^2 + CC\cdot r_3^2 + DD\cdot r_4^2\right),$$

$$C_1 = 0.5\left(AA\cdot R_1^2 + BB\cdot R_2^2 + CC\cdot R_3^2 + DD\cdot R_4^2\right).$$

2. *Estimation of technical stability measure*

$$\left[\frac{dV_B}{dt}\right]_{\max} < \frac{C1 - C0}{T}; \quad FT = \frac{C1 - C0}{T} - \left[\frac{dV_B}{dt}\right]_{\max}$$

The numerical results are reported in Figure 5.28, where the stability factors are reported as a function of the analysed quantities.

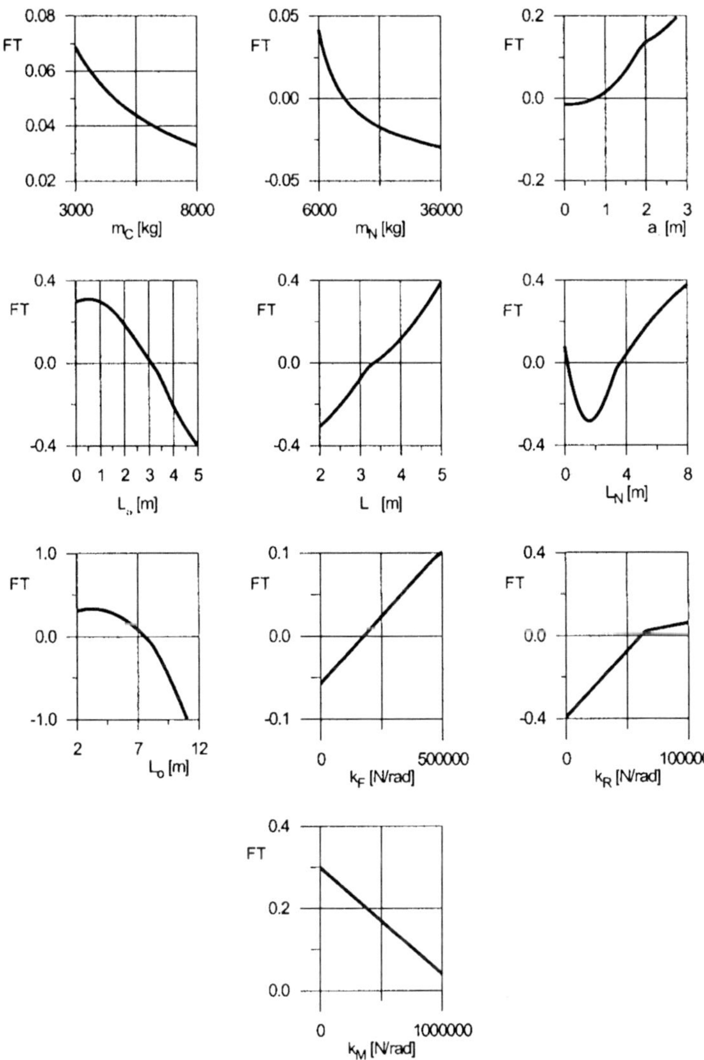

Figure 5.28. Influence of the unit vehicle on the stability in Bogusz sense.

The following conclusions are yielded from stability analysis on a basis of Examples 5.1 and 5.2:

(i) the system stability is conserved for the investigated tractor mass m_C;

(ii) an influence of the semi-trailer mass on the vehicle stability is significant. The stability decreases with an increase of the semi-trailer mass and for

$m_N > 9000\,kg$ the technical stability in the Bogusz sense is not preserved (see Figure 5.28);

(iii) an influence of position of the vehicle centre (parameter "a" in Figure 5.20); in this case for the Lyapunov stability the maximum of FT occurs for the interval $1.8 - 1.9\,m$; the system is technically unstable for $a < 0.8$;

(iv) influence of the joint position (the L_S parameter - Figure 5.20): the system is strongly stable in the sense of Lyapunov for $L_S = 1.4 - 1.6$, whereas the maximal technical stability is achieved for $L_S = 1.0\,m$ (for $L_S > 3\,m$ the system looses its stability);

(v) influence of the tractor wheel base L (see Figure 5.20): an increase of L increases both the Lyapunov and technical stability (in the latter case the system is unstable for $L > 3.3\,m$);

(vi) influence of the semi-trailer mass centre L_N (see Figure 5.20): the maximal system stability is achieved for $L_N = 1.8\,m$, and starts to decrease for $L_N > 1.8\,m$; whereas stability in the Bogusz sense is conserved for $L_N < 0.2\,m$ and $L_N > 3.5\,m$;

(vii) influence of the parameter L_0 (Figure 5.20): the Lyapunov stability increases with increase of L_0, whereas the maximal technical stability is achieved for $L_0 = 3\,m$ (for $L_0 > 7.4\,m$ the system becomes unstable);

(viii) influence of the parameters k_F, k_R and k_M: larger values of k_F, k_R and k_M correspond to stability increases in the Lyapunov sense (for $k_R < 200000\,N/rad$ the system loses stability); the same holds for the Bogusz sense stability, but $FT > 0$ for $k_F > 300000\,N/rad$ and $k_R > 600000\,N/rad$ (for $k_F < 300000\,N/rad$ and $k_R < 600000\,N/m$ the system is unstable in the Bogusz sense); FT decreases when k_M is increased (the system is stable for $k_M = 0 \div 1000000\,N/m$).

Let us briefly comment on joint friction influence on stability (see Figure 5.29). Note that the high value friction forces are generated by a special brake, which is applied to stabilize the system tractor-semi-trailer. The obtained increase of the technical stability is referred to as Coulomb friction $M_{sS} = M_{sS} \cdot \text{sign}(\dot{\varphi})$, where $\dot{\varphi}$ is the relative tractor and semi-trailer displacement.

Finally, an influence of perturbations on the tractor-semi-trailer stability is studied. In the vehicle model shown in Figures 5.21–5.25 the perturbations acting on the system are represented by the rotation moments M_{zC} (acting on the tractor) and M_{zN} (acting on the semi-trailer). In Figure 5.30 the numerical results of stability investigations are reported. The system is more sensitive to perturbations acting on the tractor, for example already $M_{zC} >$

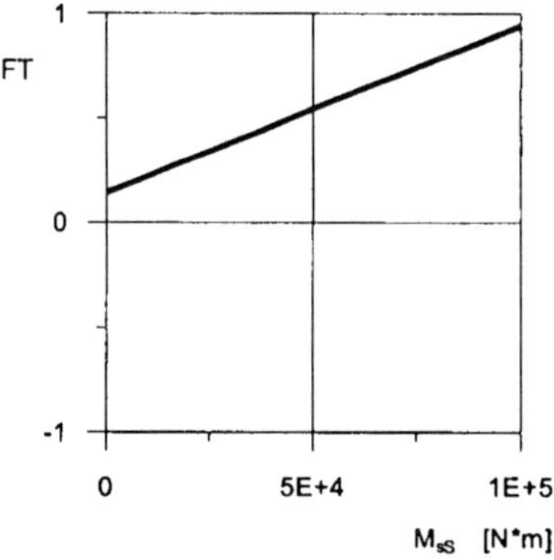

Figure 5.29. Influence of the rotational moment of friction in the joint on the technical stability.

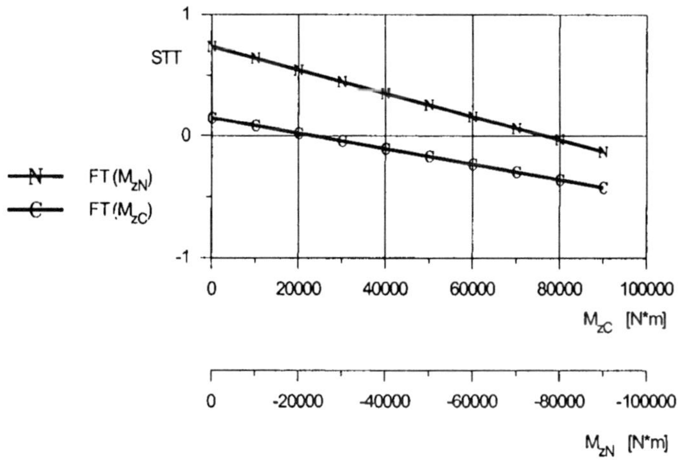

Figure 5.30. Influence of perturbations M_{zC} and M_{zN} on the technical stability.

$22000\,N \cdot m$ causes the instability, whereas the same result is obtained for $M_{zN} > 75000\,N \cdot m$.

To conclude, the effective methods of stability investigation introduced in this chapter allowed for analysis of the parameters influence on stability and the choice of optimal parameters with regard to stability. Both stability in the Lyapunov and Bogusz sense are applied. In particular, an application of stability in the technical sense yields the stability diagrams for the given norms, tolerances and deviation zones occurring in real objects and real conditions.

2.4 Stability of the System: Driver-Vehicle

It is sure that a universal human body model can not be satisfactorily designed. In addition, brain and nerve systems behaviour can not be predicted. Also psychological and emotional changes depending on the time of day can lead to complex modelling of the brain. A driver's reaction on a brake pedal or on a steering gear may depend strongly on the current driver's psychological state, alcohol, drugs, or a general physical condition. The circumstances mentioned in the above indicate a practical impossibility for obtaining a proper *driver model*. Mainly the constructed models are limited to interaction modelling for given conditions and includes mainly a driver kinematic reaction on a steering gear. As initial data usually the quantities associated with a transversal car kinematics are applied, i.e.

$$\delta_K = \Phi\left(\psi, \dot{\psi}, \ddot{\psi},\ y, \dot{y}, \ddot{y}, .\tau_\psi, \tau_y, \dots\right),$$

where: δ_K is the rotation angle of the steering gear wheel; $\psi, \dot{\psi}$ and $\ddot{\psi}$ are the angle rotational velocity and the rotational acceleration of a car; $y, \dot{y}$ and $\ddot{y}$ are the transversal displacement velocity and acceleration of a car; τ_ψ and τ_y are delays of the driver's reaction.

In what follows the so-called *anticipative* driver model with a delay is considered. The following main model hypothesis is taken into account [1]. A driver always observes a point on a road. The *observable point* lies on a required car trajectory in a distance of L_0 from the driver. L_0 depends on the car velocity and the trajectory curvature. The steering drivers quantities are the visual ones: a sight angle of the observable point and the rotational velocity $\dot{\sigma}$ (Figure 5.31).

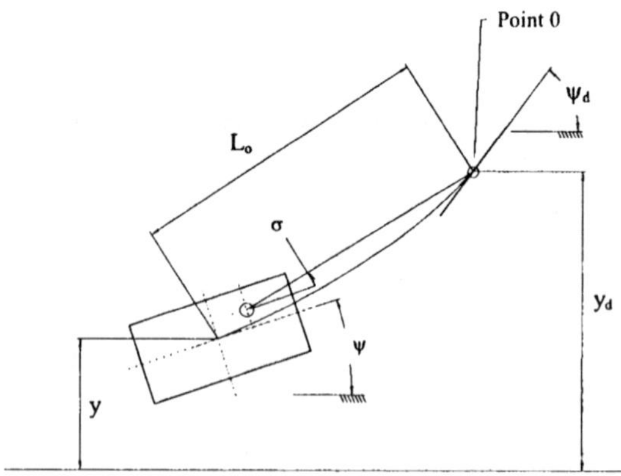

Figure 5.31. The parameters of the anticipative driver model.

The *steering function* is governed by the relation

$$\delta_K(t) = A_\sigma \cdot [\sigma(t - \tau_\sigma)] + A_{\dot\sigma} \cdot [\dot\sigma(t - \tau_{\dot\sigma})],$$

where: δ_K is the rotation angle of the steering wheel; $A_{\sigma,\dot\sigma}$ are the coefficients and $\tau_{\sigma,\dot\sigma}$ are time delays.

Taking into account the kinematical relations (see Figure 5.31) the steering function of the front car wheels roads

$$\delta_F(t) = -\frac{w_y}{L_o} \cdot y(t-\tau_y) + \frac{w_y}{L_d} \cdot y_p(t-\tau_y) - w_\psi \cdot \Psi(t-\tau_\psi) + w_\psi \cdot \Psi_d(t-\tau_\psi), \tag{5.60}$$

where: y is the transversal car position; y_d is the required position (the transversal displacement) in the point 0; ψ is the actual car angle position; ψ_d is the required car angle position in the point 0; $w_{y,\psi}$ are constant coefficients.

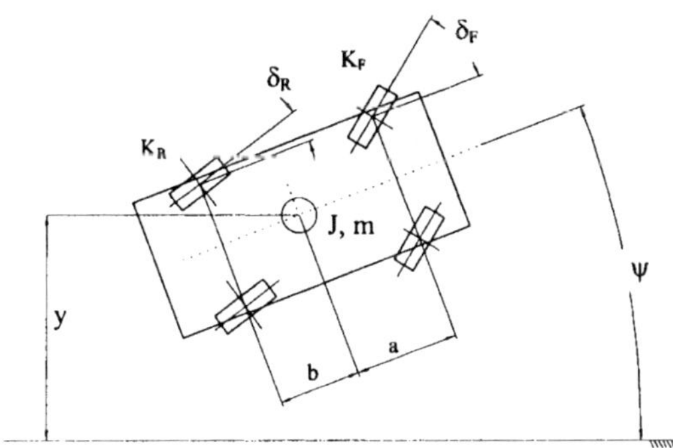

Figure 5.32. Coordinates of the car models.

The car equations of motion (see Figure 5.32) are

$$m \cdot \frac{d^2 y}{dt^2} = 2 \cdot K_F \cdot \left(\delta_F - \frac{1}{V} \cdot \frac{dy}{dt} - \frac{a}{V} \cdot \frac{d\Psi}{dt}\right) + 2 \cdot K_R \cdot \left(\delta_R - \frac{1}{V} \cdot \frac{dy}{dt} + \frac{b}{V} \cdot \frac{d\Psi}{dt}\right), \tag{5.61}$$

$$J \cdot \frac{d^2 \Psi}{dt^2} = 2 \cdot K_F \cdot \left(\delta_F - \frac{1}{V} \cdot \frac{dy}{dt} - \frac{a}{V} \cdot \frac{d\Psi}{dt}\right) \cdot a - 2 \cdot K_R \cdot \left(\delta_R - \frac{1}{V} \cdot \frac{dy}{dt} + \frac{b}{V} \cdot \frac{d\Psi}{dt}\right) \cdot b, \tag{5.62}$$

where: δ_F (δ_R) is the front (respectively, rear) wheels turn angle; Ψ is yaw angle; $K_{F,R}$ are the resistance coefficients against the transversal front (respectively, rear) tire slip; $K_{F,R} = \left.\frac{dF_{y\,F,R}}{d\alpha_{F,R}}\right|_{\alpha_{F,R}=0}$; J is the yawing moment of inertia; m is the car mass; v is the car velocity. Assuming $\delta_R = \delta_F \cdot i_{pk}$, where i_{pk} is the relative ratio of the end steering gear, and taking into account the equations (5.61) and (5.62) one gets

$$\ddot{y} = \frac{2\cdot(K_F+K_R\cdot i_{pk})}{m}\cdot\delta_F - \frac{2}{m\cdot V}\cdot(K_F+K_R)\cdot\dot{y} + \frac{2}{m\cdot V}\cdot(K_R\cdot b - K_F\cdot a)\cdot\dot{\Psi}, \quad (5.63)$$

$$\ddot{\Psi} = \frac{2\cdot(K_F\cdot a - K_R\cdot b\cdot i_{pk})}{J}\cdot\delta_F + \frac{2}{J\cdot V}\cdot(K_R\cdot b - K_F\cdot a)\cdot\dot{y} - \frac{2}{J\cdot V}\cdot\left(K_R\cdot b^2 + K_F\cdot a^2\right)\cdot\dot{\Psi}. \quad (5.64)$$

Let us introduce the notation

$$a_1 = \frac{2\cdot(K_F+K_R\cdot i_{pk})}{m}, \quad a_3 = -\frac{2\cdot(K_F+K_R)}{m\cdot V},$$

$$a_4 = \frac{2\cdot(K_R\cdot b - K_F\cdot a)}{m\cdot V}, \quad b_1 = \frac{2\cdot(K_F\cdot a - K_R\cdot b\cdot i_{pk})}{J},$$

$$b_3 = \frac{2\cdot(K_R\cdot b - K_F\cdot a)}{J\cdot V}, \quad b_4 = -\frac{2\cdot\left(K_R\cdot b^2 + K_F\cdot a^2\right)}{J\cdot V}.$$

Using the Laplace transformations in equation (5.60), (5.63) and (5.64) and assuming the straight line motion ($y_d = \psi_d = 0$) one obtains

$$\delta_F(p) = -\frac{w_y}{L_o}\cdot\left[y(p)\cdot e^{-\tau_y\cdot p}\right] - w_\psi\cdot\left[\Psi(p)\cdot e^{-\tau_\psi\cdot p}\right], \quad (5.65)$$

$$p^2\cdot y(p) = a_1\cdot\delta_F(p) + p\cdot a_3\cdot y(p) + p\cdot a_4\cdot\Psi(p), \quad (5.66)$$

$$p^2\cdot\Psi(p) = b_1\cdot\delta_F(p) + p\cdot b_3\cdot y(p) + p\cdot b_4\cdot\Psi(p). \quad (5.67)$$

The transition functions for (5.65) are denoted by

$$K_y^{\delta_F}(p) = \frac{\delta_F(p)}{y(p)}, \quad K_\psi^{\delta_F}(p) = \frac{\delta_F(p)}{\psi(p)},$$

and they read

$$K_y^{\delta_F}(p) = \frac{w_y}{L_o}\cdot e^{-\tau_y\cdot p}, \quad K_\psi^{\delta_F}(p) = w_\psi\cdot e^{-\tau_\psi\cdot p}.$$

The spectrum transmittances have the forms

$$K_y^{\delta_F}(j\cdot\omega) = \frac{w_y}{L_o}\cdot e^{-j\cdot\tau_y\cdot\omega} = \frac{w_y}{L_o}\cdot(\cos(\tau_y\cdot\omega) - j\cdot\sin(\tau_y\cdot\omega)), \quad (5.68)$$

$$K_\psi^{\delta_F}(j\cdot\omega) = \frac{w_\psi}{L_o}\cdot e^{-j\cdot\tau_\psi\cdot\omega} = \frac{w_\psi}{L_o}\cdot(\cos(\tau_\psi\cdot\omega) - j\cdot\sin(\tau_\psi\cdot\omega)), \quad (5.69)$$

and their real and imaginary parts read

$$\mathrm{Re}\left[K_y^{\delta_F}(j\cdot\omega)\right] = \frac{w_y}{L_o}\cdot\cos(\tau_y\cdot\omega),$$

$$\text{Im}\left[K_y^{\delta_F}(j\cdot\omega)\right] = \frac{w_y}{L_o}\cdot\left(-\sin\left(\tau_y\cdot\omega\right)\right), \tag{5.70}$$

$$\text{Re}\left[K_\psi^{\delta_F}(j\cdot\omega)\right] = \frac{w_\psi}{L_o}\cdot\cos\left(\tau_\psi\cdot\omega\right),$$

$$\text{Im}\left[K_\psi^{\delta_F}(j\cdot\omega)\right] = \frac{w_\psi}{L_o}\cdot\left(-\sin\left(\tau_\psi\cdot\omega\right)\right), \tag{5.71}$$

where:

$$K_y^{\delta_F}(j\cdot\omega) = \text{Re}\left[K_y^{\delta_F}(j\cdot\omega)\right] + j\cdot\text{Im}\left[K_y^{\delta_F}(j\cdot\omega)\right],$$

$$K_\psi^{\delta_F}(j\cdot\omega) = \text{Re}\left[K_\psi^{\delta_F}(j\cdot\omega)\right] + j\cdot\text{Im}\left[K_\psi^{\delta_F}(j\cdot\omega)\right]. \tag{5.72}$$

The operator transmittances of the car model obtained from (5.66) and (5.67) are denoted by

$$P_{\delta_F}^{y}(p) = \frac{y(p)}{\delta_F(p)}, \quad P_{\delta_F}^{\Psi}(p) = \frac{\Psi(p)}{\delta_F(p)},$$

and they read

$$P_{\delta_F}^{y}(p) = \frac{p\cdot a_1 + (a_4\cdot b_1 - a_1\cdot b_4)}{p^3 - p^2\cdot(a_3+b_4) + p\cdot(a_3\cdot b_4 - a_4\cdot b_3)}, \tag{5.73}$$

$$P_{\delta_F}^{\Psi}(p) = \frac{p\cdot b_1 + (a_1\cdot b_3 - a_3\cdot b_1)}{p^3 - p^2\cdot(a_3+b_4) + p\cdot(a_3\cdot b_4 - a_4\cdot b_3)}. \tag{5.74}$$

Using the notation $A = a_1$, $B = a_4\cdot b_1 - a_1\cdot b_4$, $D = a_3 + b_4$, $E = a_3\cdot b_4 - a_4\cdot b_3$, $G = b_1$, $H = a_1\cdot b_3 - a_3\cdot b_1$, we get

$$P_{\delta_F}^{y}(p) = \frac{p\cdot A + B}{p^3 - p^2\cdot D + p\cdot E}, \tag{5.75}$$

$$P_{\delta_F}^{\Psi}(p) = \frac{p\cdot G + H}{p^3 - p^2\cdot D + p\cdot E}. \tag{5.76}$$

The following spectral transmittance is yielded by equation (5.75)

$$P_{\delta_F}^{y}(j\cdot\omega) = \frac{j\cdot\omega\cdot A + B}{-j\cdot\omega^3 + \omega^2\cdot D + j\cdot\omega\cdot E}, \tag{5.77}$$

whereas the equation (5.76) yields the following spectral transmittance

$$P_{\delta_F}^{\Psi}(j\cdot\omega) = \frac{j\cdot\omega\cdot G + H}{-j\cdot\omega^3 + \omega^2\cdot D + j\cdot\omega\cdot E}. \tag{5.78}$$

The transmittance (5.77) and (5.78) have the following real and imaginary parts

$$\mathrm{Re}\left[P_{\delta_F}^{y}(j\cdot\omega)\right] = \frac{\omega^2\cdot(A\cdot E + B\cdot D) - \omega^4\cdot A}{\omega^4\cdot D^2 + \omega^2\cdot(\omega^2 - E)^2},$$

$$\mathrm{Im}\left[P_{\delta_F}^{y}(j\cdot\omega)\right] = \frac{\omega^3\cdot(A\cdot D + B) - \omega\cdot B\cdot E}{\omega^4\cdot D^2 + \omega^2\cdot(\omega^2 - E)^2}, \tag{5.79}$$

$$\mathrm{Re}\left[P_{\delta_F}^{\Psi}(j\cdot\omega)\right] = \frac{\omega^2\cdot(D\cdot H + E\cdot G) - \omega^4\cdot G}{\omega^4\cdot D^2 + \omega^2\cdot(\omega^2 - E)^2},$$

$$\mathrm{Im}\left[P_{\delta_F}^{\Psi}(j\cdot\omega)\right] = \frac{\omega^3\cdot(D\cdot G + H) - \omega\cdot E\cdot H}{\omega^4\cdot D^2 + \omega^2\cdot(\omega^2 - E)^2}, \tag{5.80}$$

where:

$$P_{\delta_F}^{y}(j\cdot\omega) = \mathrm{Re}\left[P_{\delta_F}^{y}(j\cdot\omega)\right] + j\cdot\mathrm{Im}\left[P_{\delta_F}^{y}(j\cdot\omega)\right],$$

$$P_{\delta_F}^{\psi}(j\cdot\omega) = \mathrm{Re}\left[P_{\delta_F}^{\psi}(j\cdot\omega)\right] + j\cdot\mathrm{Im}\left[P_{\delta_F}^{\psi}(j\cdot\omega)\right]. \tag{5.81}$$

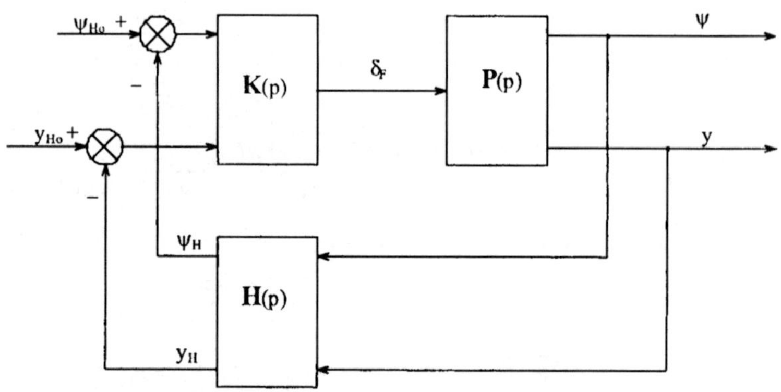

Figure 5.33. The driver-vehicle control system.

A two dimensional driver-vehicle control system is shown in Figure 5.33, where the following notation is used: $\mathbf{K}(p)$ is the transmittance of the controller (driver); $\mathbf{P}(p)$ is the matrix of the vehicle transmittance; $\mathbf{H}(p)$ is the transducers matrix. The mentioned matrices have the forms

$$\mathbf{K}(p) = \begin{bmatrix} K_y^{\delta_F}(p) & K_\psi^{\delta_F}(p) \end{bmatrix}; \quad \mathbf{P}(p) = \begin{bmatrix} P_{\delta_F}^{y}(p) \\ P_{\delta_F}^{\psi}(p) \end{bmatrix}; \quad \mathbf{H}(p) = \begin{bmatrix} 1 & 0 \\ 0 & 1 \end{bmatrix}. \tag{5.82}$$

A transfer function of the opened system (Figure 5.33) is defined by

$$\mathbf{J}(p) = \mathbf{P}(p) \cdot \mathbf{K}(p) \cdot \mathbf{H}(p), \tag{5.83}$$

and taking into account (5.82) and (5.83) we get

$$\mathbf{J}(p) = \begin{bmatrix} P_{\delta_F}^{y} \cdot K_{y}^{\delta_F} & P_{\delta_F}^{y} \cdot K_{\psi}^{\delta_F} \\ P_{\delta_F}^{\psi} \cdot K_{y}^{\delta_F} & P_{\delta_F}^{\psi} \cdot K_{\psi}^{\delta_F} \end{bmatrix}. \tag{5.84}$$

The determinant

$$\det\left(\mathbf{J}(p) - r(p) \cdot \mathbf{1}\right), \tag{5.85}$$

and the equation (5.84) yields

$$r^2(p) - r(p) \cdot \left(P_{\delta_F}^{y} \cdot K_{y}^{\delta_F} + P_{\delta_F}^{\psi} \cdot K_{y}^{\delta_F}\right) = 0, \tag{5.86}$$

where: $r(p)$ is the transmittance or the $\mathbf{J}(p)$ matrix eigenvalue. The equation (5.86) has two solutions:

1°.

$$r(p) = 0,$$

2°.

$$r(p) = P_{\delta_F}^{y} \cdot K_{y}^{\delta_F} + P_{\delta_F}^{\psi} \cdot K_{y}^{\delta_F}. \tag{5.87}$$

From (5.87) one gets

$$\begin{aligned} r(j \cdot \omega) = & \left[\mathrm{Re}\left[P_{\delta_F}^{y}(j \cdot \omega)\right] + j \cdot \mathrm{Im}\left[P_{\delta_F}^{y}(j \cdot \omega)\right]\right] \\ & \cdot \left[\mathrm{Re}\left[K_{y}^{\delta_F}(j \cdot \omega)\right] + j \cdot \mathrm{Im}\left[K_{y}^{\delta_F}(j \cdot \omega)\right]\right] \\ & + \left[\mathrm{Re}\left[P_{\delta_F}^{\psi}(j \cdot \omega)\right] + j \cdot \mathrm{Im}\left[P_{\delta_F}^{\psi}(j \cdot \omega)\right]\right] \\ & \cdot \left[\mathrm{Re}\left[K_{\psi}^{\delta_F}(j \cdot \omega)\right] + j \cdot \mathrm{Im}\left[K_{\psi}^{\delta_F}(j \cdot \omega)\right]\right], \end{aligned}$$

and separating real and imaginary parts, the following is obtained:

$$\begin{aligned} \mathrm{Re}\left[r(j \cdot \omega)\right] = & \frac{-\omega^4 \cdot A + \omega^2 \cdot (A \cdot E + B \cdot D)}{\omega^4 \cdot D^2 + \omega^2 \cdot (\omega^2 - E)^2} \cdot \frac{w_y}{L_o} \cdot \cos(\tau_y \cdot \omega) \\ & - \frac{\omega^3 \cdot (A \cdot D + B) - \omega \cdot B \cdot E}{\omega^4 \cdot D^2 + \omega^2 \cdot (\omega^2 - E)^2} \cdot \frac{w_y}{L_o} \cdot (-\sin(\tau_y \cdot \omega)) \\ & + \frac{-\omega^4 \cdot G + \omega^2 \cdot (D \cdot H + E \cdot G)}{\omega^4 \cdot D^2 + \omega^2 \cdot (\omega^2 - E)^2} \cdot \frac{w_\psi}{L_o} \cdot \cos(\tau_\psi \cdot \omega) \end{aligned}$$

$$-\frac{\omega^3 \cdot (D \cdot G + H) - \omega \cdot E \cdot H}{\omega^4 \cdot D^2 + \omega^2 \cdot (\omega^2 - E)^2} \cdot \frac{w_\psi}{L_o} \cdot (-\sin(\tau_\psi \cdot \omega)), \tag{5.88}$$

$$\mathrm{Im}\left[r(j \cdot \omega)\right] = \frac{-\omega^4 \cdot A + \omega^2 \cdot (A \cdot E + B \cdot D)}{\omega^4 \cdot D^2 + \omega^2 \cdot (\omega^2 - E)^2} \cdot \frac{w_y}{L_o} \cdot (-\sin(\tau_y \cdot \omega))$$
$$+ \frac{\omega^3 \cdot (A \cdot D + B) - \omega \cdot B \cdot E}{\omega^4 \cdot D^2 + \omega^2 \cdot (\omega^2 - E)^2} \cdot \frac{w_y}{L_o} \cdot \cos(\tau_y \cdot \omega)$$
$$+ \frac{-\omega^4 \cdot G + \omega^2 \cdot (D \cdot H + E \cdot G)}{\omega^4 \cdot D^2 + \omega^2 \cdot (\omega^2 - E)^2} \cdot \frac{w_\psi}{L_o} \cdot (-\sin(\tau_\psi \cdot \omega))$$
$$+ \frac{\omega^3 \cdot (D \cdot G + H) - \omega \cdot E \cdot H}{\omega^4 \cdot D^2 + \omega^2 \cdot (\omega^2 - E)^2} \cdot \frac{w_\psi}{L_o} \cdot \cos(\tau_\psi \cdot \omega). \tag{5.89}$$

The frequency characteristics $r(\omega)$ of the investigated system: driver (5.60) – vehicle $((5.61)$ and $(5.62))$ is shown in Figure 5.34.

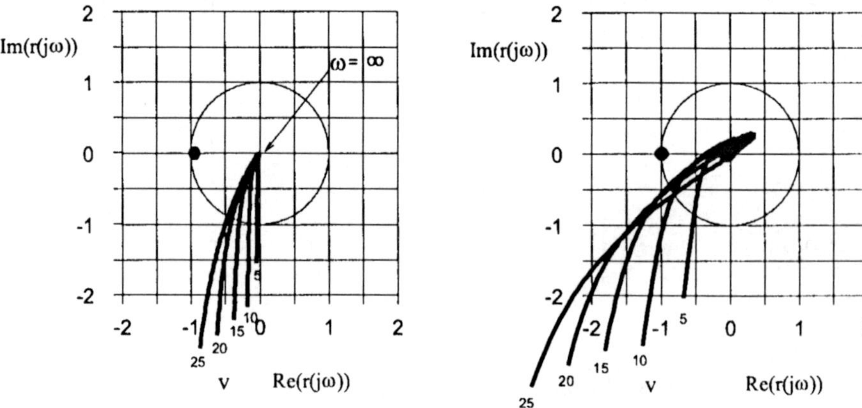

Figure 5.34. Frequency characteristics of the driver-vehicle system for five vehicle velocity values $V = 5, 10, 15, 20, 25\, m/s$ and for two sets of delay: (a) $\tau_\psi = 0$, $\tau_y = 0$; (b) $\tau_\psi = 0.3\, s$, $\tau_y = 0.3\, s$. ($L_p = 10\, m$, $w_y = 0.5\, rad \cdot m/s$, $w_\psi = 0.5$, $K_F = 60000\, N/rad$, $K_R = 60000\, N/rad$, $i_{pk} = -0.5\, rad/rad$, $J = 2000\, kg \cdot m^2$, $m = 1300\, kg$, $a = 1.25\, m$, $b = 1.25\, m$).

One may conclude that for the considered parameters the investigated system is stable. All given frequency characteristics are located far from the $(-1, j0)$ point. Although they possess one intersection point with the unit circle radius, this point is different from $(-1, j0)$. According to Nyquists' criterion the system is stable.

Our next aim is to find the critical delay values $\tau_y = \tau_\psi = \tau_{cr}$, when the characteristic $r(j\omega)$ passes through the point $(-1, j0)$. This requirement is achieved satisfying the following steps:

1°. Find ω_1, for which the frequency characteristics pass through the point $(-1, j0)$;

2°. Find the argument φ_1 corresponding to ω_1;

3°. Find the critical driver delay τ_{cr}.

Note that a transitional function with a delay has the following general form:

$$r(p) = r_0(p) \cdot e^{-p \cdot \tau},$$

where: τ is the delay; $r_0(p)$ is the transitional function of the system, when $\tau = 0$. The spectral transitional function, reads

$$r(j \cdot \omega) = r_0(j \cdot \omega) \cdot e^{-j \cdot \omega \cdot \tau},$$

and owing to (5.87)

$$r(j \cdot \omega) = P^y \cdot K_y + P^\psi \cdot K_\psi.$$

Taking into account (5.77), (5.68), (5.78) and (5.69) the following relations are obtained:

$$r(j \cdot \omega) = \frac{j \cdot \omega \cdot A + B}{-j \cdot \omega^3 + \omega^2 \cdot D + j \cdot \omega \cdot E} \cdot \frac{w_y}{L_o} \cdot e^{-j \cdot \tau_y \cdot \omega} + \frac{j \cdot \omega \cdot G + H}{-j \cdot \omega^3 + \omega^2 \cdot D + j \cdot \omega \cdot E} \cdot \frac{w_\psi}{L_o} \cdot e^{-j \cdot \tau_\psi \cdot \omega},$$

$$r_0(j \cdot \omega) = \frac{j \cdot \omega \cdot A + B}{-j \cdot \omega^3 + \omega^2 \cdot D + j \cdot \omega \cdot E} \cdot \frac{w_y}{L_o} + \frac{j \cdot \omega \cdot G + H}{-j \cdot \omega^3 + \omega^2 \cdot D + j \cdot \omega \cdot E} \cdot \frac{w_\psi}{L_o},$$

$$|r_0(j \cdot \omega)| = |\mathrm{Re}\,[r_0(j \cdot \omega)] + j \cdot \mathrm{Im}\,[r_0(j \cdot \omega)]| = \sqrt{\{\mathrm{Re}\,[r_0(j \cdot \omega)]\}^2 + \{\mathrm{Im}\,[r_0(j \cdot \omega)]\}^2}.$$

Owing to (5.77) and (5.78) the real and imaginary parts of $r_0(\omega)$ read

$$\mathrm{Re}\,[r_0(j \cdot \omega)] = \frac{-\omega^4 \cdot A + \omega^2 \cdot (A \cdot E + B \cdot D)}{\omega^4 \cdot D^2 + \omega^2 \cdot (\omega^2 - E)^2} \cdot \frac{w_y}{L_o} + \frac{-\omega^4 \cdot G + \omega^2 \cdot (D \cdot H + E \cdot G)}{\omega^4 \cdot D^2 + \omega^2 \cdot (\omega^2 - E)^2} \cdot \frac{w_\psi}{L_o},$$

$$\mathrm{Im}\,[r_0(j \cdot \omega)] = \frac{\omega^3 \cdot (A \cdot D + B) - \omega \cdot B \cdot E}{\omega^4 \cdot D^2 + \omega^2 \cdot (\omega^2 - E)^2} \cdot \frac{w_y}{L_o}$$

$$+\frac{\omega^3 \cdot (D \cdot G + H) - \omega \cdot E \cdot H}{\omega^4 \cdot D^2 + \omega^2 \cdot (\omega^2 - E)^2} \cdot \frac{w_\psi}{L_o}. \tag{5.90}$$

The value ω_1 is obtained from the relation

$$|r_0 (j \cdot \omega)| = 1 \tag{5.91}$$

and for the considered set of parameters and $v = 25\, m/s$ we get $\omega_1 = 1.1\, rad/s$. The corresponding argument is defined via the ratio

$$\tan(\varphi) = \frac{\mathrm{Im}\, [r_0 (j \cdot \omega)]}{\mathrm{Re}\, [r_0 (j \cdot \omega)]} \tag{5.92}$$

and substituting (5.92) into (5.90) one gets

$$\tan(\varphi_1) = \frac{\left[\omega_1^3 \cdot (A \cdot D + B) - \omega_1 \cdot B \cdot E\right] \cdot w_y + \left[\omega_1^3 \cdot (D \cdot G + H) - \omega_1 \cdot H \cdot E\right] \cdot w_\psi}{\left[-\omega_1^4 \cdot A + \omega_1^2 \cdot (A \cdot E + B \cdot D)\right] \cdot w_y + \left[-\omega_1^4 \cdot G + \omega_1^2 \cdot (D \cdot H + E \cdot G)\right] \cdot w_\psi}.$$

For the already mentioned parameters $\varphi_1 = 1.429\, rad$. Finally, the critical delay is given by

$$\tau_{cr} = \frac{\pi + \varphi_1}{\omega_1}, \tag{5.93}$$

and for the considered parameters $\tau_{cr} = 4.16\, s$.

One may also investigate the influence of other parameters on stability using the Nyquist criterion and the formulas (5.90)–(5.93). Some exemplary results are reported in Figure 5.35 on a basis of the formulas (5.60)–(5.62).

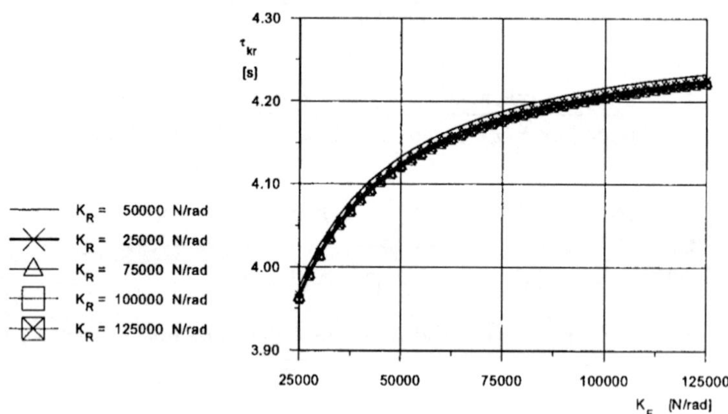

Figure 5.35. The relation $k_F(\tau_{cr})$ for different k_R.

An increase of the front axle wheel drift k_F implies an increase of the driver's delay τ_{cr}. In other words, an increase of k_F improves stability of the system: driver-vehicle.

3. Shimmy [82]

Recall that the word shimmy describes the phenomenon of self-excited wheel car vibrations around a relieving axle. It occurs mainly when a vehicle moves on a non-homogeneous surface. The road surface non-homogeneity causes the perturbations. A sudden "turn" can occur, and a wheel can move in a transversal direction. In Figure 5.36 the flexible system composed of a road wheel and a vehicle body is shown.

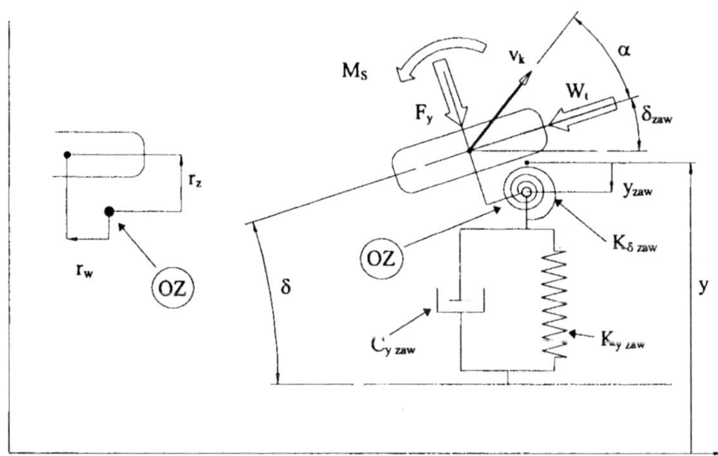

Figure 5.36. Mechanical model of the system road wheel-vehicle used to explain the shimmy phenomenon.

On the stick between a road wheel and a road surface the lateral force F_y and the stabilizing moment M_S appear. Since a vehicle suspension is flexible, both linear displacement y_{sus} and angle rotation δ_{sus} of the wheel appear. The lateral force action forces the velocity vector to turn through the slip angle α value. The new displacement deformation state can be the source for vibrations.

Following the model in Figure 5.36, stability of the system road wheel-vehicle is further studied.

The following equation governs the road wheel motion in the Y direction

$$m_k \cdot (\ddot{y}_{sus} - \ddot{y}) - F_y + K_{y\,sus} \cdot y_{sus} + C_{y\,sus} \cdot \dot{y}_{sus} = 0, \qquad (5.94)$$

where: m_k is the road wheel mass; $k_{y\,sus}$ is the vehicle suspension stiffness in the Y direction, and $C_{y\,sus}$ is the suspension damping in the Y direction.

The equation of motion governing the road wheel dynamics around the relieving axle (point OZ) reads

$$J_k \cdot (\ddot{\delta}_{sus} + \ddot{\delta}) - M_S - F_y \cdot r_w - W_t \cdot r_z + K_{\delta\,sus} \cdot \delta_{sus} = 0, \qquad (5.95)$$

where: J_k is the inertial moment of the road wheel and the associated elements (steering gear) with respect to the relieving axle OZ; $K_{\delta\, sus}$ is the stiffness of the suspension elements in the horizontal plane.

The tire characteristics, i.e. relations $F_y(\alpha)$ and $M_S(\alpha)$, are assumed to be linear and of the form

$$F_y = K_\alpha \cdot \alpha + C_\alpha \cdot \dot{\alpha}, \tag{5.96}$$

$$M_S = K_{M\alpha} \cdot \alpha + C_{M\alpha} \cdot \dot{\alpha}, \tag{5.97}$$

where: k_α is the lateral force coefficient representing the cornering tire stiffness; C_α is the lateral force coefficient representing the tire rotational damping; $K_{M\alpha}$ is the stability moment coefficient representing the tire stiffness; $C_{M\alpha}$ is the stabilizing moment coefficient representing the tire rotational damping.

Owing to both Figure 5.36 and relation $v_{ky} = \dot{y} - \dot{y}_{sus}$ one gets

$$\alpha + \delta_{sus} + \delta = \frac{1}{v_{kx}} \cdot (\dot{y} - \dot{y}_{sus}) \,. \tag{5.98}$$

Applying the Laplace transformation (5.94)-(5.98) the following relations are obtained

$$m_k \cdot s^2 \cdot y_{sus}(s) - m_k \cdot s^2 \cdot y(s) - F_y(s) + K_{y\, sus} \cdot y_{sus}(s) + C_{y\, sus} \cdot s \cdot y_{sus}(s) = 0, \tag{5.99}$$

$$J_k \cdot s^2 \cdot \delta_{sus}(s) + J_k \cdot s^2 \cdot \delta(s) - M_S(s) - F_y(s) \cdot r_w - \frac{r_z}{s} \cdot W_t(s) + K_{\delta sus} \cdot \delta_{sus}(s) = 0, \tag{5.100}$$

$$F_y(s) = K_\alpha \cdot \alpha(s) + C_\alpha \cdot s \cdot \alpha(s), \tag{5.101}$$

$$M_S(s) = K_{M\alpha} \cdot \alpha(s) + C_{M\alpha} \cdot s \cdot \alpha(s), \tag{5.102}$$

$$\alpha(s) + \delta_{sus}(s) + \delta(s) = \frac{1}{v_{kx}} \cdot s \cdot y(s) - \frac{1}{v_{kx}} \cdot s \cdot y_{sus}(s). \tag{5.103}$$

Transforming (5.99)–(5.103) the following two equations of motions are yielded:

$$\begin{aligned} y_{sus}(s) = {} & \frac{J_k \cdot s^3}{J_k \cdot s^3 + K_{\delta sus} \cdot s} \cdot y(s) - \frac{v_{kx} \cdot J_k \cdot s^2}{J_k \cdot s^3 + K_{\delta sus} \cdot s} \cdot \alpha(s) + \frac{v_{kx} \cdot \frac{1}{s} \cdot W_t(s) \cdot r_z}{J_k \cdot s^3 + K_{\delta sus} \cdot s} \\ & - \frac{v_{kx} \cdot (K_{M\alpha} + K_\alpha \cdot r_w)}{J_k \cdot s^3 + K_{\delta sus} \cdot s} \cdot \alpha(s) - \frac{v_{kx} \cdot (C_{M\alpha} + C_\alpha \cdot r_w) \cdot s}{J_k \cdot s^3 + K_{\delta sus} \cdot s} \cdot \alpha(s) \\ & + \frac{K_{\delta sus} \cdot s}{J_k \cdot s^3 + K_{\delta sus} \cdot s} \cdot y(s) - \frac{v_{kx} \cdot K_{\delta sus}}{J_k \cdot s^3 + K_{\delta sus} \cdot s} \cdot \delta(s) - \frac{v_{kx} \cdot K_{\delta sus}}{J_k \cdot s^3 + K_{\delta sus} \cdot s} \cdot \alpha(s) \end{aligned}$$

$$\begin{aligned} & m_k \cdot s^2 \cdot y_{sus}(s) - m_k \cdot s^2 \cdot y(s) - K_\alpha \cdot \alpha(s) - C_\alpha \cdot s \cdot \alpha(s) \\ & + K_{y\, sus} \cdot y_{sus}(s) + C_{y\, sus} \cdot s \cdot y_{sus}(s) = 0 \,. \end{aligned} \tag{5.104}$$

Investigating the system of equations (5.104) the following road wheel properties are studied:

(i) slip angle α dependence on the turn angle δ and the transitional function $G_{\delta}^{\alpha}(s)$;

(ii) slip angle α dependence on the transversal displacement y and the transitional function $G_{y}^{\alpha}(s)$;

(iii) stability condition.

In order to define the operator transmittance $G_{\delta}^{\alpha}(s)$ the terms with $y(s)$ and w_t are removed, and the following relation is obtained

$$\left\{\left[1+\frac{C_{\alpha}}{m_k\cdot v_{kx}^2}\right]\cdot s^4+\left[\frac{K_{\alpha}}{m_k\cdot v_{kx}^2}+\frac{C_{y\,sus}}{m_k}+\frac{C_{M\alpha}+C_{\alpha}\cdot r_w}{J_k}\right]\cdot s^3\right.$$
$$+\left[\frac{C_{\alpha}\cdot K_{\delta sus}}{m_k\cdot v_{kx}^2\cdot J_k}+\frac{K_{M\alpha}+K_{\alpha}\cdot r_w+K_{\delta sus}}{J_k}+\frac{K_{y\,sus}}{m_k}+\frac{C_{y\,sus}\cdot(C_{M\alpha}+C_{\alpha}\cdot r_w)}{m_k\cdot J_k}\right]\cdot s^2$$
$$+\left[\frac{K_{\alpha}\cdot K_{\delta sus}}{m_k\cdot v_{kx}^2\cdot J_k}+\frac{K_{y\,sus}\cdot(C_{M\alpha}+C_{\alpha}\cdot r_w)+C_{y\,sus}\cdot(K_{M\alpha}+K_{\alpha}\cdot r_w+K_{\delta sus})}{m_k\cdot J_k}\right]\cdot s$$
$$\left.+\left[\frac{K_{y\,sus}\cdot(K_{M\alpha}+K_{\alpha}\cdot r_w+K_{\delta sus})}{m_k\cdot J_k}\right]\right\}\cdot\alpha(s)$$
$$=\left\{\left[-\frac{K_{\delta sus}}{J_k}\right]\cdot s^2+\left[-\frac{C_{y\,sus}\cdot K_{\delta sus}}{m_k\cdot J_k}\right]\cdot s+\left[-\frac{K_{y\,sus}\cdot K_{\delta sus}}{m_k\cdot J_k}\right]\right\}\cdot\delta(s), \tag{5.105}$$

and hence

$$G_{\delta}^{\alpha}(s)=\frac{a_2\cdot s^2+a_1\cdot s+a_0}{b_4\cdot s^4+b_3\cdot s^3+b_2\cdot s^2+b_1\cdot s+b_0} \tag{5.106}$$

where:

$$a_2=\left[\frac{K_{\delta sus}}{J_k}\right];\quad a_1=\left[-\frac{C_{y\,sus}\cdot K_{\delta sus}}{J_k\cdot m_k}\right];\quad a_0=\left[-\frac{K_{y\,sus}\cdot K_{\delta sus}}{J_k\cdot m_k}\right];$$

$$b_4=\left[1+\frac{C_{\alpha}}{m_k\cdot v_{kx}^2}\right];\quad b_3=\left[\frac{K_{\alpha}}{m_k\cdot v_{kx}^2}+\frac{C_{M\alpha}+C_{\alpha}\cdot r_w}{J_k}+\frac{C_{y\,sus}}{m_k}\right];$$

$$b_2=\left[\frac{K_{y\,sus}}{m_k}+\frac{K_{\delta sus}+K_{M\alpha}+K_{\alpha}\cdot r_w}{J_k}+\frac{C_{\alpha}\cdot K_{\delta sus}}{J_k\cdot m_k\cdot v_{kx}^2}\right.$$
$$\left.+\frac{C_{y\,sus}\cdot(C_{M\alpha}+C_{\alpha}\cdot r_w)}{J_k\cdot m_k}\right];\quad b_1=\left[\frac{K_{\alpha}\cdot K_{\delta sus}}{J_k\cdot m_k\cdot v_{kx}^2}\right.$$

$$+\frac{K_{y\,sus}\cdot(C_{M\alpha}+C_{\alpha}\cdot r_w)+C_{y\,sus}\cdot(K_{M\alpha}+K_{\delta sus}+K_{\alpha}\cdot r_w)}{J_k\cdot m_k}\Bigg];$$

$$b_0=\left[\frac{K_{y\,sus}\cdot(K_{M\alpha}+K_{\delta sus}+K_{\alpha}\cdot r_w)}{J_k\cdot m_k}\right].$$

Furthermore, the transitional function is transformed to the following more suitable form

$$G_{\delta}^{\alpha}(s)=\frac{L_6\cdot s^6+L_5\cdot s^5+L_4\cdot s^4+L_3\cdot s^3+L_2\cdot s^2+L_1\cdot s+L_0}{m_8\cdot s^8+m_6\cdot s^6+m_4\cdot s^4+m_2\cdot s^2+m_0} \tag{5.107}$$

where:

$$L_6=b_4\cdot a_2,\ \ L_5=(b_4\cdot a_1-b_3\cdot a_2),\ \ L_4=(b_2\cdot a_2+b_4\cdot a_0-b_3\cdot a_1),$$

$$L_3=(b_2\cdot a_1-b_3\cdot a_0-b_1\cdot a_2),\ \ L_2=(b_2\cdot a_0+b_0\cdot a_2-b_1\cdot a_1),$$

$$L_1=(b_0\cdot a_1-b_1\cdot a_0),\ \ L_0=b_0\cdot a_0,\ \ m_8=(b_4^2),\ \ m_6=(2\cdot b_2\cdot b_4-b_3^2),$$

$$m_4=(2\cdot b_0\cdot b_4+b_2^2-2\cdot b_1\cdot b_3),\ \ m_2=(2\cdot b_0\cdot b_2-b_1^2),\ \ m_0=b_0^2.$$

The frequency transitional function defined via (5.107) reads

$$G_{\delta}^{\alpha}(j\cdot\omega)=\frac{L_6\cdot(j\cdot\omega)^6+L_5\cdot(j\cdot\omega)^5+L_4\cdot(j\cdot\omega)^4+L_3\cdot(j\cdot\omega)^3+L_2\cdot(j\cdot\omega)^2+L_1\cdot(j\cdot\omega)+L_0}{m_8\cdot(j\cdot\omega)^8+m_6\cdot(j\cdot\omega)^6+m_4\cdot(j\cdot\omega)^4+m_2\cdot(j\cdot\omega)^2+m_0}$$

and hence

$$G_{\delta}^{\alpha}(j\omega)=\frac{-L_6\cdot\omega^6+L_5\cdot\omega^5\cdot j+L_4\cdot\omega^4-L_3\cdot\omega^3\cdot j-L_2\cdot\omega^2+L_1\cdot\omega\cdot j+L_0}{m_8\cdot\omega^8-m_6\cdot\omega^6+m_4\cdot\omega^4-m_2\cdot\omega^2+m_0}. \tag{5.108}$$

The real and imaginary parts of G_{δ}^{α} follow

$$\mathrm{Re}\,(G_{\delta}^{\alpha}(j\cdot\omega))=\frac{-L_6\cdot\omega^6+L_4\cdot\omega^4-L_2\cdot\omega^2+L_0}{m_8\cdot\omega^8-m_6\cdot\omega^6+m_4\cdot\omega^4-m_2\cdot\omega^2+m_0}, \tag{5.109}$$

$$\mathrm{Im}\,(G_{\delta}^{\alpha}(j\cdot\omega))=\frac{L_5\cdot\omega^5-L_3\cdot\omega^3+L_1\cdot\omega}{m_8\cdot\omega^8-m_6\cdot\omega^6+m_4\cdot\omega^4-m_2\cdot\omega^2+m_0}. \tag{5.110}$$

We are going to find the frequency ω satisfying the equation

$$\mathrm{Im}\,(G_{\delta}^{\alpha}(j\cdot\omega))=0,$$

and hence the following algebraic equation is obtained

$$L_5\cdot\omega^5-L_3\cdot\omega^3+L_1\cdot\omega=0\,.$$

It has the following solutions:

(i)

$$\omega_1 = 0;$$

(ii)

$$\omega_2 = \sqrt{\frac{L_3 - \sqrt{L_3^2 - 4 \cdot L_1 \cdot L_5}}{2 \cdot L_5}};$$

(iii)

$$\omega_3 = \sqrt{\frac{L_3 + \sqrt{L_3^2 - 4 \cdot L_1 \cdot L_5}}{2 \cdot L_5}};$$

(iv)

$$\omega_4 = -\sqrt{\frac{L_3 - \sqrt{L_3^2 - 4 \cdot L_1 \cdot L_5}}{2 \cdot L_5}};$$

(v)

$$\omega_5 = -\sqrt{\frac{L_3 + \sqrt{L_3^2 - 4 \cdot L_1 \cdot L_5}}{2 \cdot L_5}}.$$

According with the Nyquist criterion, the following stability condition is obtained:

$$\mathrm{Re}\left(G_\delta^\alpha(j \cdot \omega_1)\right) > -1 \ \wedge \ \mathrm{Re}\left(G_\delta^\alpha(j \cdot \omega_2)\right) > -1 \ \wedge \ \mathrm{Re}\left(G_\delta^\alpha(j \cdot \omega_3)\right) > -1$$
$$\wedge \, \mathrm{Re}\left(G_\delta^\alpha(j \cdot \omega_4)\right) > -1 \ \wedge \ \mathrm{Re}\left(G_\delta^\alpha(j \cdot \omega_5)\right) > -1,$$

or equivalently

$$\frac{-L_6 \cdot \omega^6 + L_4 \cdot \omega^4 - L_2 \cdot \omega^2 + L_0}{m_8 \cdot \omega^8 - m_6 \cdot \omega^6 + m_4 \cdot \omega^4 - m_2 \cdot \omega^2 + m_0} > -1\,. \tag{5.111}$$

In order to define $G_y^\alpha(s)$, the terms with $\delta(s)$ and W_t are removed from the equations (5.104), and the following relation is obtained:

$$\begin{aligned}
&\Bigg\{\left[1 + \frac{C_\alpha}{m_k \cdot v_{kx}}\right] \cdot s^4 + \left[\frac{K_\alpha}{m_k \cdot v_{kx}} + \frac{(C_{M\alpha} + C_\alpha \cdot r_w)}{J_k} + \frac{C_{y\,sus}}{m_k}\right] \cdot s^3 \\
&+ \left[\frac{K_{\delta sus} \cdot C_\alpha}{J_k \cdot m_k \cdot v_{kx}} + \frac{(K_{M\alpha} + K_{\delta sus} + K_\alpha \cdot r_w)}{J_k} + \frac{K_{y\,sus}}{m_k} + \frac{C_{y\,sus} \cdot (C_{M\alpha} + C_\alpha \cdot r_w)}{J_k \cdot m_k}\right] \cdot s^2 \\
&+ \left[\frac{K_{\delta sus} \cdot K_\alpha}{J_k \cdot m_k \cdot v_{kx}} + \frac{K_{y\,sus} \cdot (C_{M\alpha} + C_\alpha \cdot r_w) + C_{y\,sus} \cdot (K_{M\alpha} + K_\alpha \cdot r_w + K_{\delta sus})}{J_k \cdot m_k}\right] \cdot s \\
&\left[\frac{K_{y\,sus} \cdot (K_{M\alpha} + K_\alpha \cdot r_w + K_{\delta sus})}{J_k \cdot m_k}\right]\Bigg\} \cdot \alpha(s) \\
&= \left\{\left[\frac{C_{y\,sus}}{m_k \cdot v_{kx}}\right] \cdot s^4 + \left[\frac{K_{y\,sus}}{m_k \cdot v_{kx}}\right] \cdot s^3 + \left[\frac{C_{y\,sus} \cdot K_{\delta sus}}{J_k \cdot m_k \cdot v_{kx}}\right] \cdot s^2 + \left[\frac{K_{y\,sus} \cdot K_{\delta sus}}{J_k \cdot m_k \cdot v_{kx}}\right] \cdot s\right\} \cdot y(s),
\end{aligned} \tag{5.112}$$

and hence

$$G_y^\alpha(s) = \frac{c_4 \cdot s^4 + c_3 \cdot s^3 + c_2 \cdot s^2 + c_1 \cdot s}{d_4 \cdot s^4 + d_3 \cdot s^3 + d_2 \cdot s^2 + d_1 \cdot s + d_0}, \tag{5.113}$$

where:

$$c_4 = \left[\frac{C_{y\,sus}}{m_k \cdot v_{kx}}\right], \quad c_3 = \left[\frac{K_{y\,sus}}{m_k \cdot v_{kx}}\right], \quad c_2 = \left[\frac{C_{y\,sus} \cdot K_{\delta sus}}{J_k \cdot m_k \cdot v_{kx}}\right],$$

$$c_1 = \left[\frac{K_{y\,sus} \cdot K_{\delta sus}}{J_k \cdot m_k \cdot v_{kx}}\right], \quad d_4 = \left[1 + \frac{C_\alpha}{m_k \cdot v_{kx}}\right],$$

$$d_3 = \left[\frac{K_\alpha}{m_k \cdot v_{kx}} + \frac{(C_{M\alpha} + C_\alpha \cdot r_w)}{J_k} + \frac{C_{y\,sus}}{m_k}\right],$$

$$d_2 = \left[\frac{K_{\delta sus} \cdot C_\alpha}{J_k \cdot m_k \cdot v_{kx}} + \frac{(K_{M\alpha} + K_{\delta sus} + K_\alpha \cdot r_w)}{J_k}\right.$$

$$\left. + \frac{K_{y\,sus}}{m_k} + \frac{C_{y\,sus} \cdot (C_{M\alpha} + C_\alpha \cdot r_w)}{J_k \cdot m_k}\right], \quad d_1 = \left[\frac{K_{\delta sus} \cdot K_\alpha}{J_k \cdot m_k \cdot v_{kx}}\right.$$

$$\left. + \frac{K_{y\,sus} \cdot (C_{M\alpha} + C_\alpha \cdot r_w) + C_{y\,sus} \cdot (K_{M\alpha} + K_\alpha \cdot r_w + K_{\delta sus})}{J_k \cdot m_k}\right],$$

$$d_0 = \left[\frac{K_{y\,sus} \cdot (K_{M\alpha} + K_\alpha \cdot r_w + K_{\delta sus})}{J_k \cdot m_k}\right].$$

The obtained transitional function (5.113) is transformed to the more suitable form

$$G_y^\alpha(s) = \frac{n_8 \cdot s^8 + n_7 \cdot s^7 + n_6 \cdot s^6 + n_5 \cdot s^5 + n_4 \cdot s^4 + n_3 \cdot s^3 + n_2 \cdot s^2 + n_1 \cdot s}{o_8 \cdot s^8 + o_6 \cdot s^6 + o_4 \cdot s^4 + o_2 \cdot s^2 + o_0}, \tag{5.114}$$

where:

$$n_8 = c_4 \cdot d_4, \quad n_7 = (c_3 \cdot d_4 - c_4 \cdot d_3), \quad n_6 = (c_2 \cdot d_4 + c_4 \cdot d_2 - c_3 \cdot d_3),$$

$$n_5 = (c_1 \cdot d_4 + c_3 \cdot d_2 - c_2 \cdot d_3 - c_4 \cdot d_1), \quad n_4 = (c_2 \cdot d_2 + c_4 \cdot d_0 - c_3 \cdot d_1 - c_1 \cdot d_3),$$

$$n_3 = (c_1 \cdot d_2 + c_3 \cdot d_0 - c_2 \cdot d_1), \quad n_2 = (c_2 \cdot d_0 - c_1 \cdot d_1), \quad n_1 = (c_1 \cdot d_0),$$

$$o_2 = (2 \cdot d_0 \cdot d_2 - d_1^2), \quad o_8 = d_4^2, \quad o_6 = (2 \cdot d_2 \cdot d_4 - d_3^2),$$

$$o_4 = (2 \cdot d_0 \cdot d_4 + d_2^2 - 2 \cdot d_1 \cdot d_3), \quad o_0 = d_0^2.$$

The frequency transitional function (see (5.114)) has the form

$$G_y^\alpha(j\omega) = \left(n_8 \cdot \omega^8 - n_7 \cdot \omega^7 \cdot j - n_6 \cdot \omega^6 + n_5 \cdot \omega^5 \cdot j + n_4 \cdot \omega^4 - n_3 \cdot \omega^3 \cdot j\right.$$

$$\left. - n_2 \cdot \omega^2 + n_1 \cdot \omega \cdot j\right) / \left(o_8 \cdot \omega^8 - o_6 \cdot \omega^6 + o_4 \cdot \omega^4 - o_2 \cdot \omega^2 + o_0\right), \tag{5.115}$$

and hence its real and imaginary parts read

$$\mathrm{Re}\left(G_y^\alpha(j \cdot \omega)\right) = \frac{n_8 \cdot \omega^8 - n_6 \cdot \omega^6 + n_4 \cdot \omega^4 - n_2 \cdot \omega^2}{o_8 \cdot \omega^8 - o_6 \cdot \omega^6 + o_4 \cdot \omega^4 - o_2 \cdot \omega^2 + o_0},$$

Table 5.5. The parameters associated with Figure 5.37.

$\omega_i\|_{\mathrm{Im}\left[G_\delta^\alpha(j\omega_i)\right]=0}$		$\mathrm{Re}\left[G_\delta^\alpha(j\omega_i)\right]$ $[rad/rad]$	
$i=1$	0	$i=1$	-0.869
$i=2$	72.26	$i=2$	-9.46
$i=3$	141.42	$i=3$	0.013
$i=4$	-72.26	$i=4$	-9.46
$i=5$	141.42	$i=5$	0

$$\mathrm{Im}\left(G_y^\alpha(j\cdot\omega)\right) = \frac{-n_7\cdot\omega^7 + n_5\cdot\omega^5 - n_3\cdot\omega^3 + n_1\cdot\omega}{o_8\cdot\omega^8 - o_6\cdot\omega^6 + o_4\cdot\omega^4 - o_2\cdot\omega^2 + o_0}.$$

We are going to find the frequency for which

$$\mathrm{Im}\left(G_y^\alpha(j\cdot\omega)\right) = 0,$$

and hence

$$-n_7\cdot\omega^7 + n_5\cdot\omega^5 - n_3\cdot\omega^3 + n_1\cdot\omega = 0\,. \tag{5.116}$$

Since the first solution $\omega_1 = 0$, the six others can be found from the cubic algebraic equation of the form

$$-n_7\cdot x^3 + n_5\cdot x^2 - n_3\cdot x + n_1 = 0,$$

where: $x=\omega^2$. The stability condition reads

$$\mathrm{Re}\left(G_y^\alpha(j\cdot\omega)\right) > -1,$$

or equivalently

$$\text{for } \omega=\omega_{1,2,3,4,5,6,7} \quad \frac{n_8\cdot\omega^8 - n_6\cdot\omega^6 + n_4\cdot\omega^4 - n_2\cdot\omega^2}{o_8\cdot\omega^8 - o_6\cdot\omega^6 + o_4\cdot\omega^4 - o_2\cdot\omega^2 + o_0} > -1\,. \tag{5.117}$$

EXAMPLE 5.3 *Analyse stability of the system road wheel-vehicle for the shimmy model shown in Figure 5.36 and the following data:* $v_{kx} = 20\,m/s$, $m_k = 5\,kg$, $J_k = 2\,kg{\cdot}m^2$, $r_w = 0.05\,m$, $r_z = 0.03\,m$, $K_{y\,sus} = 100000\,N/m$, $C_{y\,sus} = 10\,N\cdot s/m$, $K_{\delta\,sus} = 10000\,N/rad$, $K_\alpha = 10000\,N/rad$, $C_\alpha = 10\,N\cdot s/rad$, $K_{M\alpha} = 1000\,N\cdot m/rad$, $C_{M\alpha} = 1N\cdot m\cdot s/rad$.

In Figure 5.37 the frequency characteristic of the investigated system for small damping coefficients $C_{y\,sus} = 1\,N\cdot s/m$, $C_\alpha = 1\,N\cdot s/rad$, $C_{M\,\alpha} = 0.1\,N\cdot m\cdot s/rad$, *is reported.*

In Figure 5.38 the frequency characteristic of the system with small stiffness $K_{y\,sus} = 10000\,N/m$ *is shown.*

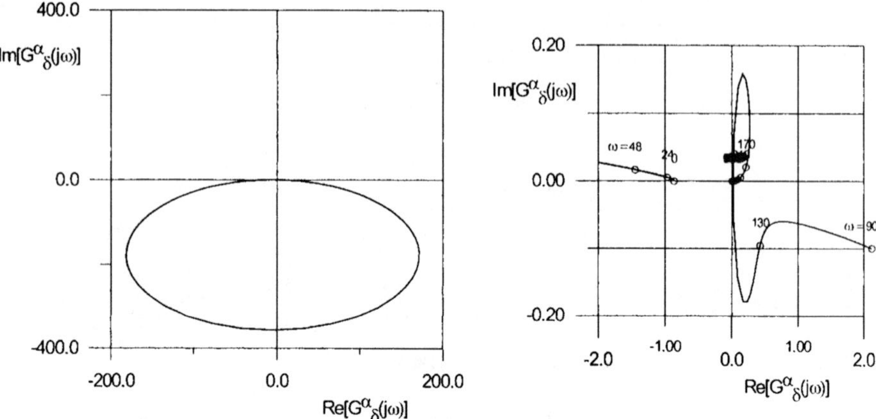

Figure 5.37. The characteristic $G_\delta^\alpha(j\omega)$ exhibiting the system instability for $C_{y\,sus} = 1\,N \cdot s/m$, $C_\alpha = 1\,N \cdot s/rad$, $C_{M\,\alpha} = 0.1\,N \cdot m \cdot s/rad$.

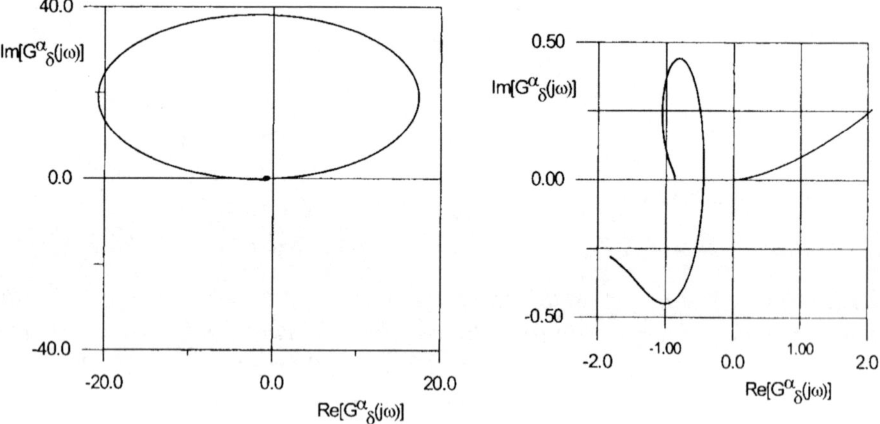

Figure 5.38. The characteristic $G_\delta^\alpha(j\omega)$ exhibiting the system instability for $K_{y\,sus} = 10000\,N/m$.

Table 5.6. The parameters associated with Figure 5.38.

$\omega_i\vert_{\text{Im}[G_\delta^\alpha(j\omega_i)]=0}$		Re $[G_\delta^\alpha(j\omega_i)]$ $[rad/rad]$	
$i = 1$	0	$i = 1$	-0.869
$i = 2$	44.77	$i = 2$	-0.445
$i = 3$	67.87	$i = 3$	-4.35
$i = 4$	-44.77	$i = 4$	-0.445
$i = 5$	-67.87	$i = 5$	0

Table 5.7. The parameters associated with Figure 5.39.

$\omega_i\vert_{\text{Im}[G_\delta^\alpha(j\omega_i)]=0}$		Re $[G_\delta^\alpha(j\omega_i)]$ $[rad/rad]$	
$i = 1$	0	$i = 1$	-0.869
$i = 2$	83.39	$i = 2$	4.165
$i = 3$	141.56	$i = 3$	9.635
$i = 4$	-83.39	$i = 4$	4.165
$i = 5$	-141.56	$i = 5$	0

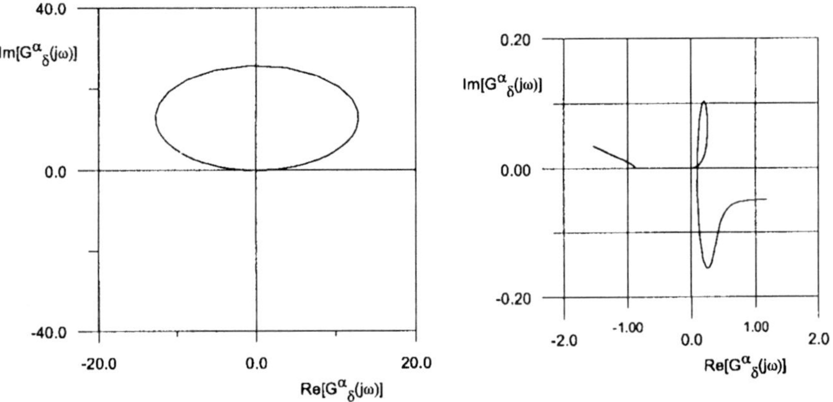

Figure 5.39 The characteristic $G_\delta^\alpha(j\omega)$ of the stable system.

Since in both Figures 5.37 and 5.38 the associated frequency characteristics contain the point $(-1, j_0)$, the system is unstable in the sense of Lyapunov with respect to Nyquist criterion. The associated system parameters are reported in Tables 5.5 and 5.6.

The frequency characteristics of the stable system ($K_{y\,ouo} = 100000\,N/m$, $C_{y\,sus} = 10\,N \cdot s/m$, $C_\alpha = 10\,N \cdot s/rad$, $C_{M\,\alpha} = 1\,N \cdot m \cdot s/rad$) is shown in Figure 5.39, and the associated parameters are reported in Table 5.6.

Chapter 6

VERTICAL DYNAMICS

In this chapter we study the dynamics of a suspension system model consisting of two bodies vibrating with a passive and an active constraint. A stability investigation is carried out and Poincaré maps and bifurcation diagrams are reported.

1. Two Degrees-of-Freedom System Dynamics [5, 44, 48, 63, 71, 79]

The active suspension shown in Figure 6.1 is analysed. The vibrating system is composed of two bodies: the road wheel with its associated elements has mass m_1, whereas the vehicle car body possesses mass m_2.

Two constraint systems are applied:

(i) *passive system* with the following stiffnesses: k_1 - stiffness of the constraints between a road surface and a road wheel body; k_2 - stiffness of the constraints between a road wheel and a vehicle body; k_{22} - stiffness of the additional constraints (bumper) between a road wheel and a vehicle body.

(ii) *active system* including the block U with stiffness and damping controlled via an electronic system applying signals from the acceleration transducers (velocities and displacements) of the vehicle body z_2 and the road wheel body z_1.

The passive (conventional) and the active suspension differ by the block U (or damping element) and by the elastic parameters.

The suspension system dynamics is governed by the differential equations (see Figure 6.1)

$$m_1 \cdot \ddot{z}_1 + U(t) - k_2 \cdot (z_2 - z_1) + (m_1 + m_2) \cdot g - F_{z1} = 0, \qquad (6.1)$$

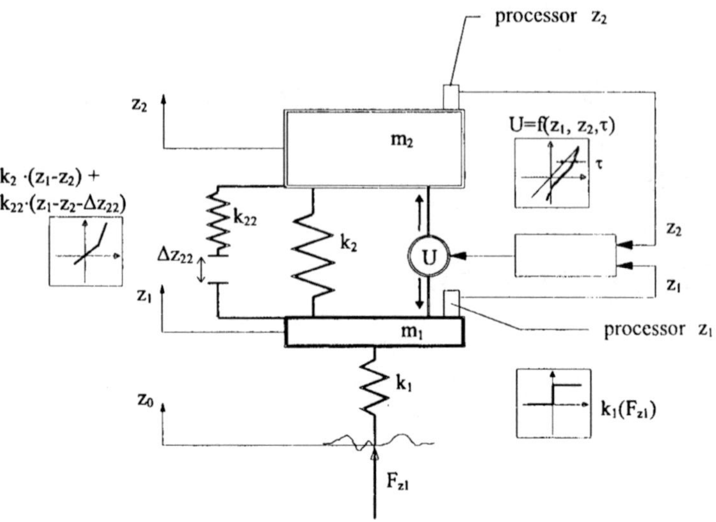

Figure 6.1. Model of the active suspension. (m_1 - road wheel and the associated elements mass; m_2 - body vehicle mass; z_1 - displacement of the mass m_1; z_2 - displacement of the mass m_2; Δz_{22} - static distance between m_2 and the bumper, i.e. additional elastic element with the stiffness k_{22}; $U(t)$ - controlling force generated via the active system and acting on the road wheel bodies and the vehicle body; F_{z1} - vertical force of the mutual interaction of the road wheel and the road surface).

$$m_2 \cdot \ddot{z}_2 - U(t) + k_2 \cdot (z_2 - z_1) = 0. \tag{6.2}$$

Depending on free and external system dynamics the following cases are separated:

1°. The road wheel is always in contact with the road surface

$$F_{z1} = (m_1 + m_2) \cdot g + k_1 \cdot (z_0 - z_1) > 0.$$

2°. The road wheel looses its contact with the road surface

$$F_{z1} = (m_1 + m_2) \cdot g + k_1 \cdot (z_0 - z_1) < 0 \Rightarrow F_{z0} = 0$$

3°. The vehicle suspension is so large that further displacement is bounded by the bumper

$$(z_1 - z_2) > \Delta z_{22}.$$

In the latter case, the body m_2 is supported via the stiffness $k_2 + k_{22}$.

The controlling force $U = U(t)$ acting on the road wheel and the vehicle body is generated by the final control unit of the active suspension. A so-called optimal control functional is applied. In our case, the vertical vehicle body acceleration $\ddot{z}_2$ and the relative bodies displacement $z_1 - z_2$ as a comfortable driving feature are taken, and the relative road wheel displacement $(z_1 - z_0)$ as a safe driving feature are taken, and the associated costs function is defined via the formula

$$J = \frac{1}{2} \int_0^t \left(\rho_1 \cdot (\ddot{z}_2)^2 + \rho_2 \cdot (z_1 - z_2)^2 + \rho_3 \cdot (z_1 - z_0)^2 \right) dt,$$

where: $\rho_{1,2,3}$ are the weight coefficients.

The optimalisation of the active suspension should satisfy the following criterion: $J_{t \to \infty} \to \min$.

Depending on the assumed vehicle requirements the weight coefficients will have various values. If driving comfort is the most important, then the coefficients φ_1 and φ_2 achieve relatively large values (particularly φ_1). If the most important factor will be the running safety, then the relative displacements $(z_0 - z_1)$ should be the smallest ones, and hence φ_3 achieves a relatively large value. In both cases considered, the choice and the way of control of the force $U(t)$ in order to realize the assumed criterion is important. In general the control function possesses the arguments $z_1, z_2, \dot{z}_1$ and $\dot{z}_2$ with time delay τ between control quantities time measurement and the time instant of the force $U(t)$ generation. The exemplary function reads

$$U(t) = k_{A1} \cdot z_1 (t - \tau) + k_{A2} \cdot z_2 (t - \tau) + k_{B2} \cdot z_2^3 (t - \tau)$$
$$+ C_{A1} \cdot \dot{z}_1 (t - \tau) + C_{A2} \cdot \dot{z}_2 (t - \tau),$$

where: τ - time delay; k_{A1} - stiffness of the unit proportional to the wheel displacement z_1; k_{A2} - stiffness of the unit proportional to the vehicle body displacement z_2; k_{B2} - stiffness of the unit proportional to the third power of the car body displacement z_2; C_{A1} - damping of the unit proportional to the velocity of the wheel displacement z_1; C_{A2} - damping of the unit proportional to the velocity of the vehicle body displacement z_2.

In order to illustrate an action of the *active suspension* the simulation results are shown in Figures 6.2–6.5. It is assumed that the road wheel moving on a flat surface approaches, in a certain time instant, a single sinusoidal barrier. The control function $U(t)$ has the mathematical relation introduced earlier, and stiffness and damping are chosen in a way to secure a high security running, i.e. via achieving the smallest vertical load changes of the road wheel F_{z1}.

In Figure 6.2, dynamics of the *conventional suspension* with the parameters: $K_1 = 200000\ N/m$, $K_2 = 10000\ N/m$, $C_2 = 1000\ N \cdot s/m$ is reported. Running through the barrier leads to essential changes of the road wheel vertical load F_{z1}. However, this vertical dynamics is essentially improved via introduction of the *active suspension*. In Figure 6.3 the time history of the passive road force F_{z1} and the control force $U(t)$ as well as the vertical kinematics of the road wheel $\dot{z}_1(z_1)$ for the active suspension are shown, the instantaneous decrease of the passive road force, when the road wheel leaves the surface irregularity, achieves the value of $1500\,N$, whereas in the case of the *conventional* suspension it achieved the value of $0\,N$ (see Figure 6.2).

Improper choice of the control parameters can yield to "stiff" dynamics exhibited by the road wheel reflection phenomenon. In Figure 6.4 running dynamics through the surface sinusoidal irregularity with $k_{B2} = -6.5 \cdot 10^7\,N/m$ is reported. The investigated model exhibits irregular vibrations, and the road wheel undergoes jumps by losing contact with the road surface.

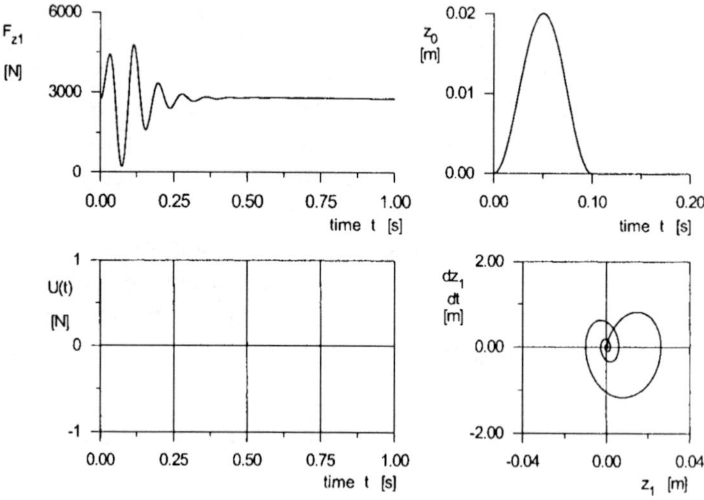

Figure 6.2. Running dynamics through the sinusoidal road surface for conventional suspension ($K_1 = 200000\,N/m$, $K_2 = 10000\,N/m$, $C_2 = 1000\,N/m$).

Improper choice of the parameters of control $U(t)$ can lead to instability of the control process with higher vibration frequency, when a delay is relatively large ($\tau > \tau_{cr}$). In Figure 6.5 the running dynamics through the sinusoidal road irregularity is shown for $\tau = 0.0044\,s$. The investigated system exhibits the undamped oscillation of $U(t)$, which excites the vertical road wheel vibrations.

The given examples indicate a validity of investigation of stability of control units of the active car suspensions.

In the next step, a stability of an active suspension system with the non-linear control function

$$U(t) = k_{A1} \cdot z_1\,(t-\tau) + k_{A2} \cdot z_2\,(t-\tau) + k_{B2} \cdot z_2^2\,(t-\tau)$$
$$+C_{A1} \cdot \dot{z}_1\,(t-\tau) + C_{A2} \cdot \dot{z}_2\,(t-\tau) \qquad (6.3)$$

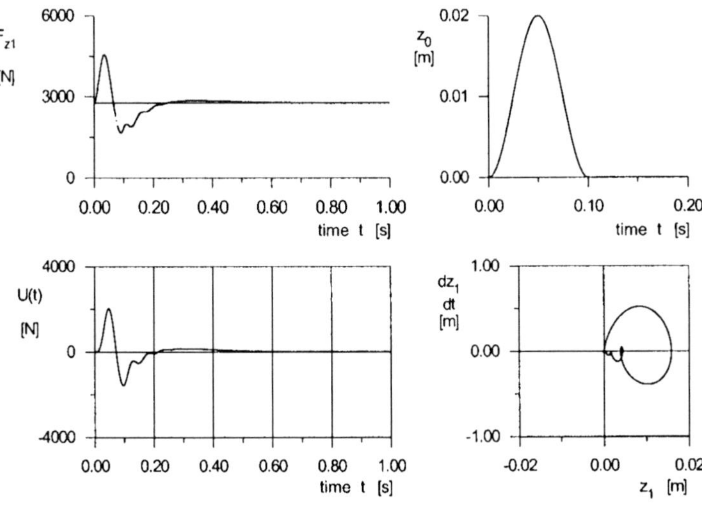

Figure 6.3. Running dynamics through the sinusoidal road surface for the active suspension ($k_1 = 200000\,N/m$, $k_2 = 9000\,N/m$, $k_{A1} = 250000\,N/m$, $k_{A2} = -50000\,N/m$, $k_{B2} = -500000\,N/m$, $C_{A1} = 1000\,N \cdot s/m$, $C_{A2} = -10000\,N \cdot s/m$, $t = 0.0021s$).

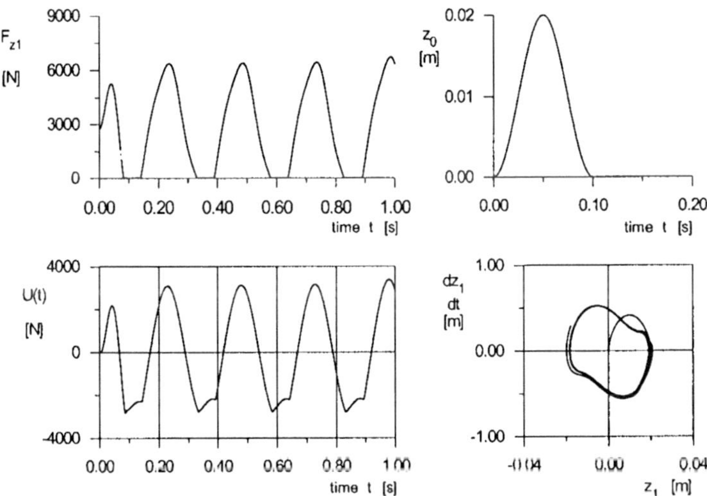

Figure 6.4. Running dynamics through the sinusoidal road surface for the active suspension ($k_{B2} = -650000000\,N/m$, $k_1 = 200000\,N/m$, $k_2 = 9000\,N/m$, $k_{A1} = 25000\,N/m$, $k_{A2} = -25000\,N/m$, $C_{A1} = 10000\,N \cdot s/m$, $C_{A2} = -10000\,N \cdot s/m$, $t = 0.0021s$).

is studied. The investigated system is governed by the relations (6.1)–(6.3) of Case 1°. The function (6.3) is developed into the Taylor series including only

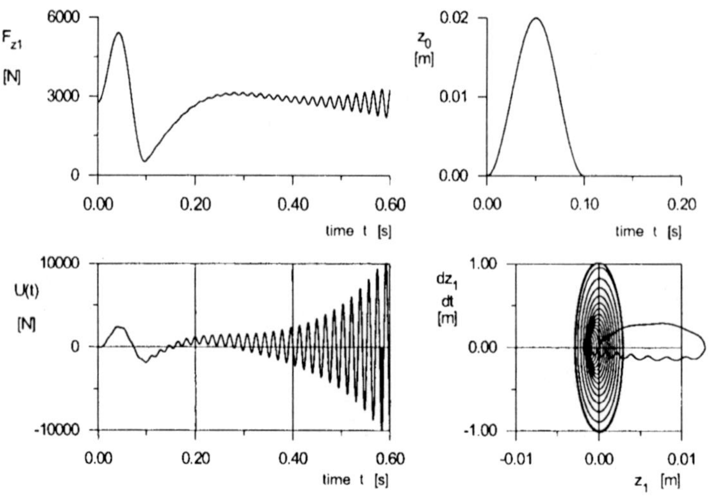

Figure 6.5. Running dynamics through the sinusoidal surface with active suspension ($k_1 = 200000\,N/m,\ k_2 = 5000\,N/m,\ k_{A1} = 25000\,N/m,\ k_{A2} = -25000\,N/m,\ k_{B2} = -2000000\,N/m, C_{A1} = 10000\,N \cdot s/m, C_{A2} = -10000\,N \cdot s/m, t = 0.0044\,s$).

the first term of the function $U(t)$:

$$U(t) = k_{A1} \cdot z_1(t) + k_{A2} \cdot z_2(t) + k_{B2} \cdot z_2^2(t) + C_{A1} \cdot \dot{z}_1(t) + C_{A2} \cdot \dot{z}_2(t)$$

$$+k_{A1} \cdot \tau \cdot \dot{z}_1(t) + k_{A2} \cdot \tau \cdot \dot{z}_2(t) + 2 \cdot k_{B2} \cdot \tau \cdot z_2(t) \cdot \dot{z}_2(t) + C_{A1} \cdot \tau \cdot \ddot{z}_1(t) + C_{A2} \cdot \tau \cdot \ddot{z}_2(t). \tag{6.4}$$

Substituting (6.4) into (6.1) one gets

$$(m_1 + C_{A1} \cdot \tau) \cdot \ddot{z}_1 = -(k_{A1} + k_2 + k_1) \cdot z_1 - (k_{A2} - k_2) \cdot z_2 - (C_{A1} + k_{A1} \cdot \tau) \cdot \dot{z}_1$$

$$-(C_{A2} + k_{A2} \cdot \tau) \cdot \dot{z}_2 - C_{A2} \cdot \tau \cdot \ddot{z}_2 + (-k_{B2} \cdot z_2^2 - 2k_{B2} \cdot \tau \cdot z_2 \cdot \dot{z}_2 + k_1 \cdot z_0), \tag{6.5}$$

whereas substituting (6.4) into (6.2) yields

$$(m_2 - C_{A2} \cdot \tau) \cdot \ddot{z}_2 = +(k_{A1} + k_2) \cdot z_1 + (k_{A2} - k_2) \cdot z_2 + (C_{A1} + k_{A1} \cdot \tau) \cdot \dot{z}_1$$

$$+(C_{A2} + k_{A2} \cdot \tau) \cdot \dot{z}_2 + C_{A1} \cdot \tau \cdot \ddot{z}_1 + (k_{B2} \cdot z_2^2 + 2k_{B2} \cdot \tau \cdot z_2 \cdot \dot{z}_2). \tag{6.6}$$

Substituting (6.5) and (6.6) and introducing the notation

$$a_1 = \frac{k_1 \cdot C_{A2} \cdot \tau - m_2 \cdot (k_1 + k_2 + k_{A1})}{m_1 \cdot m_2 - m_1 \cdot C_{A2} \cdot \tau + m_2 \cdot C_{A1} \cdot \tau};$$

$$a_2 = \frac{(k_2 - k_{A2}) \cdot m_2}{m_1 \cdot m_2 - m_1 \cdot C_{A2} \cdot \tau + m_2 \cdot C_{A1} \cdot \tau};$$

$$a_3 = \frac{-(C_{A1} + k_{A1} \cdot \tau) \cdot m_2}{m_1 \cdot m_2 - m_1 \cdot C_{A2} \cdot \tau + m_2 \cdot C_{A1} \cdot \tau};$$

$$a_4 = \frac{-(C_{A2} + k_{A2} \cdot \tau) \cdot m_2}{m_1 \cdot m_2 - m_1 \cdot C_{A2} \cdot \tau + m_2 \cdot C_{A1} \cdot \tau};$$

$$a_5 = \frac{k_1 \cdot z_0 \cdot (m_2 - C_{A2} \cdot \tau) - k_{B2} \cdot m_2 \cdot (z_2^2 + 2\tau \cdot z_2 \cdot \dot{z}_2))}{m_1 \cdot m_2 - m_1 \cdot C_{A2} \cdot \tau + m_2 \cdot C_{A1} \cdot \tau};$$

one gets

$$\ddot{z}_1 = a_1 \cdot z_1 + a_2 \cdot z_2 + a_3 \cdot \dot{z}_1 + a_4 \cdot \dot{z}_2 + a_5 \,. \tag{6.7}$$

Whereas substituting (6.5) into (6.6) and introducing the notation

$$b_1 = \frac{(k_{A1} + k_2) \cdot m_1 - C_{A1} \cdot \tau \cdot k_1}{m_1 \cdot m_2 - m_1 \cdot C_{A2} \cdot \tau + m_2 \cdot C_{A1} \cdot \tau};$$

$$b_2 - \frac{(k_{A2} - k_2) \cdot m_1}{m_1 \cdot m_2 - m_1 \cdot C_{A2} \cdot \tau + m_2 \cdot C_{A1} \cdot \tau};$$

$$b_3 = \frac{(C_{A1} + k_{A1} \cdot \tau) \cdot m_1}{m_1 \cdot m_2 - m_1 \cdot C_{A2} \cdot \tau + m_2 \cdot C_{A1} \cdot \tau};$$

$$b_4 = \frac{(C_{A2} + k_{A2} \cdot \tau) \cdot m_1}{m_1 \cdot m_2 - m_1 \cdot C_{A2} \cdot \tau + m_2 \cdot C_{A1} \cdot \tau};$$

$$b_5 = \frac{k_{B2} \cdot m_1 \cdot (z_2^2 + 2\tau \cdot z_2 \cdot \dot{z}_2) + C_{A1} \cdot \tau \cdot k_1 \cdot z_0}{m_1 \cdot m_2 - m_1 \cdot C_{A2} \cdot \tau + m_2 \cdot C_{A1} \cdot \tau};$$

one obtains

$$\ddot{z}_2 = b_1 \cdot z_1 + b_2 \cdot z_2 + b_3 \cdot \dot{z}_1 + b_4 \cdot \dot{z}_2 + b_5 \,. \tag{6.8}$$

To sum up, the obtained equation (6.7) and (6.8), governs the active suspension dynamics

$$\ddot{z}_1 = a_1 \cdot z_1 + a_2 \cdot z_2 + a_3 \cdot \dot{z}_1 + a_4 \cdot \dot{z}_2 + a_5;$$

$$\ddot{z}_2 = b_1 \cdot z_1 + b_2 \cdot z_2 + b_3 \cdot \dot{z}_1 + b_4 \cdot \dot{z}_2 + b_5 \,. \tag{6.9}$$

Since the non-homogeneous equations stability is preserved when the stability of homogeneous equations holds, then the latter are further investigated:

$$\dot{x}_1 = x_3;$$

$$\dot{x}_2 = x_4;$$

$$\dot{x}_3 = a_1 \cdot x_1 + a_2 \cdot x_2 + a_3 \cdot x_3 + a_4 \cdot x_4;$$

$$\dot{x}_4 = b_1 \cdot x_1 + b_2 \cdot x_2 + b_3 \cdot x_3 + b_4 \cdot x_4; \tag{6.10}$$

where:

$$x_1 = z_1,\ \ x_2 = z_2,\ \ x_3 = \dot{z}_1,\ \ x_4 = \dot{z}_2\ .$$

The characteristic determinant associated with the equations (6.10) has the form

$$W = \begin{vmatrix} -\lambda & 0 & 1 & 0 \\ 0 & -\lambda & 0 & 1 \\ a_1 & a_2 & a_3 - \lambda & a_4 \\ b_1 & b_2 & b_3 & b_4 - \lambda \end{vmatrix} \tag{6.11}$$

and hence the following characteristic form is obtained:

$$W = \lambda^4 - (a_3 + b_4) \cdot \lambda^3 - a_1 \cdot \lambda^2 + (b_2 + a_1 \cdot b_4 - b_1 \cdot a_4) \cdot \lambda$$

$$+ a_2 \cdot b_3 - b_2 \cdot a_3 + a_1 \cdot b_2 - b_1 \cdot a_2\ . \tag{6.12}$$

Introducing the notation

$$A_3 = -(a_3 + b_4),\ \ A_2 = -a_1,\ \ A_1 = b_2 + a_1 \cdot b_4 - b_1 \cdot a_4,$$

$$A_0 = a_2 \cdot b_3 - b_2 \cdot a_3 + a_1 \cdot b_2 - b_1 \cdot a_2,$$

the following characteristic equation is obtained:

$$\lambda^4 + A_3 \cdot \lambda^3 + A_2 \cdot \lambda^2 + A_1 \cdot \lambda^1 + A_0 = 0\ . \tag{6.13}$$

The necessary condition for stability of the system (6.10) is the positiveness of all coefficients of the characteristic equation (6.13), i.e.

$$A_0 > 0 \wedge A_1 > 0 \wedge A_2 > 0 \wedge A_3 > 0\ . \tag{6.14}$$

The sufficient stability conditions of the system (6.10) require the positiveness of the determinants

$$D_1 = A_3,\quad D_2 = \begin{vmatrix} A_3 & 1 \\ A_1 & A_2 \end{vmatrix},$$

$$D_3 = \begin{vmatrix} A_3 & 1 & 0 \\ A_1 & A_2 & A_3 \\ 0 & A_0 & A_1 \end{vmatrix},\quad D_4 = \begin{vmatrix} A_3 & 1 & 0 & 0 \\ A_1 & A_2 & A_3 & 1 \\ 0 & A_0 & A_1 & A_2 \\ 0 & 0 & 0 & A_0 \end{vmatrix}.$$

The condition $D_1 > 0$ yields

$$A_3 > 0, \tag{6.15}$$

the condition $D_2 > 0$ yields

$$A_3 \cdot A_2 - A_1 > 0\ . \tag{6.16}$$

On the other hand, the condition $D_3 > 0$ yields

$$A_1 \cdot A_2 \cdot A_3 - A_0 \cdot A_3^2 - A_1^2 > 0, \tag{6.17}$$

whereas $D_4 > 0$ and $D_3 > 0$ is equivalent to the inequality

$$A_0 > 0\,. \tag{6.18}$$

The sufficient stability conditions of the system (6.10) are defined by the inequality

$$A_0 > 0;\;\; A_3 > 0;\;\; A_3 \cdot A_2 - A_1 > 0;\;\; A_1 \cdot A_2 \cdot A_3 - A_0 \cdot A_3^2 - A_1^2 > 0\,. \tag{6.19}$$

For the fixed parameters $m_1 = 33\,kg$, $m_2 = 200\,kg$, $k_1 = 200000\,N/m$, $k_2 = 9000\,N/m$, $k_{A1} = 250000\,N/m$, $k_{A2} = -50000\,N/m$, $k_{B2} = -500000\,N/m$, $C_{A1} = 1000\,N \cdot s/m$, $C_{A2} = -10000\,N \cdot s/m$, $\tau = 0.0021\,s$ the stability thresholds are reported in Figure 6.6 (the necessary condition (6.14)) and in Figure 6.7 (the sufficient condition (6.19)).

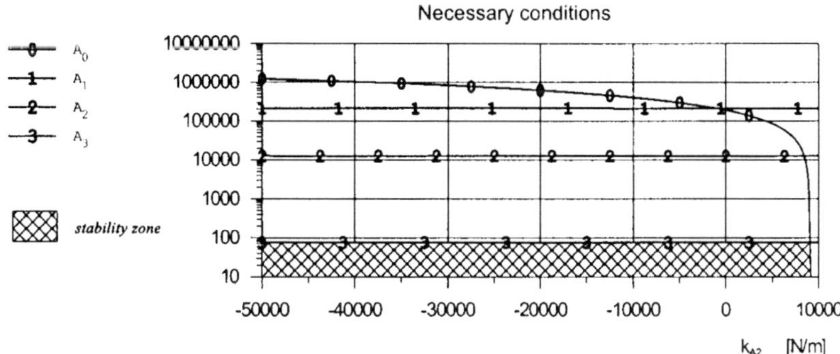

Figure 6.6. Influence of stiffness proportional to the displacement z_2 on the active suspension stability in the sense of Lyapunov (the necessary condition (6.14)).

As it has been shown in Figure 6.6 and 6.7, the investigated system with active suspension is stable in the Lyapunov sense for $k_{A2} < 9000\,N/m$.

Now our system is investigated in the sense of technical stability [37] (see (6.10)).

The initial conditions zone ω and the admissible solutions zone have the forms

$$\omega \equiv \{C_1 \cdot x_1^2 + C_2 \cdot x_2^2 + C_3 \cdot x_3^2 + C_4 \cdot x_4^2 < r^2\},$$

$$\Omega \equiv \{C_1 \cdot x_1^2 + C_2 \cdot x_2^2 + C_3 \cdot x_3^2 + C_4 \cdot x_4^2 \leqslant R^2\}, \;\text{ where: } R < r,\; C_{1,2,3,4} > 0,$$

or equivalently

$$\omega \equiv \left\{x_1^2 < r_1^2, x_2^2 < r_2^2, x_3^2 < r_3^2, x_4^2 < r_4^2\right\},$$

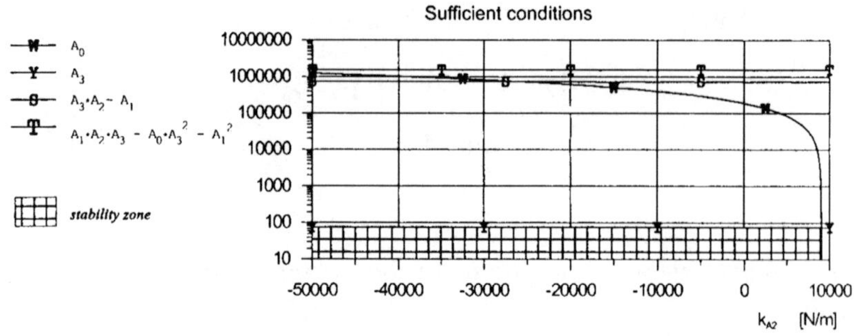

Figure 6.7. Influence of stiffness proportional to the displacement z_2 on the active suspension stability in the sense of Lyapunov (the sufficient condition (6.19)).

$$\Omega \equiv \left\{x_1^2 \leqslant R_1^2, x_2^2 \leqslant R_2^2, x_3^2 \leqslant R_3^2, x_4^2 \leqslant R_4^2\right\}.$$

The scalar function $V_B(x_1, x_2, x_3, x_4)$ for the system (6.10) reads

$$V_B(x_1, x_2, x_3, x_4) = \frac{1}{2}\left(AA \cdot x_1^2 + BB \cdot x_2^2 + CC \cdot x_3^2 + DD \cdot x_4^2\right).$$

It is assumed that the coefficients satisfy the inequalities

$$AA > 0, \quad BB > 0, \quad CC > 0, \quad DD > 0.$$

If there exists a number C_0 satisfying the inequality $C_0 \geq V_B(x_1, x_2, x_3, x_4)$ for $(x_1, x_2, x_3, x_4) \in \omega$, $C_0 > 0$, i.e.

$$C_0 = \sup_{x_i \in \omega} V_B(x_1, x_2, x_3, x_4) = \frac{1}{2} \cdot \left(AA \cdot r_1^2 + BB \cdot r_2^2 + CC \cdot r_3^2 + DD \cdot r_4^2\right), \tag{6.20}$$

and the number C_1 satisfying the inequality $C_1 < V_B(x_1, x_2, x_3, x_4)$ for $(x_1, x_2, x_3, x_4) \in \Omega$, $C_1 > 0$, i.e.

$$C_1 = \inf_{x_i \notin \Omega} V_B(x_1, x_2, x_3, x_4) = \frac{1}{2} \cdot \left(AA \cdot R_1^2 + BB \cdot R_2^2 + CC \cdot R_3^2 + DD \cdot R_4^2\right), \tag{6.21}$$

then for

$$\frac{dV_B}{dt} = +CC \cdot a_3 \cdot x_3^2 + DD \cdot b_4 \cdot x_4^2 + DD \cdot b_1 \cdot x_1 \cdot x_4 + (AA + CC \cdot a_1) \cdot x_1 \cdot x_3$$
$$+ CC \cdot a_2 \cdot x_2 \cdot x_3 + (BB + DD \cdot b_2) \cdot x_2 \cdot x_4 + (CC \cdot a_4 + DD \cdot b_3) \cdot x_3 \cdot x_4 \tag{6.22}$$

the stability condition in the Bogusz sense reads

$$\sup_{x_i \in \Omega/\omega \; t_1 \leqslant t \leqslant t_1 + T} \left\{\frac{dV_B}{dt}\right\} < \frac{C_1 - C_0}{T}. \tag{6.23}$$

Satisfaction of the inequality (6.23) means, that the system governed by the equations (6.10) is technically stable with respect to zones ω and Ω and perturbation R. The condition (6.23) can be easy verified via the numerical simulations. The problem is reduced to the numerical analysis of $\dot{V}_B$ in $\{x_1, x_2, x_3, x_4\} \in \Omega/\omega$, where

$$\Omega/\omega \equiv \left\{r_1^2 < x_1^2 \leqslant R_1^2 \wedge r_2^2 < x_2^2 \leqslant R_2^2 \wedge r_3^2 < x_3^2 \leqslant R_3^2 \wedge r_4^2 < x_4^2 \leqslant R_4^2\right\}.$$

The technical stability factor is defined by the dependence

$$FT = \frac{C_1 - C_0}{T} - \frac{dV_B}{dt}.$$

Taking into account the relations (6.21), (6.22) and (6.23) the stability factor is defined in the following way:

$$FT = \frac{C_1 - C_0}{T} - CC \cdot a_3 \cdot x_3^2 + DD \cdot b_4 \cdot x_4^2 - DD \cdot b_1 \cdot x_1 \cdot x_4$$
$$-(AA + CC \cdot a_1) \cdot x_1 \cdot x_3 - CC \cdot a_2 \cdot x_2 \cdot x_3 - (BB + DD \cdot b_2) \cdot x_2 \cdot x_4$$
$$-(CC \cdot a_4 + DD \cdot b_3) \cdot x_3 \cdot x_4 \,. \tag{6.24}$$

Recall that if FT is positive then the system is stable. Let us investigate the influence of the chosen vehicle parameters governed by the system of differential equations (6.10) on the stability for some arbitrarily chosen parameters. Zones ω and Ω are taken intuitively. The following initial conditions are taken:

$$z_1 = 0 \pm 0.005\,m, \quad z_2 = 0 \pm 0.01\,m, \quad \dot{z}_1 = 0.01\,m/s, \quad \dot{z}_2 = 0.02\,m/s\,.$$

The admissible solutions are as follows:

$$z_1 = 0 \pm 0.010\,m, \; z_2 = 0 \pm 0.02\,m, \; \dot{z}_1 = 0 \pm 0.01\,m/s, \; \dot{z}_2 = 0 \pm 0.25\,m/s\,.$$

The numerical results and the stability factors are reported in Figure 6.8. It is worth noticing that the FT dependence on the stiffness k_{A2} qualitatively overlaps with the Lyapunov dependence (see Figure 6.6). The technical stability depends on the other system parameters. In particular, its dependence on time delay τ is remarkable. In words, the stability factor is improved for small delay values, and the essential control improvement is achieved for $\tau < 0.05\,s$.

2. Analysis of Suspension Vibrations

Consider first dynamics of the passive non-linear suspension shown in Figure 6.1, where instead of the unit U the damping element C_2 is used. Its vibrations are governed by the equations

$$m_1 \cdot \ddot{z}_1 - C_2 \cdot (\dot{z}_2 - \dot{z}_1) - k_2^* \cdot (z_2 - z_1) + (m_1 + m_2) \cdot g - F_{z1} = 0, \tag{6.25}$$

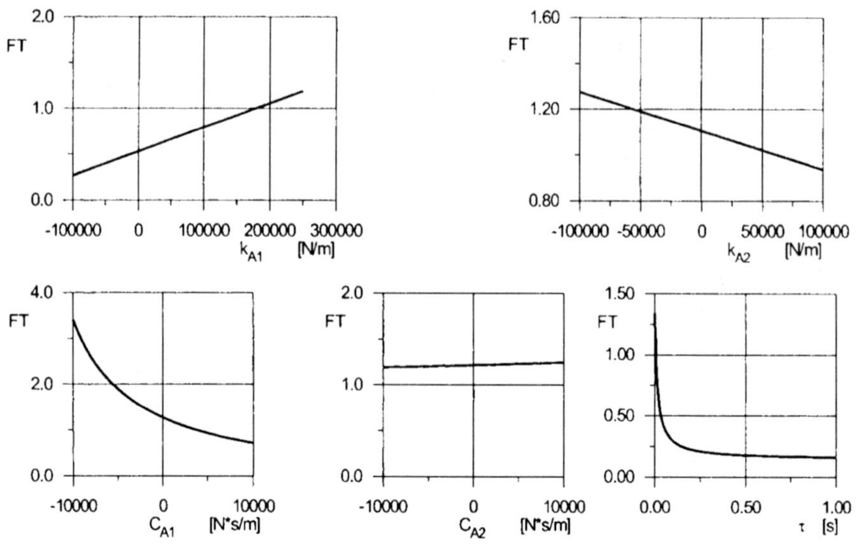

Figure 6.8. Influence of the active suspension parameters on the technical stability in the Bogusz sense.

$$m_2 \cdot \ddot{z}_2 + C_2 \cdot (\dot{z}_2 - \dot{z}_1) + k_2^* \cdot (z_2 - z_1) = 0. \tag{6.26}$$

It is assumed that the stiffness k_2^* is treated either as the constant or as variable parameter via the following rules:

$$(z_1 - z_2) \leqslant \Delta z_{22} \Rightarrow k_2^* = k_2 = const,$$

$$(z_1 - z_2) > \Delta z_{22} \Rightarrow k_2^* = (\Delta z_{22} - (z_1 - z_2))^2 \cdot k_{22}, \text{ where } k_{22} = const\,. \tag{6.27}$$

The following cases are considered:

1°. The road wheel is always pressed to the road surface, i.e.

$$F_{z1} = (m_1 + m_2) \cdot g + k_1 \cdot (z_0 - z_1) > 0\,. \tag{6.28}$$

2°. The lack of contact between the wheel and the road appears,

$$F_{z1} = (m_1 + m_2) \cdot g + k_1 \cdot (z_0 - z_1) < 0 \Rightarrow F_{z1} = 0\,.$$

Furthermore, it is assumed that the analysed system is excited kinematically by the road shape

$$z_0 = z_{0m} \cdot \cos(\omega \cdot t), \tag{6.29}$$

where: z_{0m} is the amplitude and ω is the frequency of the excitation. Two different excitations are analysed: (i) harmonic excitation without stochastic perturbations ($z_{0m} = const$) and (ii) harmonic excitation with stochastically perturbed amplitude ($z_{0m} \neq const$). The stochastic perturbations are simulated via the analogous procedures as in Example 4.1, using the "random" function. The stochastic variable is taken for $n = 12$ sampling in the form

$$\xi_n = \xi_{r1} + \xi_{r2} + \xi_{r3} + \ldots + \xi_{r12} - 6 \,. \tag{6.30}$$

The simple algorithm yielding the stochastic variable with the normal distribution by applying the function "random" and the formula (6.30) has the form:

```
for i=1 to 12;
   ksi(i):=random;
   sksi:=sksi+ksi(i);
next;
sksi:=sksi-6;
```

Owing to equations (6.25) and (6.26), dynamics of the passive non-linear suspension harmonically excited with randomly perturbed amplitude is governed by the equations

$$\begin{aligned} \dot{u} &= a_1 \cdot x + a_2 \cdot u + a_3 \cdot y + a_4 \cdot z + e_1, \\ \dot{z} &= u, \\ \dot{x} &= b_1 \cdot x + b_2 \cdot u + b_3 \cdot y + b_4 \cdot z, \\ \dot{y} &= x, \end{aligned} \tag{6.31}$$

where:

$$\frac{dz_2}{dt} = x, \quad z_2 = y, \quad \frac{dz_1}{dt} = u, \quad z_1 = z, \quad a_1 = \frac{C_2}{m_1}, \quad a_2 = -\frac{C_2}{m_1},$$

$$a_3 = \frac{k_2^*}{m_1}, \quad a_4 = -\frac{k_2^*}{m_1}, \quad b_1 = \frac{C_2}{m_2}, \quad b_2 = -\frac{C_2}{m_2}, \quad b_3 = \frac{k_2^*}{m_2},$$

$$b_4 = \frac{k_2^*}{m_2}, \quad e_1 = \frac{F_{z1} - (m_1 + m_2) \cdot g}{m_1},$$

$$z_0 = z_{0m} \cdot \cos(\omega \cdot t), \quad z_{0m} = z_{00} \cdot (1 + a_0 \cdot sksi) \,. \tag{6.32}$$

In the above the following notation is used: a_0 - relative value of the harmonic road surface amplitude; z_{00} - amplitude of the harmonic road surface without random perturbation; z_{0m} - amplitude of the harmonic road surface with random perturbations; $sksi$ - random variable.

EXAMPLE 6.1 *[16] This example presents theoretical analysis of excited vibrations of the vehicle in a wide range of excitation frequencies (from 1 to 80 Hz). The mathematical model and calculations were prepared for the accepted physical model of the vehicle. The model was used to simulate the excited vertical vibrations. The bifurcation figures with an excitation frequency as a bifurcation parameter were determined on the basis of the simulation results – the changes of kinematics values in time. Bifurcation diagrams give a picture of vehicle vibrations. The picture gives the possibility of identification of characteristic parameters of springs and damping elements and can be used for control, diagnostic aims, and for making technical investigations of vehicle suspension.*

The following notation is applied:

A, B - *point - front or rear arm joint,*

OF, OR - *point - front or rear axle,*

F, R - *point front or rear wheel contact plane with the road surface,*

Z - *vertical displacement of body mass centre,*

$\ddot{z}, \dfrac{d^2 z}{dt^2}$ - *second derivative of the displacement z on time t,*

ε - *rotation angle of the vehicle body,*

$\ddot{\varepsilon}, \dfrac{d^2 \varepsilon}{dt^2}$ - *second derivative of the angular displacement ε on time t,*

z_F, z_R - *vertical displacement of front or rear arm joint,*

z_{OF}, z_{OR} - *vertical displacement of front or rear wheel,*

f_{sF}, f_{sR} - *displacement of suspension front or rear elements,*

h_F, h_R - *kinematics excitation (resulting for ex. road profile),*

φ_F, φ_R - *front or rear wheel angle of rotation,*

β_F, β_R - *angular displacement of front or rear suspension spring and damping elements,*

α_F, α_R - *angular displacement of front or rear suspension arm,*

m - *body mass,*

J - *body inertia moment according to the mass centre,*

m_F, m_R - *front or rear wheel mass,*

J_F, J_R - *inertia moment of the front or rear wheel (according to the axle of rotation),*

F_{sC}, F_{sD} - *force in the spring element of the front or rear suspension,*

F_{tC}, F_{tD} - *force in the damping element of the front or rear suspension,*

F_{zF}, F_{zR} - *vertical force acting from the road surface on the front or rear wheel,*

F_{xF}, F_{xR} - *horizontal force acting from the road surface on the front or rear wheel,*

M_{TF}, M_{TR} - *friction torque of braking element of the front or rear wheel,*

R_{KF}, R_{KR} - *dynamic radius of the front or rear wheel,*

L_{AOF} - *front suspension arm length,*

L_{BOR} - *rear suspension arm length.*

The object of investigation is a vehicle model with a body connected to wheels by single longitudinal arms, in the front pushed arm, in the rear – trailed arm – Fig. 6.9 and Fig. 6.10. A suspension system with longitudinal arms give the model a general character, because the kinematics of any suspension system can be presented using the kinematics of an artificial arm with "transient" length and with a "transient" centre of rotation. The model consists of a flat system of inertia elements $((m, J), (m_F, J_F)$ and $(m_R, J_R))$, and elements without inertia $((A - OF)$ and $(B - OR))$, connected by the spring and damping elements. It was accepted that the model of tires is also non-linear.

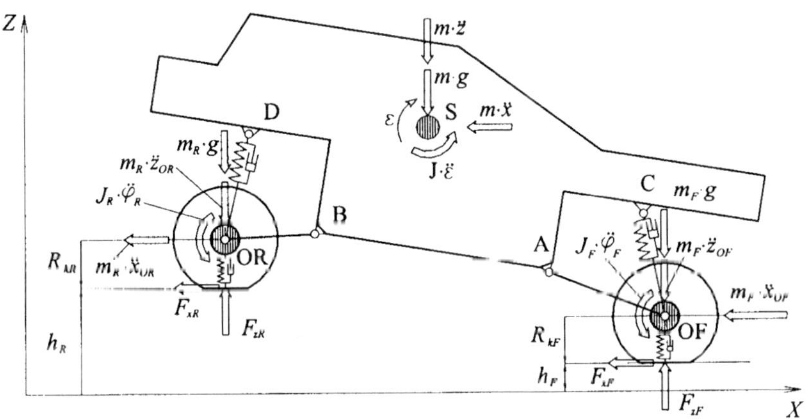

Figure 6.9. Vehicle model.

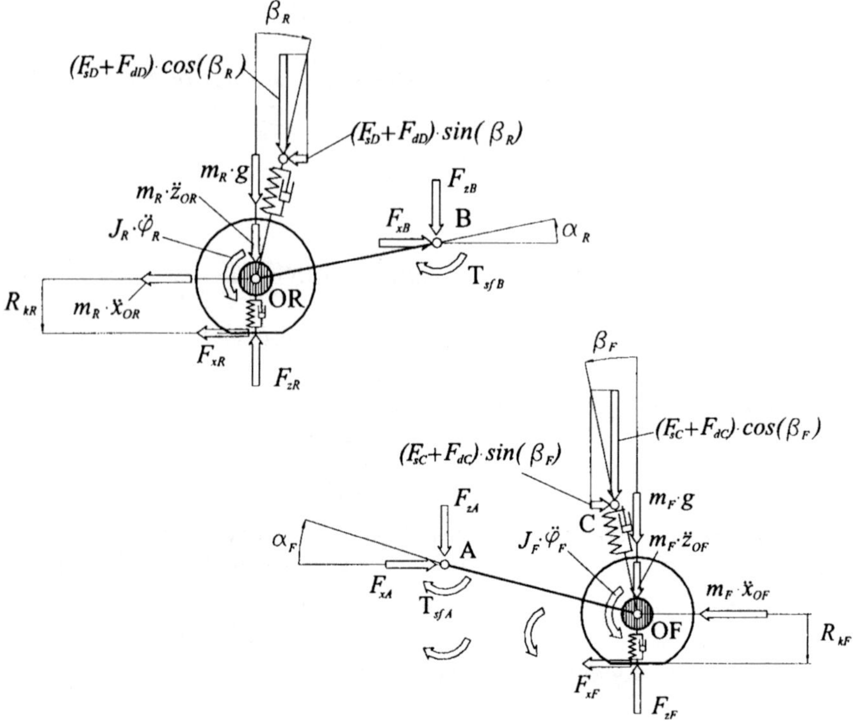

Figure 6.10. Vehicle suspension model.

Damping forces in the suspension shock absorbers are described using the following equations:

$$for\ \frac{df_s}{dt} < 0 \quad F_t = 0.6 \cdot C_t \cdot \left[\left(\frac{df_s}{dt} \right) + 0.1 \cdot \left(\frac{df_s}{dt} \right)^2 + 0.006 \cdot \left(\frac{df_s}{dt} \right)^3 \right],$$

$$for\ \frac{df_s}{dt} > 0 \quad F_t = C_t \cdot \left[\left(\frac{df_s}{dt} \right) - 0.1 \cdot \left(\frac{df_s}{dt} \right)^2 + 0.006 \cdot \left(\frac{df_s}{dt} \right)^3 \right],$$

$$for\ \frac{df_s}{dt} = 0 \quad F_t = 0,$$

where coefficient C_t for the front (F) and rear (R) axle: $C_{tF} = 2400\, N \cdot s/m$ and $C_{tR} = 3500\, N \cdot s/m$.

The considered vehicle model is a system with seven degrees of freedom, as one can see in Figs. 6.9, 6.10. The generalized co-ordinates accepted here are the following kinematic values: $\varepsilon, x_S, z_S, \alpha_F, \alpha_R, \varphi_F, \varphi_R$. These values are shown with the vehicle model in Fig. 6.11.

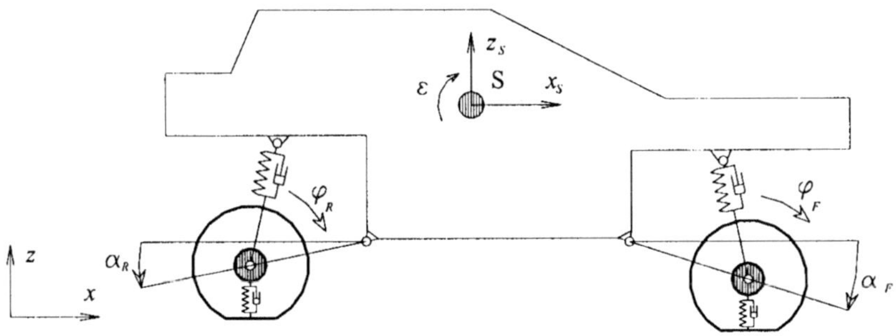

Figure 6.11. Generalized co-ordinates for vehicle model $\varepsilon, x_S, z_S, \alpha_F, \alpha_R, \varphi_F, \varphi_R$.

The following quantities were introduced:

$$q_1 = \frac{d\varepsilon}{dt}, \quad q_2 = \varepsilon, \quad q_3 = \frac{dx_S}{dt}, \quad q_4 = x_S, \quad q_5 = \frac{dz_S}{dt}, \quad q_6 = z_S,$$

$$q_7 = \frac{d\alpha_F}{dt}, \quad q_8 = \alpha_F, \quad q_9 = \frac{d\alpha_R}{dt}, \quad q_{10} = \alpha_R, \quad q_{11} = \frac{d\varphi_F}{dt},$$

$$q_{12} = \varphi_F, \quad q_{13} = \frac{d\varphi_R}{dt}, \quad q_{14} = \varphi_R.$$

Basic equations of the vehicle model motion – Figs. 6.9–6.11 – were presented as a system of derivative equations $\frac{dq_i}{dt}(q_i, \ldots)$.

$$\frac{dq_1}{dt} = \{F_{zB}\cdot(q_4 - x_B) + F_{xB}\cdot(q_6 - z_B) - [F_{zA}\cdot(x_C - q_4) + F_{xA}\cdot(q_6 - z_A)$$
$$+ [F_{sD} + F_{tD}]\cdot\cos(\beta_R)\cdot(q_4 - x_D)] - [(F_{sD} + F_{tD})\cdot\sin(\beta_R)\cdot(q_6 - z_D)$$
$$- (F_{sC} + F_{tC})\cdot\cos(\beta_F)\cdot(x_C - q_4)] + (F_{sC} + F_{tC})\cdot\sin(\beta_F)\cdot(q_6 - z_C)\}\cdot\frac{1}{J},$$

$$\frac{dq_2}{dt} = q_1,$$

$$\frac{dq_3}{dt} = \frac{-F_{xR} - F_{xF}}{m + m_F + m_R},$$

$$\frac{dq_4}{dt} = q_3,$$

$$\frac{dq_5}{dt} = [-m\cdot g + F_{zB} + F_{zA} + (F_{sD} + F_{tD})\cdot\cos(\beta_R) + (F_{sC} + F_{tC})\cdot\cos(\beta_F)]\cdot\frac{1}{m},$$

$$\frac{dq_6}{dt} = q_5,$$

$$\frac{dq_7}{dt} = \frac{m_F\cdot g - F_{zF} + F_{zA} + (F_{sC} + F_{tC})\cdot\cos(\beta_F)}{m_F\cdot\cos(q_8)\cdot L_{AOF}}$$
$$+ \frac{\frac{d^2 z_{OF}}{dt^2}}{\cos(q_8)\cdot L_{AOF}} + \frac{\sin(q_8)}{\cos(q_8)}\left(\frac{dq_8}{dt}\right)^2,$$

$$\frac{dq_8}{dt} = q_7,$$

$$\frac{dq_9}{dt} = \frac{m_R \cdot g - F_{zR} + F_{zB} + (F_{sD} + F_{tD}) \cdot \cos(\beta_R)}{m_R \cdot \cos(q_{10}) \cdot L_{BOR}}$$

$$+ \frac{\frac{d^2 z_{OR}}{dt^2}}{\cos(q_{10}) \cdot L_{BOR}} + \frac{\sin(q_{10})}{\cos(q_{10})} \cdot \left(\frac{dq_{10}}{dt}\right)^2,$$

$$\frac{dq_{10}}{dt} = q_9,$$

$$\frac{dq_{11}}{dt} = -\frac{M_{TF}}{J_{KF}} + \frac{F_{xF} \cdot R_{KF}}{J_{KF}},$$

$$\frac{dq_{12}}{dt} = q_{11},$$

$$\frac{dq_{13}}{dt} = -\frac{M_{TR}}{J_{KR}} + \frac{F_{xR} \cdot R_{KR}}{J_{KR}},$$

$$\frac{dq_{14}}{dt} = q_{13}.$$

The derivative equations system $\frac{dq_i}{dt}(q_i, \ldots)$ *was solved using the numerical procedure Runge–Kutta IV. The time step for calculation was constant at the level of* $\Delta t = 2^{-12}$ *s, it means:* $\Delta t = 0.000244140625\,s$. *The step value was chosen after many tests and numerical investigations.*

Simulation numerical tests were developed using this system model and vehicle model. Excitation of vibrations was done in two cases:

- *excitation for front wheels* $h_F(t) = a_{har} \cdot \sin(2 \cdot \pi \cdot n \cdot t)$,
- *excitation for front and end wheels simultaneously:*

$$h_F(t) = a_{har} \cdot \sin(2 \cdot \pi \cdot n \cdot t) \quad \textit{and} \quad h_R(t) = a_{har} \cdot \sin(2 \cdot \pi \cdot n \cdot t).$$

Simulations were performed for different values of excitation frequencies from $n = 1$ *to* $n = 80\,Hz$. *Amplitude of excitation had constant value for all simulations at the rate of* $a_{har} = 0.02\,m$.

Initial values of simulation

From one point of view, the vehicle body behaviour (motion, vibrations) might be interesting, form the other – behaviour of wheels (motion, displacement) is interesting when one investigates vertical vehicle vibrations. The first problems are important when one pays attention to travelling comfort, the second - when safety of travel is considered. The following initial values were accepted for these reasons:

z_A, z_B $[m]$ - *body vertical displacement for point* A, B - *Fig. 6.9,*

$\frac{dz_A}{dt}, \frac{dz_R}{dt}$ $[m/s]$ - *velocity of body vertical displacement for point* A, R,
z_{OF}, z_{OR} $[m]$ - *wheel vertical displacement for point* OF, OR,
$\frac{dz_{OF}}{dt}, \frac{dz_{OR}}{dt}$ $[m/s]$ - *velocity of wheel vertical displacement for point* OF, OR.

Vehicle parameters

The values of basic parameters for the investigated vehicle model were accepted as follows:

- *vehicle body mass* - $m = 1400\,kg$;
- *body inertial moment according to the centre of mass* - $J = 1200\,kg{\cdot}m^2$;
- *mass of the front or rear wheel with suspension elements* - $m_F = 30\,kg$, $m_R = 25\,kg$;
- *wheel inertia moment (according to the wheel axle) for front or rear wheel* - $J_F = 2\,kg{\cdot}m^2$, $J_R = 1.5\,kg{\cdot}m^2$;
- *stiffness coefficient for front or rear suspension* - $K_{sF} = 25000\,N/m$, $K_{sR} = 20000\,N/m$;
- *damping coefficient for front or rear suspension* - $C_{tF} = 2400\,N \cdot s/m$, $C_{tR} = 3500\,N \cdot s/m$;
- *suspension arm length for front or rear* - $L_{AOF} = 0.5\,m$, $L_{BOR} = 0.5\,m$;
- *wheel free radius* - $R_{KFsw} = 0.25\,m$, $R_{KRsw} = 0.25\,m$.

Other parameters (length and other dimensions of the vehicle model) are similar to those of European passenger car of lower middle class.

Series of initial values were received as the result of computer simulation done for the investigated system model. They were the basis for plotting the bifurcation diagrams. The excitation frequency - n *[Hz] was accepted as the bifurcation parameter.*

The series of bifurcation curve values were created for sampling time delay $t_0 = 0$:

$$\left\{ z_i(t)|_{t=t_0+nk}, n \right\} \quad and \quad \left\{ \frac{dz}{dt_i}(t)\bigg|_{t=t_0+nk}, n \right\}.$$

The results were recordered after receiving a steady state of vibrations in order to avoid the influence of initial transient states in the system.

The bifurcation diagrams $\left\{ \frac{dz_A}{dt}, n \right\}$, *and* $\{z_{OF}, n\}$ *are shown in Fig. 6.12, where* $\frac{dz_A}{dt}$ - *vertical displacement velocity for vehicle body point* A, z_{OF} -

vertical displacement for front wheel point OF. The curves were drawn for standard parameter values for a vehicle suspension model: $C_{tF} = 2400\, N \cdot s/m$. System vibrations of the presented curves is the following phenomenon: bifurcation branch appears at excitation frequency of approx. 35 Hz. This bifurcation disappears at a frequency of approx. 78 Hz.

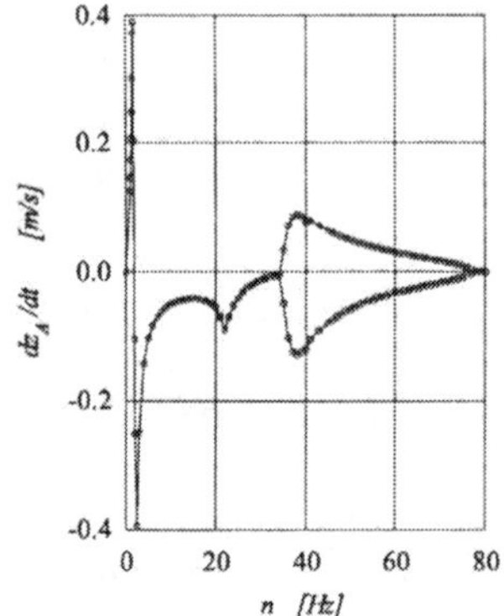

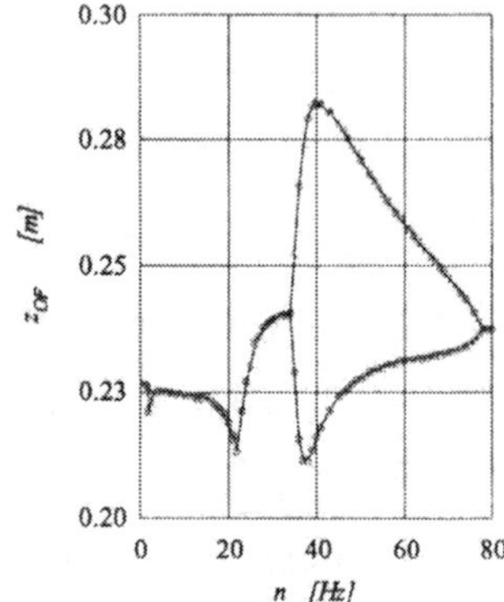

Figure 6.12. Bifurcation diagrams showing body vibrations for point A $\left\{\frac{dz_A}{dt}, n\right\}$ and showing front wheel vibrations - $\{z_{OF}, n\}$. Damping coefficient for front suspension value - $C_{tF} = 2400\, N \cdot s/m$. Excitation on front axle.

The bifurcation diagrams $\left\{\frac{dz_A}{dt}, n\right\}$, and $\{z_{OF}, n\}$ are shown in Fig. 6.13, for lower values of parameter describing damping of the front suspension shock absorber $C_{tF} = 1800\, N \cdot s/m$, which means for a lower damping of vibration. The system vibrations are of harmonic character, for the majority of vibrations frequency values. The characteristic feature of the curves is that, similarly as in Fig. 6.12, the bifurcation appears here in the range of excitation frequencies approx. $34 \div 78\, Hz$. The secondary branch - bi-bifurcation appears in the range of excitation frequencies of approx. $54 \div 65\, Hz$, for curves $\left\{\frac{dz_A}{dt}, n\right\}$, and $\{z_{OF}, n\}$.

The characteristic values of bifurcation diagrams are given in Table 6.1. These are: the range of excitation frequencies for bifurcation, bifurcation amplitude and the range of excitation frequencies for bi-bifurcation appearance.

The Poincaré maps were additionally presented in Fig. 6.15, 6.15 and 6.16 for a better explanation of bifurcation curves for vehicle model with lower front suspension damping - Fig. 6.13. These curves were drawn for such excitation

[1]amplitude values are not for curves with bifurcation, but for chaotic vibrations with maximal amplitude

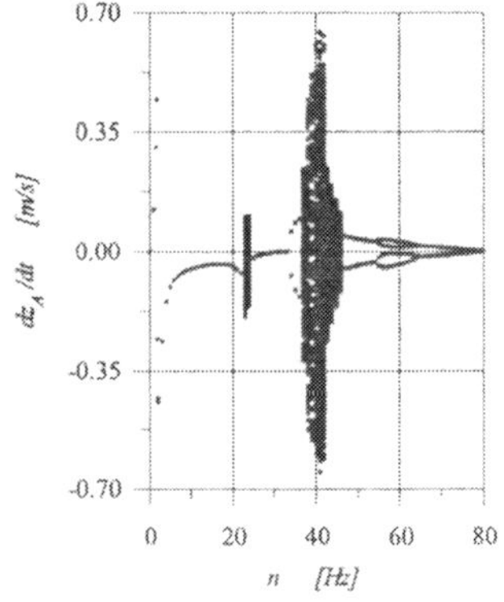

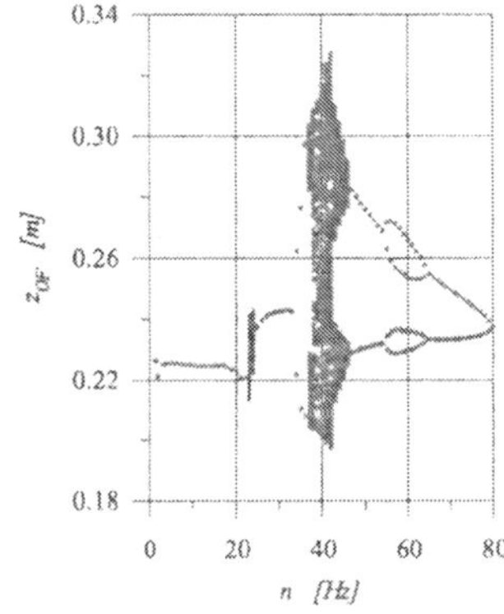

Figure 6.13. Bifurcation diagrams showing body vibrations for point A - $\left\{\frac{dz_A}{dt}, n\right\}$ and showing front wheel vibrations - $\{z_{OF}, n\}$. Damping coefficient for front suspension value - $C_{tF} = 1800\,N\cdot s/m$. Excitation on front axle.

Table 6.1. The characteristic values of bifurcation curves.

Bifurcation diagrams/ vehicle data	*The range of bifurcation appearance* n *[Hz]*	*Amplitude (dual) of bifurcation*	*The range of bi-bifurcation appearance* n *[Hz]*
$\left\{\frac{dz_{OF}}{dt}, n\right\}$ / standard data	$35 \div 78$	0.22 m/s	–
$\{z_{OF}, n\}$ / standard data	$35 \div 78$	0.071 m	–
$\left\{\frac{dz_{OF}}{dt}, n\right\}$ $/C_{tF}\langle C_{tF\,std}$	$32 \div 76$	1.3^1m	$54 \div 65$
$\{z_{OF}, n\}$ $/C_{tF}\langle C_{tF\,std}$	$32 \div 76$	0.13^1m	$54 \div 65$

frequencies at which the non-harmonic motion was presented in bifurcation curves.

Analysing the bifurcation diagrams – Fig. 6.13 – and Poincaré map – Fig. 6.14 – one can see that point A (mounting point for front arm) vertical motion has a character of quasi-periodic dual loop curves for $\frac{dz_A}{dt}(z_A)$*. For the excitation frequency* $37\,Hz$*, point A vertical motion has also a quasi-periodic character with dual loop curves* $\frac{dz_A}{dt}(z_A)$*, but the amplitude of this motion is several times higher than the previous one. At excitation frequency of* $39\,Hz$*, point A vertical motion has also quasi harmonic character with dual loop curves* $\frac{dz_A}{dt}(z_A)$*, but one can observe some chaotic behaviour. At excitation frequency* $41\,Hz$*, point A vertical motion has also chaotic character. For higher values of excitation frequency - for* $n = 43$ *and* $n = 45\,Hz$ *- system*

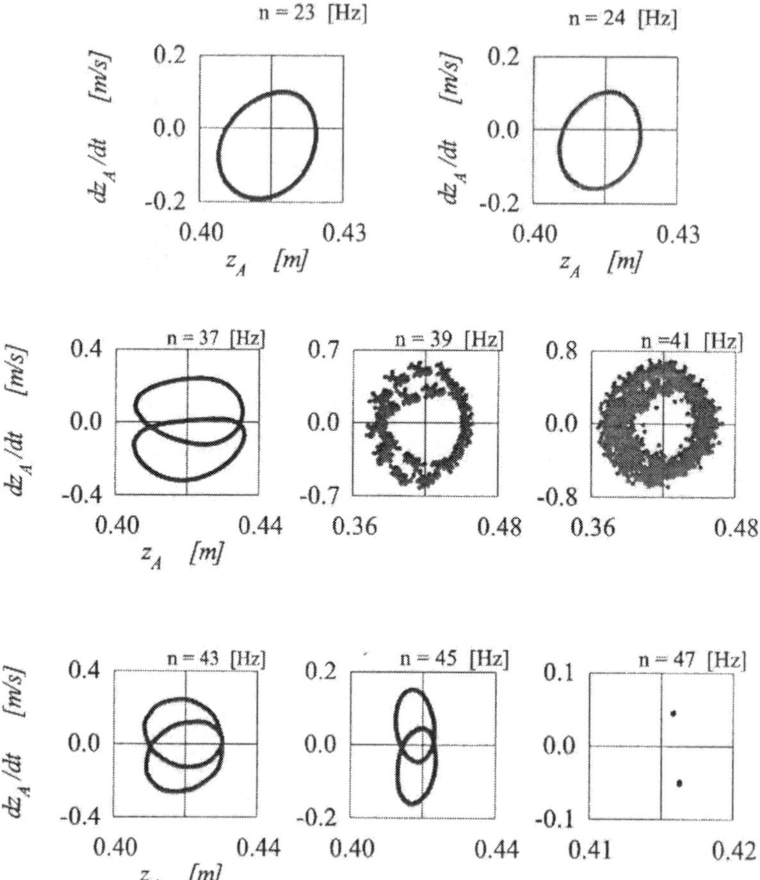

Figure 6.14. Poincaré maps showing vibrations for point $A\left\{\frac{dz_A}{dt}, z_A\right\}$. Damping coefficient for front suspension lower value - $C_{tF} = 1800\,N\cdot s/m$. Excitation on front axle for selected excitation frequencies.

motion comes back to quasi-periodic motion with dual loop curves $\frac{dz_A}{dt}(z_A)$, *and at excitation frequency* $n = 47\,Hz$ *one can see harmonic motion with bifurcation.*

Poincaré maps are shown in Fig. 6.15 $\frac{dz_{OF}}{dt}(z_{OF})$ *- describing point* OF *motion (front wheel vertical motion). These maps concern bifurcation curves* $z_{OF}(n)$*, which were drawn for lower damping* ($C_{tF} = 1800\,N\cdot s/m$) *- Fig. 6.13.*

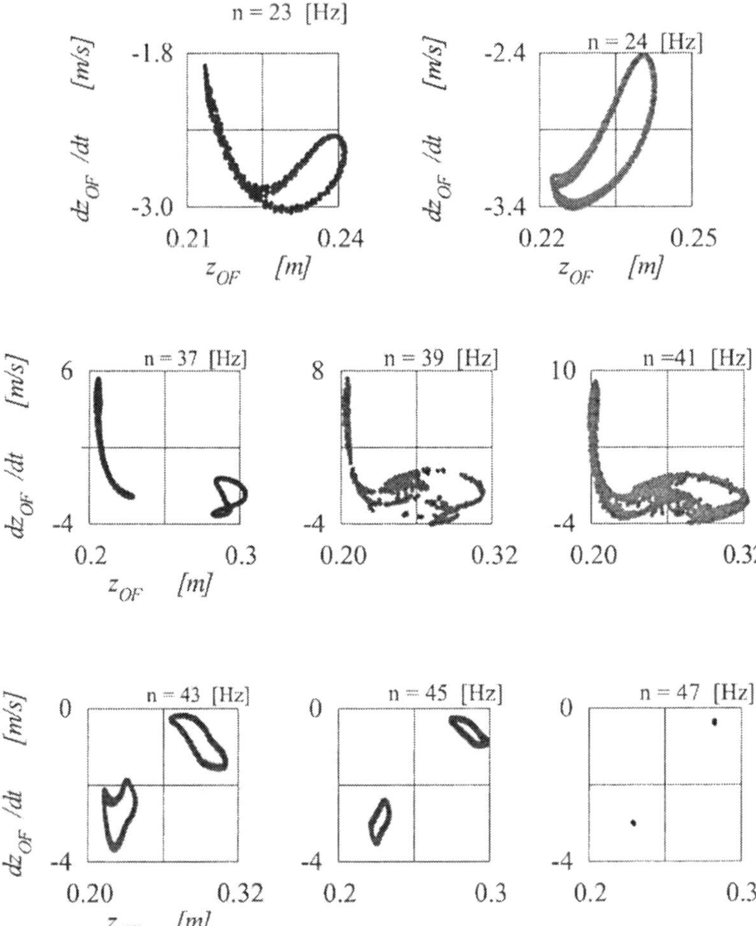

Figure 6.15. Poincaré maps showing front wheel vibrations $\left\{\frac{dz_{OF}}{dt}, z_{OF}\right\}$. Damping coefficient for front suspension lower value - $C_{tF} = 1800\, N \cdot s/m$. Excitation on front axle for selected excitation frequencies.

Basing on bifurcation diagrams - Fig. 6.13 - and Poincaré map - Fig. 6.15 - one can see that at excitation frequencies 23 *and* 24 Hz *front wheel point* OF *vertical motion is a quasi harmonic motion. At excitation frequency* 37 Hz, *point* OF *vertical motion has quasi-periodic character with dual loop curves* $\frac{dz_{OF}}{dt}(z_{OF})$ *At excitation frequency* 39 Hz *point* OF *motion maintains the character of quasi-periodic motion with dual loop curves* $\frac{dz_{OF}}{dt}(z_{OF})$, *but there are some symptoms of chaotic motion. For the following values of excitation* *frequencies - $n = 43$ and $n = 45\,Hz$ - the system motion is quasi harmonic with dual loop curves, but at excitation frequency $n = 47\,Hz$ one can see harmonic motion with bifurcation.*

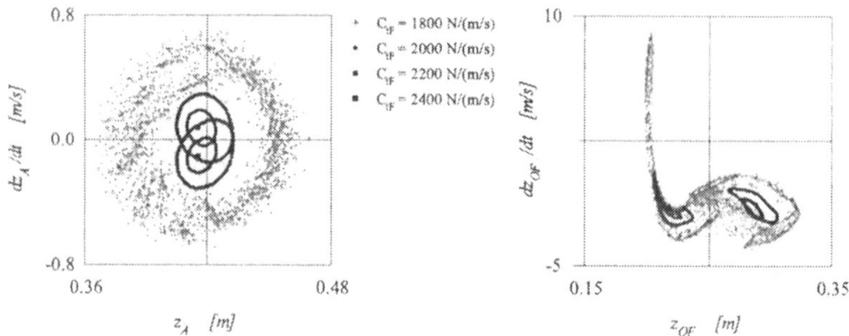

Figure 6.16. Poincaré maps showing body vibrations for point A - $\left\{\frac{dz_A}{dt}, z_A\right\}$ and showing front wheel vibrations $\left\{\frac{dz_{OF}}{dt}, z_{OF}\right\}$ for different values of front suspension damping coefficient C_{tF}. Excitation on front axle at frequency 41 Hz.

Poincaré map $\frac{dz_A}{dt}(z_A)$ is presented in Fig. 6.16 for excitation frequency 41 Hz. The influence of damping on system motion is shown in this figure. as it results from the presented map, and also from Fig. 6.14, for damping coefficient $C_{tF} = 1800\,N{\cdot}s/m$ point A motion (body motion) has a chaotic character. The increase of damping (for this kind of excitation) causes a change in vibration character - for $C_{tF} = 20000\,N\cdot s/m$ motion stops to be chaotic and it evolves quasi-periodic with two loops. Further increase of damping coefficient value results in amplitude decrease $\frac{dz_A}{dt}$ and z_A, and for $C_{tF} = 2400\,N\cdot s/m$ the motion becomes harmonic with bifurcation.

The kinematic input signals on front wheels and rear wheels are described by the following equations:

$$h_F(t) = a_{har}\cdot\sin(\omega\cdot t) \quad \textit{and} \quad h_R(t) = a_{har}\cdot\sin(\omega\cdot t).$$

Some results of investigations obtained for standard values of suspension parameters and for excitation acting simultaneously on front and rear wheels are presented in Fig. 6.17.

For the case of vibration excitation acting simultaneously on front and rear wheels, bifurcation diagrams have a different shape than that for vibration excitation acting on one wheel axle - see Fig. 6.12. In Fig. 6.17 one can see bifurcation and also pulsation impulse change of the value $\frac{dz_A}{dt}$. Character of this pulsation is shown exactly at frequency range $n = 48$ and $n = 52\,Hz$.

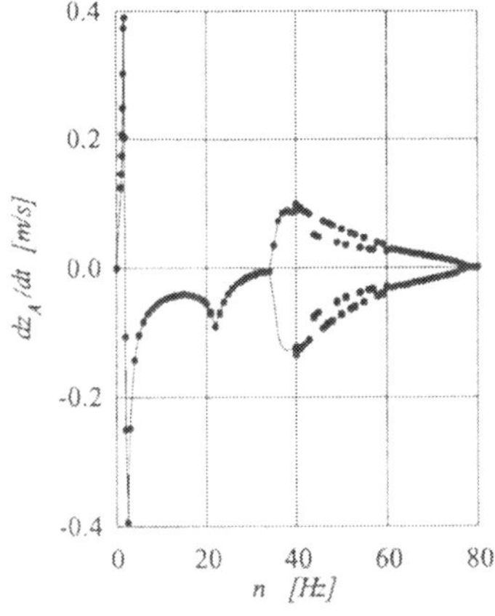

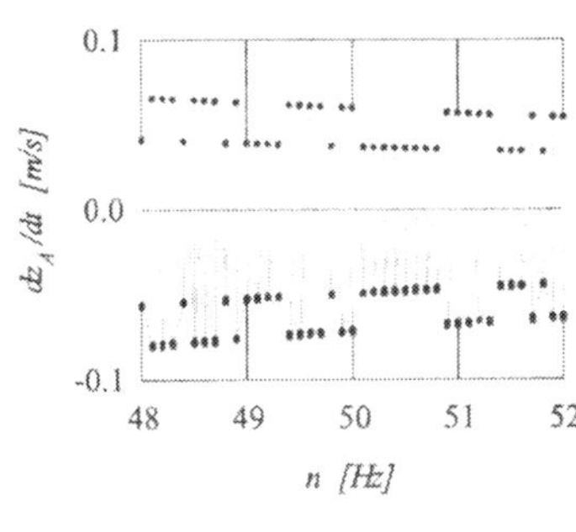

Figure 6.17. Bifurcation diagrams showing body vibrations for point A - $\left\{ \frac{dz_A}{dt}, n \right\}$. Damping coefficient for front suspension standard value - $C_{tF} = 2400\ N \cdot s/m$. Excitation on front and rear axles - simultaneously.

To conclude, in this example the results of simulation tests for vehicle model whose structure is shown in Figs. 6.9, 6.10, also with non-linear characteristics of spring and damping elements are presented. The assumed values of parameters and inertia elements for this vehicle model were similar to those of passenger car of European lower middle class. The excitation for vehicle model was harmonic. The excitation character was the same as that acting on a vehicle running on a road of certain profile or during investigation on seismic test bed. The tests were carried out for two cases:

1 *excitation acting on vehicle front wheels,*

2 *excitation acting on vehicle front and rear wheels simultaneously.*

The results of these tests are presented as bifurcation diagrams and Poincaré maps Figs. 6.12–6.17.

The following conclusions were drawn on the basis of the analysis of bifurcation diagrams:

- *In the case of excitation acting on the wheels of one axle, bifurcation diagrams* $\left\{ \frac{dz_A}{dt}, n \right\}$, $\{z_{OF}, n\}$ *the curves are* smooth (continuous) - *Fig. 6.12.*

- *Bifurcation has the shape of* snake head - *Fig. 6.12. This bifurcation appears in the range of* 34 *to* 80 Hz.

- *In the case of lower, insufficient suspension damping, one can see dual loops in bifurcation diagrams - bi-bifurcation - Fig. 6.13. Bifurcation has the shape of a* snake head with glasses.

- *In the case of lower, insufficient suspension damping, in some range of excitation frequency, some symptoms of quasi-periodic motion can be observed - see bifurcation diagrams in Fig. 6.13 and Poincaré maps in Figs. 6.14, 6.15 and 6.16.*

- *In the case of lower, insufficient suspension damping, in some range of excitation frequencies, some symptoms of chaotic motion can be observed - see bifurcation diagrams in Fig. 6.13 and Poincaré maps in Figs. 6.14, 6.15 and 6.16.*

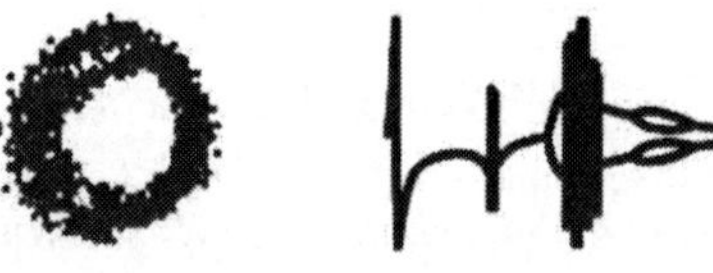

- *In the case of excitation acting simultaneously on front and rear wheels,* pulse slit *can be observed in bifurcation diagrams, values pulsation showing instability of function* - $\left\{ \frac{dz_A}{dt}, n \right\}$ - *Fig. 6.17.*

- *Bifurcation diagrams created for excitation frequencies ranging from* 1 *to* 80 Hz *are the picture of vehicle model vibration, which can be changed by introducing changes in the parameters of vehicle suspension characteristic.*
- *Change (approx.* 25%*) of parameter values for the test results presented here, connected with damping, had a serious influence on the shape of bifurcation diagrams, both in quantitative and in qualitative sense.*
- The sensivity *of shape of bifurcation diagrams with respect to the changes of values of parameters characteristics of vehicle suspension damping is so high that the bifurcation diagram can be used for* identification *of these characteristics or for estimation of suspension technical state.*

Chapter 7

TRANSVERSAL TILT DYNAMICS

In this chapter dynamics of a transversal tilt is illustrated by analysis of two fundamental cases. In the first case the mechanical system consists of a vehicle body with massless stiffness and damping, and of a transversal pendulum with friction. The pendulum models a suspended load. Small system vibrations and their stability in the sense of Lyapunov and Bogusz are analysed.

In the second case the model composed of a vehicle body and the road wheels suspension is investigated. The influence of the height of the vehicle body mass centre on both vehicle transversal and transversal tilt dynamics is analysed.

1. Dynamics of a Vehicle Body with the Pendulum Type Load

The small vibrations of the two-body system composed of the vehicle body and the self-aligningly suspended load are analysed (see Figure 7.1). The flat

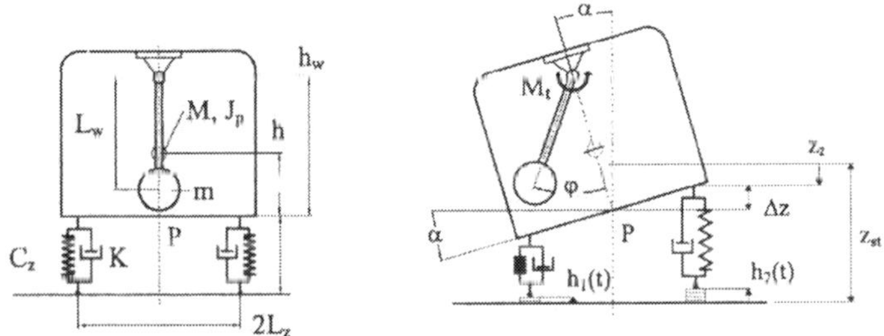

Figure 7.1. The analysed system: vehicle body-pendulum.

body is elastically supported via two lumped elements and is coupled with the simple physical pendulum. The flat body is characterized via mass m and inertial moment J_p with respect to the point P. The system kinematic energy reads

$$E = \frac{1}{2} \cdot m \cdot [(h_w - L_w) \cdot \dot{\alpha} + L_w \cdot \dot{\varphi}]^2 + \frac{1}{2} \cdot J_p \cdot \dot{\alpha}^2 + \frac{1}{2} \cdot (M + m) \cdot \dot{z}^2. \quad (7.1)$$

The pendulum potential energy associated with its successive motion paths, i.e. from A_0 to A, from A to B, from B to C, and from C to D (see Figure 7.2) reads

$$V_w = -m \cdot g \cdot z - m \cdot g \cdot (h_w - L_w) \cdot (1 - \cos(\alpha))$$

$$-m \cdot g \cdot L_w \cdot (1 - \cos(\alpha)) + m \cdot g \cdot L_w \cdot (1 - \cos(\varphi - \alpha)). \quad (7.2)$$

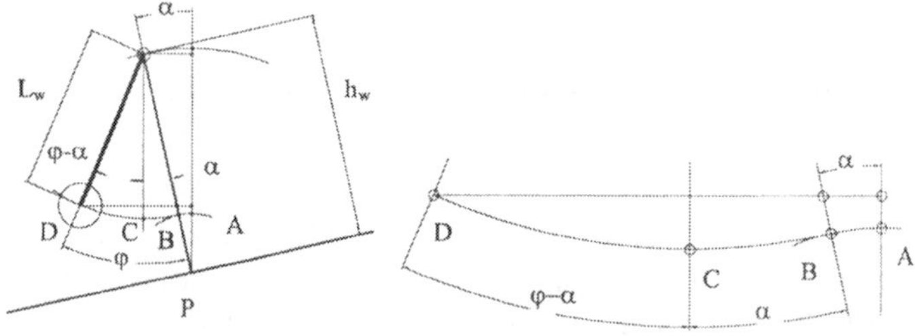

Figure 7.2. Illustration of the vertical pendulum path from the point A to the point B.

The potential energy of the vehicle body for the case $h_1 = 0$, $h_2 \neq 0$ has the form

$$V_p = -M \cdot g \cdot z + \frac{1}{2} \cdot C_z \cdot (z + \alpha \cdot L_z)^2 + \frac{1}{2} \cdot C_z \cdot (z - \alpha \cdot L_z + h_2(t))^2 \quad (7.3)$$

and the system potential energy is

$$V = -m \cdot g \cdot z - m \cdot g \cdot [(h_w - L_w) \cdot (1 - \cos(\alpha))$$

$$-m \cdot g \cdot L_w \cdot (1 - \cos(\alpha)) + m \cdot g \cdot L_w \cdot (1 - \cos(\varphi - \alpha)) - M \cdot g \cdot z$$

$$+\frac{1}{2} \cdot C_z \cdot (z + \alpha \cdot L_z)^2 + \frac{1}{2} \cdot C_z \cdot (z - \alpha \cdot L_z + h_2(t))^2. \quad (7.4)$$

Assuming viscous suspension damping (K) and the Coulomb type friction (M_F) associated with pendulum motion, the following dissipation function is obtained:

$$D = \frac{1}{2} \cdot K \cdot (\dot{z} + \dot{\alpha} \cdot L_z)^2 + \frac{1}{2} \cdot K \cdot \left(\dot{z} - \dot{\alpha} \cdot L_z + \dot{h}_2(t)\right)^2 + MT * \text{sign}(\dot{\varphi}) \cdot \varphi. \quad (7.5)$$

The work of external forces reads

$$F_2 = C_z \cdot (z - \alpha \cdot L_z + h_2(t)) + K \cdot \left(\dot{z} - \dot{\alpha} \cdot L_z + \dot{h}_2(t)\right) . \qquad (7.6)$$

In order to apply the Lagrange formalism the following derivatives are calculated:

$$\frac{d}{dt}\frac{\partial E}{\partial \dot{\alpha}} = m \cdot [(h_w - L_w) \cdot \ddot{\alpha} + L_w \cdot \ddot{\varphi}] \cdot (h_w - L_w) + J_p \cdot \ddot{\alpha},$$

$$\frac{d}{dt}\frac{\partial E}{\partial \dot{\varphi}} = m \cdot [(h_w - L_w) \cdot \ddot{\alpha} + L_w \cdot \ddot{\varphi}] \cdot L_w,$$

$$\frac{d}{dt}\frac{\partial E}{\partial \dot{z}} = (M + m) \cdot \ddot{z},$$

$$\frac{d}{dt}\frac{\partial E}{\partial \dot{h}_2} = 0, \qquad (7.7)$$

$$\frac{\partial V}{\partial \alpha} = -m \cdot g \cdot h_w \cdot \sin(\alpha) - m \cdot g \cdot L_w \cdot \sin(\varphi - \alpha)$$
$$+C_z \cdot (z + \alpha \cdot L_z) \cdot L_z - C_z \cdot (z - \alpha \cdot L_z + h_2(t)) \cdot L_z,$$

$$\frac{\partial V}{\partial \varphi} = m \cdot g \cdot L_w \cdot \sin(\varphi - \alpha),$$

$$\frac{\partial V}{\partial z} = C_z \cdot (z + \alpha \cdot L_z) + C_z \cdot (z - \alpha \cdot L_z + h_2(t)) - (M + m) \cdot g,$$

$$\frac{\partial V}{\partial h_2} = C_z \cdot (z - \alpha \cdot L_z + h_2(t)), \qquad (7.8)$$

$$\frac{\partial D}{\partial \dot{\alpha}} = K \cdot (\dot{z} + \dot{\alpha} \cdot L_z) \cdot L_z - K \cdot \left(\dot{z} - \dot{\alpha} \cdot L_z + \dot{h}_2(t)\right) \cdot L_z,$$

$$\frac{\partial D}{\partial \varphi} = MT \cdot sign(\dot{\varphi}),$$

$$\frac{\partial D}{\partial \dot{z}} = K \cdot (\dot{z} + \dot{\alpha} \cdot L_z) + K \cdot \left(\dot{z} - \dot{\alpha} \cdot L_z + \dot{h}_2(t)\right),$$

$$\frac{\partial D}{\partial \dot{h}_2} = K \cdot \left(\dot{z} - \dot{\alpha} \cdot L_z + \dot{h}_2(t)\right). \qquad (7.9)$$

Assuming small oscillations the following Lagrange equations are obtained:

$$\frac{d}{dt}\frac{\partial E}{\partial \dot{\alpha}} + \frac{\partial V}{\partial \alpha} + \frac{\partial D}{\partial \dot{\alpha}} = m \cdot [(h_w - L_w) \cdot \ddot{\alpha} + L_w \cdot \ddot{\varphi}] \cdot (h_w - L_w)$$
$$+J_p \cdot \ddot{\alpha} - m \cdot g \cdot h_w \cdot \alpha - m \cdot g \cdot L_w \cdot \varphi + m \cdot g \cdot L_w \cdot \alpha$$
$$+C_z \cdot (z + \alpha \cdot L_z) \cdot L_z - C_z \cdot (z - \alpha \cdot L_z + h_2(t)) \cdot L_z$$

$$-K \cdot (\dot{z} + \dot{\alpha} \cdot L_z) \cdot L_z - K \cdot \left(\dot{z} - \dot{\alpha} \cdot L_z + \dot{h}_2(t)\right) \cdot L,$$

$$\frac{d}{dt}\frac{\partial E}{\partial \dot{\varphi}} + \frac{\partial V}{\partial \varphi} + \frac{\partial D}{\partial \varphi} = +m \cdot (h_w - L_w) \cdot L_w \cdot \ddot{\alpha} + m \cdot L_w^2 \cdot \ddot{\varphi}$$

$$+m \cdot g \cdot L_w \cdot \varphi - m \cdot g \cdot L_w \cdot \alpha + MT \cdot sign(\dot{\varphi}) \equiv 0,$$

$$\frac{d}{dt}\frac{\partial E}{\partial \dot{z}} + \frac{\partial V}{\partial z} + \frac{\partial D}{\partial \dot{z}} = (M + m) \cdot \ddot{z} + C_z \cdot z + C_z \cdot L_z \cdot \alpha$$

$$+C_z \cdot z - C_z \cdot L_z \cdot \alpha + C_z \cdot h_2(t) + K \cdot \dot{z} + K \cdot L_z \cdot \dot{\alpha}$$

$$+K \cdot \dot{z} - K \cdot L_z \cdot \dot{\alpha} + K \cdot \dot{h}_2(t) \equiv 0,$$

$$\frac{d}{dt}\frac{\partial E}{\partial \dot{h}_2} + \frac{\partial V}{\partial h_2} + \frac{\partial D}{\partial \dot{h}_2} + \frac{\partial D}{\partial h_2} = C_z \cdot z - C_z \cdot L_z \cdot \alpha$$

$$+C_z \cdot h_2(t) + K \cdot \dot{z} - K \cdot L_z \cdot \dot{\alpha} + K \cdot \dot{h}_2(t) \equiv F_2 \,. \tag{7.10}$$

First, free vibrations are considered, without dissipation ($h_2 = K = M_F = 0$). The problem is reduced to analysis of two equations of motion

$$\ddot{\alpha} = A \cdot \alpha + B \cdot \varphi, \qquad \ddot{\varphi} = C \cdot \alpha + D \cdot \varphi, \tag{7.11}$$

where:

$$A = \left[-\frac{2}{J_p} \cdot C_z \cdot L_z^2\right], \quad C = \left[\frac{g}{L_w} + \frac{2}{J_p \cdot L_w} \cdot C_z \cdot L_z^2 \cdot (h_w - L_w)\right],$$

$$B = \left[\frac{1}{J_p} \cdot m \cdot g \cdot h_w\right], \quad D = \left[-\frac{g}{L_w} - \frac{h_w}{J_p \cdot L_w} \cdot m \cdot g \cdot (h_w - L_w)\right].$$

Substituting $\dot{\alpha} = x$, $\dot{\varphi} = y$ into (7.11), the following first order differential perturbation equations are obtained:

$$\dot{x} = 0 \cdot x + 0 \cdot y + A \cdot \alpha + B \cdot \varphi,$$

$$\dot{y} = 0 \cdot x + 0 \cdot y + C \cdot \alpha + D \cdot \varphi,$$

$$\dot{\alpha} = 1 \cdot x + 0 \cdot y + 0 \cdot \alpha + 0 \cdot \varphi,$$

$$\dot{\varphi} = 0 \cdot x + 1 \cdot y + 0 \cdot \alpha + 0 \cdot \varphi,$$

and the associated characteristic equation reads

$$s^4 + (-A - D) \cdot s^2 + (A \cdot D - C \cdot B) = 0,$$

which is transformed to the following one ($j = s^2$):

$$j^2 + (-A - D) \cdot j + (A \cdot D - C \cdot B) = 0 \,.$$

The Lyapunov stability conditions are

$$-A - D > 0, \qquad A \cdot D - C \cdot B > 0\,.$$

Considering the first inequality and accounting (7.11) the first Lyapunov stability condition reads

$$\text{for} \quad 1 - \frac{2 \cdot C_z \cdot L_z^2}{m \cdot g \cdot h_w} < 0, \quad L_w > \frac{J_p \cdot g + m \cdot g \cdot h_w^2}{m \cdot g \cdot h_w - 2 \cdot C_z \cdot L_z^2}\,. \tag{7.12}$$

Introducing the following data: $J_p = 500\,kg \cdot m^2$, $C_z = 20000\,N/m$, $L_z = 0.5\,m$, $h_w = 1\,m$ the stability zones generated by the first stability condition are reported in Figure 7.3.

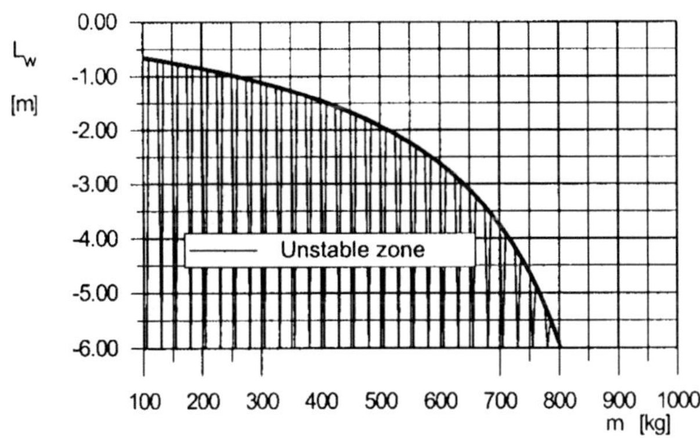

Figure 7.3. Stable and unstable zones on the (L_w, m) plane (L_w, m are pendulum length and mass, respectively).

It is clear that for a real vehicle the first stability condition is always satisfied. The stability condition reads

$$\left[-\frac{2}{J_p} \cdot C_z \cdot L_z^2\right] \cdot \left[-\frac{g}{L_w} - \frac{h_w}{J_p \cdot L_w} \cdot m \cdot g \cdot (h_w - L_w)\right]$$

$$- \left[\frac{g}{L_w} + \frac{2}{J_p \cdot L_w} \cdot C_z \cdot L_z^2 \cdot (h_w - L_w)\right] \cdot \left[\frac{1}{J_p} \cdot m \cdot g \cdot h_w\right] > 0\,,$$

and after some manipulation it achieves the form

$$h_w < \frac{2 \cdot C_z \cdot L_z^2}{m \cdot g}\,. \tag{7.13}$$

In Figure 7.4 stable and unstable zones are reported for the earlier mentioned fixed parameters.

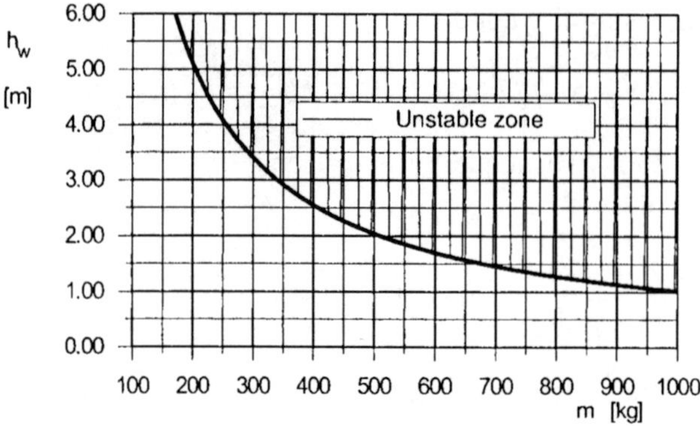

Figure 7.4. Stable and unstable zones in the (h_w, m) parameters plane (h_w, m) is the height and the pendulum mass, respectively).

One may conclude that the vehicle stability depends strongly on the pendulum hanging at h_w. For example, for the mass $m = 50\,kg$ the critical pendulum hanging at $h_w = 2\,m$.

In Figures 7.5–7.9 time histories and phase projections of the investigated system are shown. In the phase projections the dependencies are $\dot{\alpha}(\alpha)$ and $\dot{\varphi}(\varphi)$, whereas in the time histories $\alpha(t)$ and $\varphi(t)$ are shown for different pendulum mass values. The other parameters are as follows: $L_z = 0.5\,m$, $h = 1.1\,m$, $L_w = 1.5\,m$, $h_w = 1.6\,m$, $M = 2000\,kg$, $C_z = 20000\,N/m$, $J_p = 2000\,kg \cdot m^2$.

For $m = 60\,kg$ free system vibrations are quasi-periodic (see Figure 7.5). For $m = 80.3\,kg$ the vehicle body and the pendulum move periodically (Figure 7.6). The peculiar motion for $m = 395\,kg$ is reported in Figure 7.7. Increasing the mass further the system approaches a Sacker–Neimark bifurcation (see Figure 7.8 for $m = 625\,kg$). The newly born quasi-periodic motion includes the carrier wave with the frequency $\sim 0.0366\,Hz$. Further increase of the control parameter m forces the trajectories to approach infinity, i.e. a catastrophe occurs.

In general, the system vehicle body-pendulum is governed by the differential equations (7.11). Recall that the coordinate describing vertical road surface irregularities $h_z(t)$ (see Figure 7.1) has been treated as the generalized coordinate. However, now it is treated as the *kinematic excitation*. It is assumed that this excitation is of the harmonic type

$$h_2 = h_{2m} \cdot \sin(2 \cdot \pi \cdot n \cdot t), \quad \dot{h}_2 = 2 \cdot \pi \cdot n \cdot h_{2m} \cdot \cos(2 \cdot \pi \cdot n \cdot t),$$

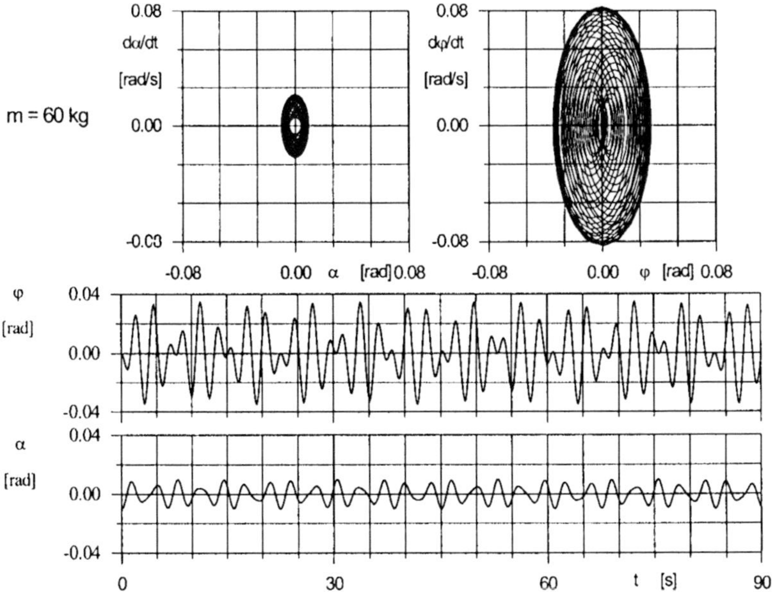

Figure 7.5. Phase portraits and time histories ($m = 60\,kg$).

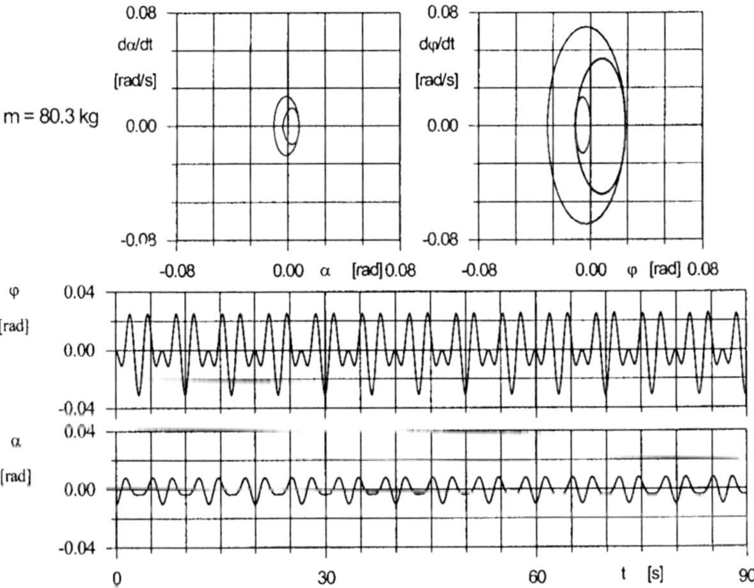

Figure 7.6. Phase portraits and time histories ($m = 80.3\,kg$).

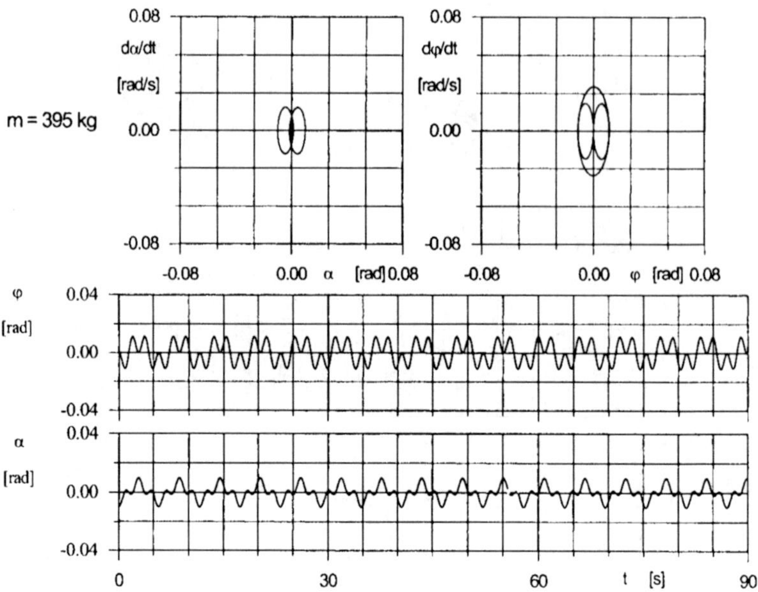

Figure 7.7. Phase portraits and time histories ($m = 595\ kg$).

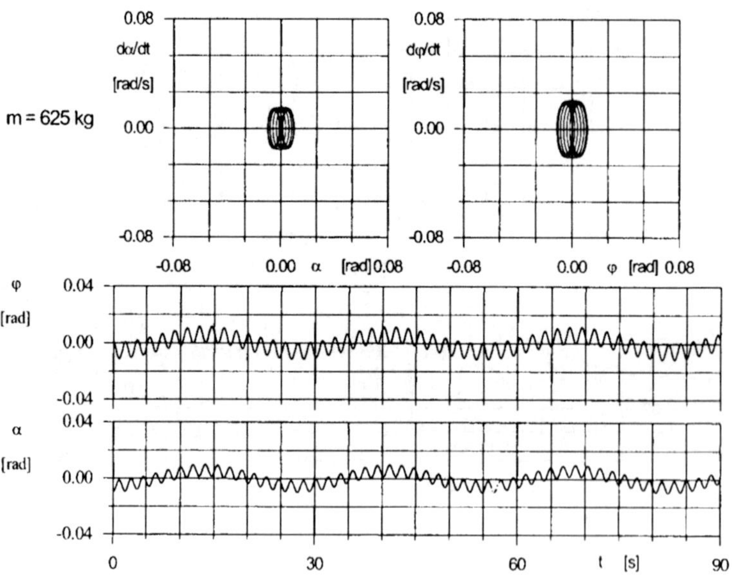

Figure 7.8. Phase portraits and time histories ($m = 625\ kg$).

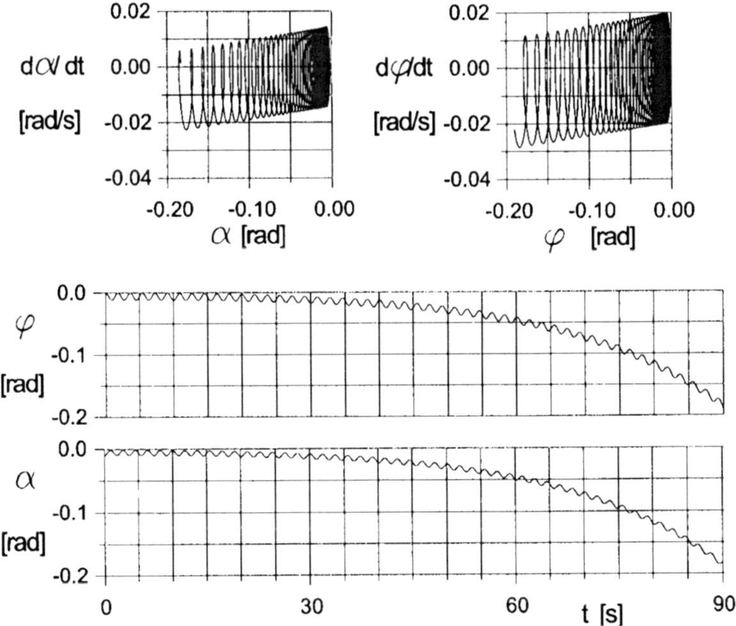

Figure 7.9. Phase portraits and time histories ($m = 637.6\,kg$).

where: n denotes frequency $[Hz]$, and h_{2m} is the amplitude $[m]$. Introducing the new variables

$$q_1 = \dot{\varphi}, \;\; q_2 = \varphi, \;\; q_3 = \dot{\alpha}, \;\; q_4 = \alpha, \;\; q_5 = \dot{z}, \;\; q_6 = z,$$

the system dynamics is governed by the following six first order differential equations

$$\begin{aligned}
\dot{q}_1 &= a_{11} \cdot q_1 + a_{12} \cdot q_3 + a_{13} \cdot q_5 + b_{11} \cdot q_2 + b_{12} \cdot q_4 + b_{13} \cdot q_6 + e_1, \\
\dot{q}_2 &= q_1, \\
\dot{q}_3 &= a_{21} \cdot q_1 + a_{22} \cdot q_3 + a_{23} \cdot q_5 + b_{21} \cdot q_2 + b_{22} \cdot q_4 + b_{23} \cdot q_6 + e_2, \\
\dot{q}_4 &= q_3, \\
\dot{q}_5 &= a_{31} \cdot q_1 + a_{32} \cdot q_3 + a_{33} \cdot q_5 + b_{31} \cdot q_2 + b_{32} \cdot q_4 + b_{33} \cdot q_6 + e_3, \\
\dot{q}_6 &= q_5,
\end{aligned} \tag{7.14}$$

where the coefficients a_{ij} read

$$a_{12} = \left[\frac{2 \cdot K \cdot L_z^2 \cdot (h_w - L_w)}{L_w \cdot J_p} \right], \quad b_{11} = \left[-\frac{g}{L_w} - \frac{m \cdot g \cdot h_w \cdot (h_w - L_w)}{L_w \cdot J_p} \right],$$

$$b_{12} = \left[\frac{g}{L_w} + \frac{2 \cdot C_z \cdot L_z^2 \cdot (h_w - L_w)}{L_w \cdot J_p}\right],$$

$$e_1 = \left[-\frac{K \cdot L_z \cdot (h_w - L_w)}{L_w \cdot J_p} \cdot \dot{h}_2(t) - \frac{C_z \cdot L_z \cdot (h_w - L_w)}{L_w \cdot J_p} \cdot h_2(t)\right.$$

$$\left. - \frac{m \cdot (h_w - L_w)^2 + J_p}{m \cdot L_w^2 \cdot J_p} \cdot MT \cdot sign(\dot{\varphi})\right]$$

$$a_{22} = \left[-\frac{2 \cdot K \cdot L_z^2}{L_w \cdot J_p}\right], \quad b_{21} = \left[\frac{m \cdot g \cdot h_w}{J_p}\right], \quad b_{22} = \left[-\frac{2 \cdot C_z \cdot L_z^2}{J_p}\right],$$

$$e_2 = \left[\frac{K \cdot L_z}{J_p} \cdot \dot{h}_2(t) + \frac{C_z \cdot L_z}{J_p} \cdot h_2(t) + \frac{(h_w - L_w)}{J_p \cdot L_w} \cdot MT \cdot sign(\dot{\varphi})\right],$$

$$a_{33} = \left[-\frac{2 \cdot K}{(m + M)}\right], \quad b_{33} = \left[-\frac{2 \cdot C_z}{(m + M)}\right],$$

$$e_3 = \left[-\frac{K}{(m + M)} \cdot \dot{h}_2(t) - \frac{C_z}{(m + M)} \cdot h_2(t)\right].$$

In Figures 7.10 and 7.11 phase projections $\dot{\alpha}(\alpha)$ and $\dot{\varphi}(\varphi)$ and time histories $\alpha(t)$ and $\varphi(t)$ are reported for two different damping values. The mentioned characteristics have been obtained via numerical solution of the equation (7.14) using a fourth order Runge-Kutta method for the following fixed parameters: $L_z = 0.5\,m$, $h = 1.1\,m$, $L_w = 1.5\,m$, $h_w = 1.6\,m$, $m = 500\,kg$, $M = 2000\,kg$, $C_z = 20000\,N/m$, $J_p = 600\,kg \cdot m^2$.

It occurred that the dynamics of the system is very sensitive to the pendulum damping. Time histories $\alpha(t)$ and $\varphi(t)$ shown in Figure 7.10 without damping ($M = 0$) display long term oscillations. However, introduction of friction acting on the pendulum dynamics ($MT = 20\,N \cdot m$) yields a few times shorter period of vibrations.

The kinematically excited vibrations in the form of graphs $\dot{\alpha}(\alpha)$, $\dot{\varphi}(\varphi)$, $\alpha(t)$ and $\varphi(t)$ are shown in Figures 7.12 and 7.13 for two excitation frequencies: $h_2(t) - n = 0.34\,Hz$ and $n = 0.70\,Hz$. The following parameters are fixed: $L_z = 0.5\,m$, $h = 1.1\,m$, $L_w = 1.5\,m$, $h_w = 1.6\,m$, $m = 100 \div 670\,kg$, $M = 1200\,kg$, $C_z = 20000\,N/m$, $K_r = 1000\,N/(m/s)$, $J_p = 600\,kg \cdot m^2$, $MT = 20\,N \cdot m$.

Two exemplary results shown in Figures 7.12 and 7.13 display qualitatively different system dynamics, and hence the bifurcation diagrams are highly required to trace the system dynamics. Taking pendulum mass as the bifurcation parameter, the bifurcation diagrams are reported in Figures 7.14 and 7.15. Analysis of the bifurcation diagrams allows us to trace high sensitivity to the frequency changes. The larger frequency value $n = 0.70\,Hz$ in the investigated mass interval $[0, 637]$ yields vibrations with smaller amplitudes:

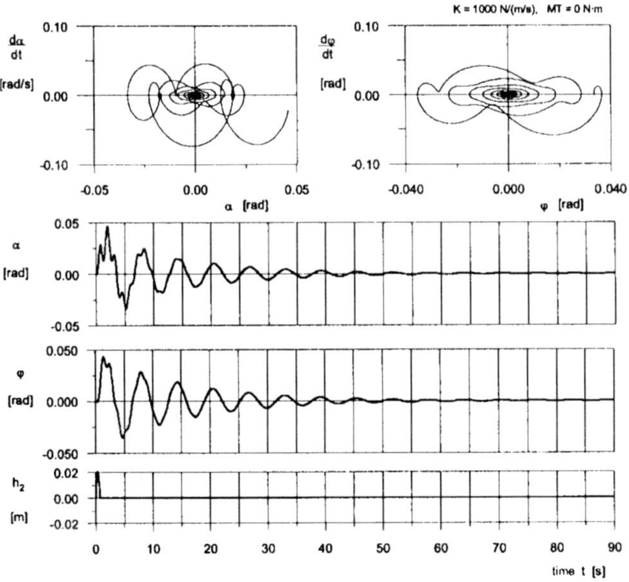

Figure 7.10. Free vibrations and sinusoidal impulse $h_2(t)$ ($K = 1000\,N/(m/s)$; $MT = 0$).

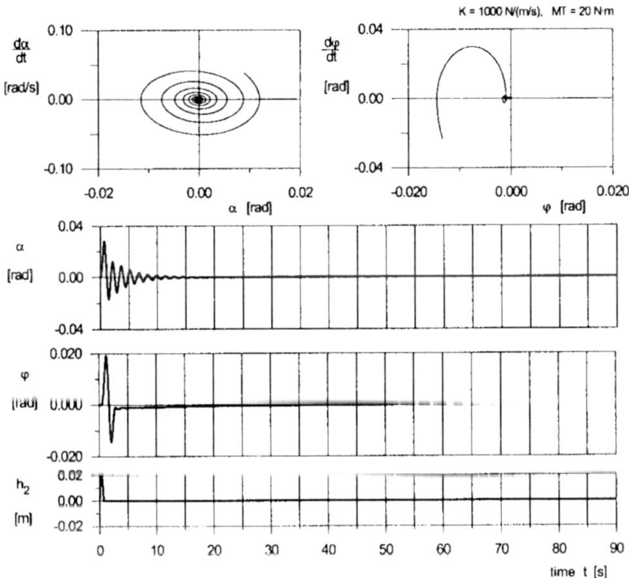

Figure 7.11. Free vibrations and sinusoidal impulse $h_2(t)$ ($K = 1000\,N/(m/s)$; $MT = 20\,N \cdot m$).

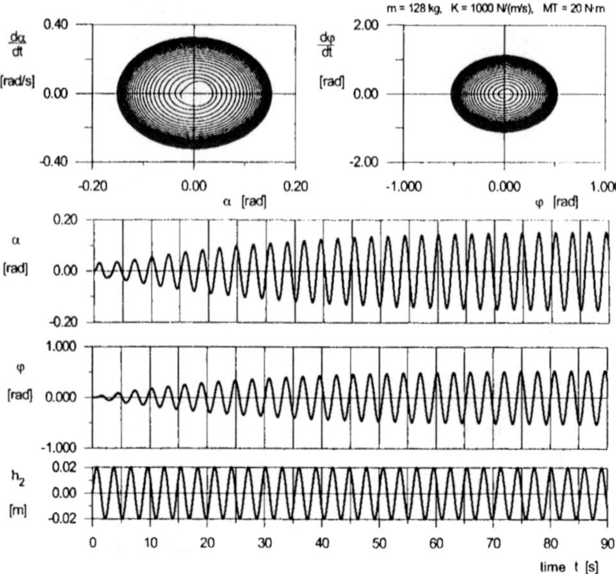

Figure 7.12. Phase portraits and time histories. Forced sinusoidal vibration $n = 0.34\,Hz$. Damped motion of the body and the pendulum.

(i) for $n = 0.70\,Hz \Rightarrow A_{\alpha\ \max} = -0.12\,rad$ and $A_{\varphi\ \max} = +0.075\,rad$,

(ii) for $n = 0.34\,Hz \Rightarrow A_{\alpha\ \max} = -0.70\,rad$ and $A_{\varphi\ \max} = -0.58\,rad$.

Dynamics of the analysed system depends essentially on the pendulum mass. Two qualitatively different dynamics can be distinguished. From $m = 133\,kg$ the so-called *semicritical dynamics* occurs (in particular for $n = 0.34\,Hz$). It is characterized by a major increase of amplitudes of the vehicle body (α) and the pendulum (φ). For larger pendulum mass the vibration amplitude decreases. However, this observation is true until the mass threshold (the so-called critical mass $m_{cr} = 637\,kg$). For $m > m_{cr}$ the motion is unstable and unbounded. The semicritical pendulum mass $m_{scr} = 133\,kg$.

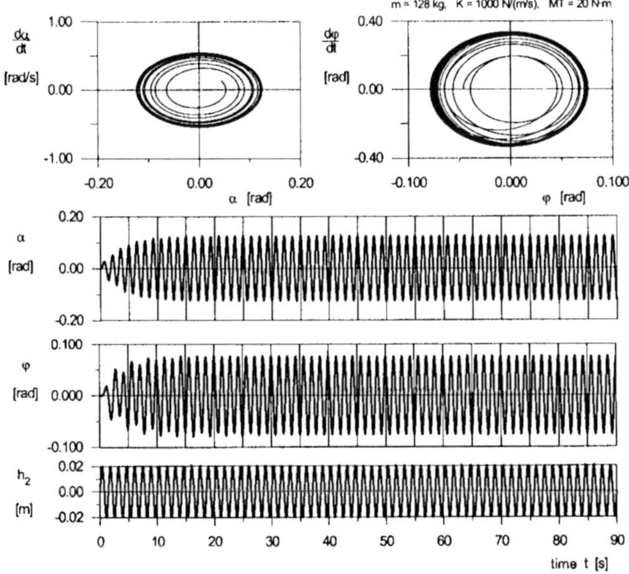

Figure 7.13. Phase portraits and time histories. Forced sinusoidal vibration $n = 0.70\,Hz$. Damped motion of the body and the pendulum.

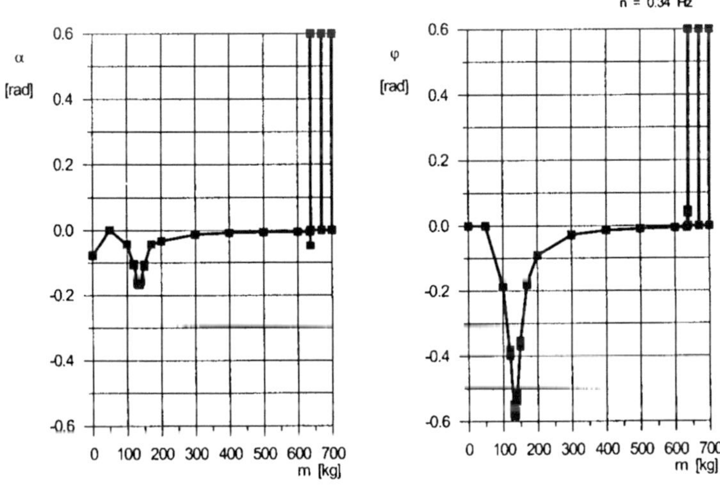

Figure 7.14. Bifurcation diagrams $\alpha(m)$ and $\varphi(m)$ of sinusoidally driven system with the frequency $n = 0.34\,Hz$.

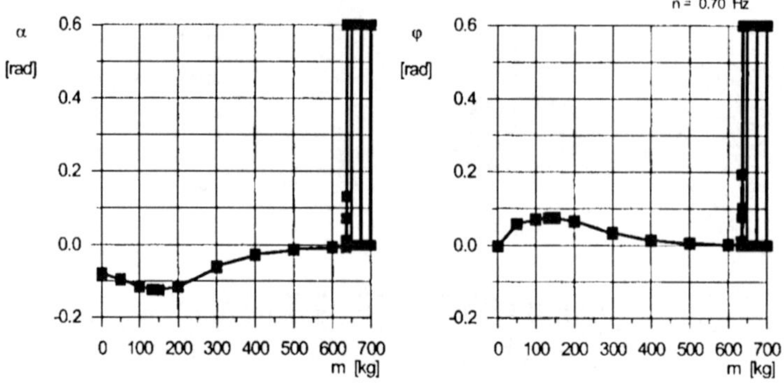

Figure 7.15. Bifurcation diagrams $\alpha(m)$ and $\varphi(m)$ of sinusoidally driven system with the frequency $n = 0.70\,Hz$.

2. Roll of a Vehicle Body, Instantaneous Roll Centre and Stability [43, 52, 72, 76, 94, 101]

A curvilinear vehicle trajectory accounting for a vehicle body displacement, roll and the motion around an instantaneous roll centre is now considered. The forces acting on the vehicle body and road wheels are shown in Figure 7.16.

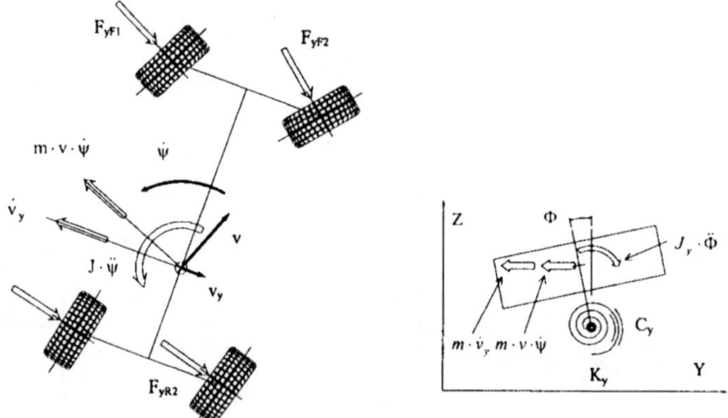

Figure 7.16. The forces acting on the car body and the road wheels when the vehicle moves on the curvilinear trajectory (the vehicle body is supported via rotational stiffness K_y and damping C_y).

The equations of motion in direction Y and ψ read

$$m \cdot \dot{v}_y + m \cdot v \cdot \dot{\psi} - F_{yF1} - F_{yF2} - F_{yR1} - F_{yR2} = 0, \tag{7.15}$$

$$J \cdot \ddot{\psi} - F_{yF1} \cdot a - F_{yF2} \cdot a + F_{yR1} \cdot b + F_{yR2} \cdot b = 0, \tag{7.16}$$

where: a (b) are distances between the vehicle centre and the front (rear) axle.

Owing to Figure 7.16 it is assumed that the instantaneous rotation vehicle body centre lies on the vertical axle of symmetry of the vehicle body at the distance h from the vehicle body mass centre. The vehicle roll dynamics is governed by the equation

$$J_y \cdot \ddot{\Phi} + C_y \cdot \dot{\Phi} + K_y \cdot \Phi - m \cdot \dot{v}_y \cdot h - m \cdot v \cdot \dot{\psi} \cdot h = 0, \qquad (7.17)$$

where: h - vehicle mass centre position; K_y, C_y - stiffness and damping in Φ direction, respectively. It is assumed that the vehicle moves with a constant velocity v, and that

$$v_y = \beta \cdot v_s \; . \qquad (7.18)$$

Since the angle displacement of the vehicle body generates the angle displacements of the guiding elements of suspension, the road wheels move in both Y direction and angularly in the ZY plane.

It is assumed that the friction force generated in tires is proportional to the slip angle. In fact, this is only true for small angles. The assumed dependence of the lateral force F_y versus tire slip angle α and versus sloping of the wheel γ is shown in Figure 7.17 (see the tire characteristics reported in Figure 3.15). Owing to the mentioned assumptions the following relations are obtained:

$$F_{yF1} = K_{F1} \cdot \alpha_{F1} + A_{kF} \cdot \gamma_{F1}; \quad F_{yF2} = K_{F2} \cdot \alpha_{F2} + A_{kF} \cdot \gamma_{F2};$$

$$F_{yR1} = K_{R1} \cdot \alpha_{R1} + A_{kR} \cdot \gamma_{R1}; \quad F_{yR2} = K_{R2} \cdot \alpha_{R2} + A_{kR} \cdot \gamma_{R2}; \qquad (7.19)$$

where: $A_{kF,R} > 0$.

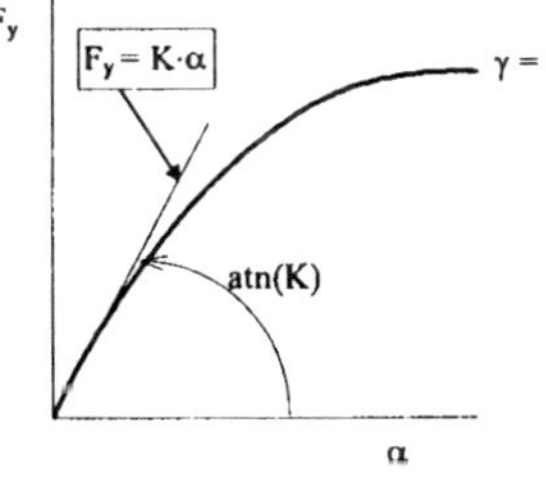

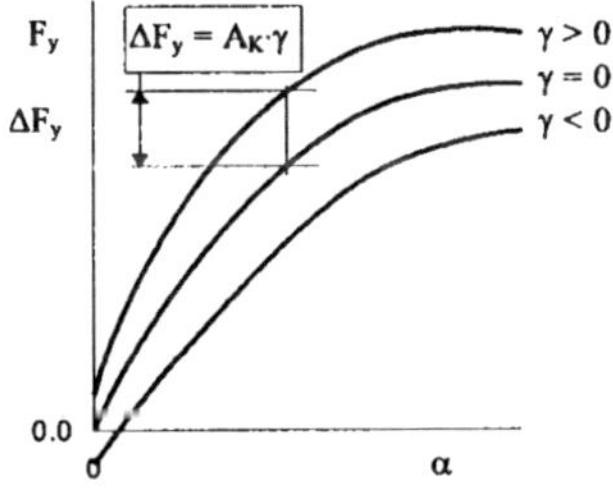

Figure 7.17. A transversal force F_y versus the tire slip angle α and the wheel slope angle γ.

Owing to relations (7.19), when $\gamma > 0$ then the increment ΔF_y is also positive and $F_y(\alpha, \gamma > 0) > F_y(\alpha, \gamma = 0)$. On the other hand, when $\gamma < 0$ then the increment $\Delta F_y < 0$ and $F_y(\alpha, \gamma < 0) < F_y(\alpha, \gamma = 0)$. Recall that if a road wheel is sloped contrary to the F_y sense, then γ is negative. In the case, when a road wheel is sloped in agreement with the vector force F_y sense, then the γ sign is positive.

It is obvious that a dependence of γ_i on the angular vehicle body displacement Φ depends on the suspension design. In Figure 7.17 three possible cases of the system composed of vehicle body-guiding elements of suspension-road wheel, subjected to transversal force action are reported (for instance, during run on a turn; see also 3).

The mentioned dependence is here simplified tc the linear one of the form

$$\gamma_{F1} = (B_\gamma)_{F1} \cdot \Phi \cdot \text{sgn}(\Phi); \quad \gamma_{F2} = (B_\gamma)_{F2} \cdot \Phi \cdot \text{sgn}(\Phi);$$
$$\gamma_{R1} = (B_\gamma)_{R1} \cdot \Phi \cdot \text{sgn}(\Phi); \quad \gamma_{R2} = (B_\gamma)_{R2} \cdot \Phi \cdot \text{sgn}(\Phi). \tag{7.20}$$

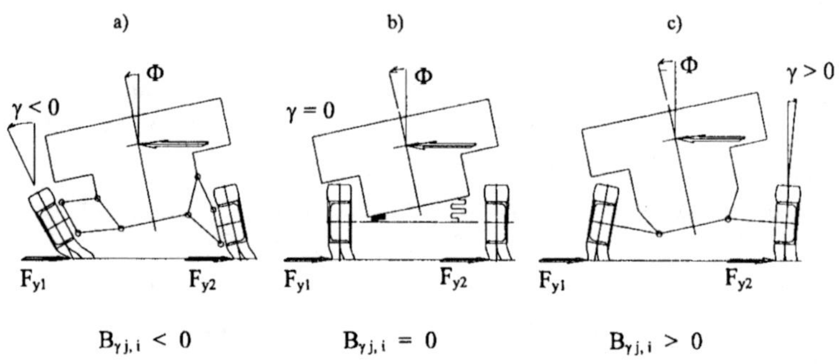

Figure 7.18. Road wheel plane slope γ_i versus angular body car displacement Φ (roll angle) for various suspension design.

Observe that the B_γ value depends on road wheel suspension properties. Figure 7.18 displays three following possibilities: (i) $B_{\gamma1,2} < 0$, $\gamma_{1,2} < 0$ (Fig. 7.18a); (ii) $B_{\gamma1,2} = 0$, $\gamma_{1,2} = 0$ (Fig. 7.18b); (iii) $B_{\gamma1,2} > 0$, $\gamma_{1,2} > 0$ (Fig. 7.18c). In our further considerations the phenomenon of a road wheel position change and its displacement in the transversal direction with respect to the longitudinal vehicle axle owing to angular body vehicle and suspension guiding elements displacements is accounted (see Figure 7.19).

Assuming small angular vehicle body displacements the following simple relations are applied:

$$\dot{y}_{F1} = A_{yF1} \cdot \dot{\Phi}; \quad \dot{y}_{F2} = A_{yF2} \cdot \dot{\Phi};$$
$$\dot{y}_{R1} = A_{yR1} \cdot \dot{\Phi}; \quad \dot{y}_{R2} = A_{yR2} \cdot \dot{\Phi}. \tag{7.21}$$

The tire slip angles of the successive road wheels are defined via the relation (3.4) and the velocity planes shown in Figure 3.28. Since for small angles the following approximation holds, $v_s \pm \dot{\psi} \cdot Y_{F1,2,3,4} \approx v_s$, the relation defining the tire slip angles are simplified to the forms

$$\alpha_{F1} = \delta_{F1} - \beta - \dot{y}_{F1} \cdot \frac{1}{v} - \dot{\psi} \cdot \frac{a}{v}; \quad \alpha_{F2} = \delta_{F2} - \beta - \dot{y}_{F2} \cdot \frac{1}{v} - \dot{\psi} \cdot \frac{a}{v};$$

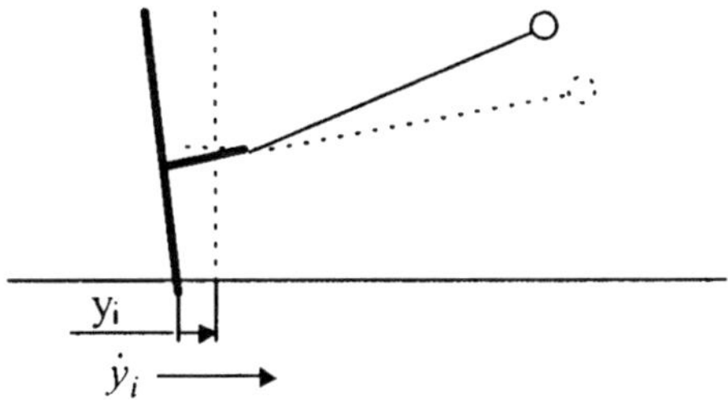

Figure 7.19. A transversal road wheel displacement during run on a turn.

$$\alpha_{R1} = \delta_{R1} - \beta - \dot{y}_{R1} \cdot \frac{1}{v} + \dot{\psi} \cdot \frac{b}{v}; \quad \alpha_{R2} = \delta_{R2} - \beta - \dot{y}_{R2} \cdot \frac{1}{v} + \dot{\psi} \cdot \frac{b}{v}. \quad (7.22)$$

The equations (7.15)–(7.22) govern a road vehicle dynamics running on a curvilinear trajectory and possessing a flexible vehicle body support. Note that the mathematical model obtained accounts for both angular and linear road wheels displacement and the changes of tire characteristics generated by them. Owing to the mentioned relations, the road vehicle dynamics in Y direction is governed by the equation

$$\begin{aligned} \dot{\beta} = & \left[\left(K_{F1} + K_{F2} + K_{R1} \cdot i_{R/F} + K_{R2} \cdot i_{R/F} \right) \cdot \frac{1}{m \cdot v} \right] \cdot \delta \\ & + \left[\frac{-1}{m \cdot v} \cdot (K_{F1} + K_{F2} + K_{R1} + K_{R2}) \right] \cdot \beta \\ & + \left[\frac{-1}{m \cdot v^2} \cdot (K_{F1} \cdot A_{yF1} + K_{F2} \cdot A_{yF2} + K_{R1} \cdot A_{yR1} + K_{R2} \cdot A_{yR2}) \right] \cdot \dot{\Phi} \\ & + \left[-(K_{F1} \cdot a + K_{F2} \cdot a - K_{R1} \cdot b - K_{R2} \cdot b) \cdot \frac{1}{m \cdot v^2} - 1 \right] \cdot \dot{\psi} \\ & + \left[\left(A_{kF} \cdot \left((B_\gamma)_{F1} + (B_\gamma)_{F2} \right) + A_{kR} \cdot \left((B_\gamma)_{R1} + (B_\gamma)_{R2} \right) \right) \cdot \frac{1}{m \cdot v} \right] \cdot \Phi \cdot \mathrm{sgn}(\Phi), \end{aligned} \quad (7.23)$$

where: δ is the "averaged" steering angle of the front road wheels; $i_{R/F}$ is transmission ratio in the steering gear between steering angles of the rear and front wheels.

The equation of road vehicle motion in ψ direction reads

$$\begin{aligned} \ddot{\psi} = & \left[\left(K_{F1} \cdot a + K_{F2} \cdot a - K_{R1} \cdot b \cdot i_{R/F} - K_{R2} \cdot b \cdot i_{R/F} \right) \cdot \frac{1}{J} \right] \cdot \delta \\ & + \left[(-K_{F1} \cdot a - K_{F2} \cdot a + K_{R1} \cdot b + K_{R2} \cdot b) \cdot \frac{1}{J} \right] \cdot \beta \end{aligned}$$

$$+\left[(-K_{F1}\cdot A_{yF1}\cdot a - K_{F2}\cdot A_{yF2}\cdot a + K_{R1}\cdot A_{yR1}\cdot b + K_{R2}\cdot A_{yR2}\cdot b)\cdot\frac{1}{v\cdot J}\right]\dot{\Phi}$$

$$+\left[(-K_{F1}\cdot a^2 - K_{F2}\cdot a^2 - K_{R1}\cdot b^2 - K_{R2}\cdot b^2)\cdot\frac{1}{v\cdot J}\right]\cdot\dot{\psi}$$

$$+\left[(A_{kF}\cdot B_{\gamma_{F1}}\cdot a + A_{kF}\cdot B_{\gamma_{F2}}\cdot a - A_{kR}\cdot B_{\gamma_{R1}}\cdot b - A_{kR}\cdot B_{\gamma_{R2}}\cdot b)\cdot\frac{1}{J}\right]\cdot\Phi\cdot\mathrm{sgn}(\Phi). \tag{7.24}$$

Finally, the real vehicle motion in Φ direction is governed by the equation

$$\ddot{\Phi} = -\frac{C_y}{J_y}\cdot\dot{\Phi} - \frac{K_y}{J_y}\cdot\Phi + \frac{m\cdot v\cdot h}{J_y}\cdot\dot{\beta} + \frac{m\cdot v\cdot h}{J_y}\cdot\dot{\psi}\,. \tag{7.25}$$

Introducing notation

$$a_1 = \left[\left(K_{F1} + K_{F2} + K_{R1}\cdot i_{R/F} + K_{R2}\cdot i_{R/F}\right)\cdot\frac{1}{m\cdot v}\right],$$

$$a_2 = \left[-(K_{F1} + K_{F2} + K_{R1} + K_{R2})\cdot\frac{1}{m\cdot v}\right],$$

$$a_3 = \left[-(K_{F1}\cdot A_{yF1} + K_{F2}\cdot A_{yF2} + K_{R1}\cdot A_{yR1} + K_{R2}\cdot A_{yR2})\cdot\frac{1}{m\cdot v^2}\right],$$

$$a_4 = \left[-(K_{F1}\cdot a + K_{F2}\cdot a - K_{R1}\cdot b - K_{R2}\cdot b)\cdot\frac{1}{m\cdot v^2} - 1\right],$$

$$a_5 = \left[\left(A_{kF}\cdot\left((B_\gamma)_{F1} + (B_\gamma)_{F2}\right) + A_{kR}\cdot\left((B_\gamma)_{R1} + (B_\gamma)_{R2}\right)\right)\cdot\frac{1}{m\cdot v}\right],$$

$$b_1 = \left[\left(K_{F1}\cdot a + K_{F2}\cdot a - K_{R1}\cdot b\cdot i_{R/F} - K_{R2}\cdot b\cdot i_{R/F}\right)\cdot\frac{1}{J}\right],$$

$$b_2 = \left[(-K_{F1}\cdot a - K_{F2}\cdot a + K_{R1}\cdot b + K_{R2}\cdot b)\cdot\frac{1}{J}\right],$$

$$b_3 = \left[(-K_{F1}\cdot A_{yF1}\cdot a - K_{F2}\cdot A_{yF2}\cdot a + K_{R1}\cdot A_{yR1}\cdot b + K_{R2}\cdot A_{yR2}\cdot b)\cdot\frac{1}{v\cdot J}\right],$$

$$b_4 = \left[(-K_{F1}\cdot a^2 - K_{F2}\cdot a^2 - K_{R1}\cdot b^2 - K_{R2}\cdot b^2)\cdot\frac{1}{v\cdot J}\right],$$

$$b_5 = \left[(A_{kF}\cdot B_{\gamma_{F1}}\cdot a + A_{kF}\cdot B_{\gamma_{F2}}\cdot a - A_{kR}\cdot B_{\gamma_{R1}}\cdot b - A_{kR}\cdot B_{\gamma_{R2}}\cdot b)\cdot\frac{1}{J}\right],$$

$$c_1 = -\frac{C_y}{J_y},\quad c_2 = -\frac{K_y}{J_y},\quad c_3 = \frac{m\cdot v\cdot h}{J_y},\quad c_4 = \frac{m\cdot v\cdot h}{J_y},$$

the governing differential equations take the form

$$
\begin{aligned}
\dot{\beta} &= +a_1 \cdot \delta + a_2 \cdot \beta + a_3 \cdot \dot{\Phi} + a_4 \cdot \dot{\psi} + a_5 \cdot |\Phi| \,, \\
\ddot{\psi} &= b_1 \cdot \delta + b_2 \cdot \beta + b_3 \cdot \dot{\Phi} + b_4 \cdot \dot{\psi} + b_5 \cdot |\Phi| \,, \\
\ddot{\Phi} &= c_1 \cdot \dot{\Phi} + c_2 \cdot \Phi + c_3 \cdot \dot{\beta} + c_4 \cdot \dot{\psi} \,.
\end{aligned} \tag{7.26}
$$

Applying new variables,

$$\beta = x_1, \;\; \dot{\beta} = \dot{x}_1, \;\; \Phi = x_2, \;\; \dot{\Phi} = x_3, \;\; \ddot{\Phi} = \dot{x}_3, \;\; \dot{\psi} = x_4, \;\; \ddot{\psi} = \dot{x}_4,$$

the equations (7.26) are replaced by the following first order set of differential equations:

$$\dot{x}_1 = a_2 \cdot x_1 + 0 \cdot x_2 + a_3 \cdot x_3 + a_4 \cdot x_4 + a_5 \cdot |x_2| + a_1 \cdot \delta,$$

$$\dot{x}_2 = x_3,$$

$$\dot{x}_3 = c_3 {\cdot} a_2 {\cdot} x_1 {+} c_2 {\cdot} x_2 {+} (c_1 {+} c_3 {\cdot} a_3) {\cdot} x_3 {+} (c_3 {\cdot} a_4 {+} c_4) {\cdot} x_4 {+} c_3 {\cdot} a_5 {\cdot} |x_2| {+} c_3 {\cdot} a_1 {\cdot} \delta,$$

$$\dot{x}_4 = b_2 \cdot x_1 + 0 \cdot x_2 + b_3 \cdot x_3 + b_4 \cdot x_4 + b_1 \cdot \delta + b_5 \cdot |x_2| \,.$$

Finally, introducing the coefficients

$$cc_1 = a_2, \;\; cc_2 = 0, \;\; cc_3 = a_3, \;\; cc_4 = a_4, \;\; cc_5 = a_5 \cdot |x_2| + a_1 \cdot \delta,$$

$$bb_1 = c_3 \cdot a_2, \;\; bb_2 = c_2, \;\; bb_3 = (c_1 + c_3 \cdot a_3), \;\; bb_4 = (c_3 \cdot a_4 + c_4),$$

$$bb_5 = c_3 \cdot a_5 \cdot |x_2| + c_3 \cdot a_1 \cdot \delta,$$

$$aa_1 = b_2, \;\; aa_2 = 0, \;\; aa_3 = b_3, \;\; aa_4 = b_4, \;\; aa_5 = b_1 \cdot \delta + b_5 \cdot |x_2| \,,$$

the following system of differential equations *governing a road vehicle model during its running on a curvilinear trajectory* is obtained:

$$
\begin{aligned}
\dot{x}_1 &= cc_1 \cdot x_1 + cc_2 \cdot x_2 + cc_3 \cdot x_3 + cc_4 \cdot x_4 + cc_5, \\
\dot{x}_2 &= x_3, \\
\dot{x}_3 &= bb_1 \cdot a_2 \cdot x_1 + bb_2 \cdot x_2 + bb_3 \cdot x_3 + bb_4 \cdot x_4 + bb_5, \\
\dot{x}_4 &= aa_1 \cdot x_1 + aa_2 \cdot x_2 + aa_3 \cdot x_3 + aa_4 \cdot x_4 + aa_5.
\end{aligned} \tag{7.27}
$$

In Figures 7.20, 7.21 time histories obtained from numerical simulations of equation (7.27) are shown. The turn angle of the front wheels $\delta_F(t) = A \cdot \sin(B \cdot t)$ serves as excitation, whereas the outputs are the vehicle rotation ψ (yaw) and the angular displacement of the car body ϕ (fourth order Runge-Kutta method is applied during simulations).

The computations have been carried out for the following fixed parameters: $m = 1200\,kg$, $v = 20\,m/s$, $a = 1.1\,m$, $b = 1.4\,m$, $h = 0.5\,m$,

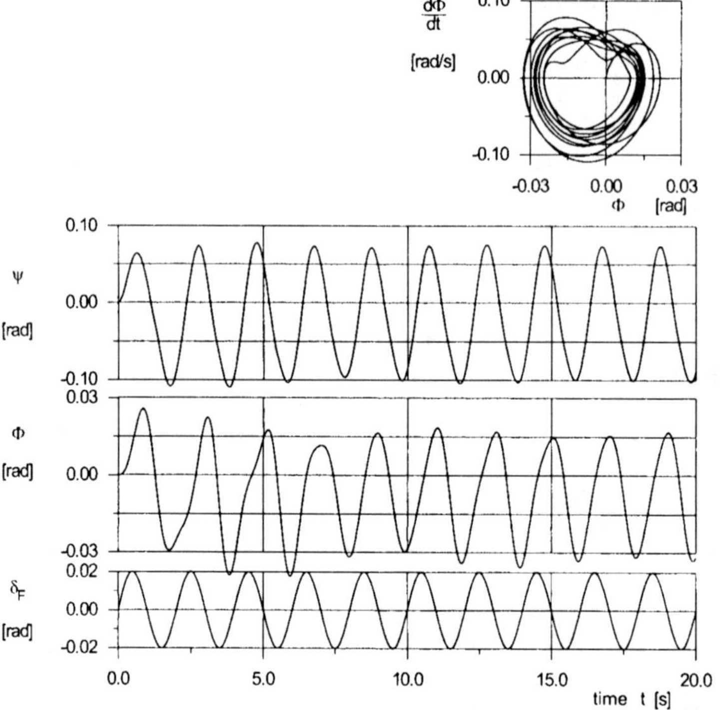

Figure 7.20. Time histories of the vehicle kinematics quantities during its curvilinear run ($K_y = 36 \cdot 10^3 N/rad$).

$K_{F1} = 15000\,N/rad$, $K_{F2} = 15000\,N/rad$, $K_{R1} = 15000\,N/rad$, $K_{R2} = 15000\,N/rad$, $A_{kF} = 2000\,N/rad$, $A_{kR} = 2400\,N/rad$, $A_{yF1} = 0.05\,m/rad$, $A_{yF2} = 0.05\,m/rad$, $A_{yR1} = 0.05\,m/rad$, $A_{yR2} = 0.05\,m/rad$, $B_{\gamma F1} = -1$, $B_{\gamma F2} = -1$, $B_{\gamma R1} = 1$, $B_{\gamma R2} = 1$, $C_y = 90\,N \cdot m \cdot s/rad$, $J_y = 1200\,kg \cdot m^2$, $J = 2200\,kg \cdot m^2$, ($i_{R/F} = 0$).

Analysis of the Figures 7.20, 7.21 yields the obvious conclusion that increasing the tilting suspension stiffness K_y decreases the transversal tilting of the vehicle body. In addition, it causes also a decrease of the yaw amplitudes (ψ). The latter means that a tendency to oversteering is decreased.

In the next step of our considerations a stability of the system equilibrium state is defined as follows: drift angle $\beta = 0$; yaw $\varphi = 0$; yaw velocity $\dot{\varphi} = 0$ and tilting velocity $\dot{\psi} = 0$. It describes the road vehicle moving is a straight line without any drift or yaw tendency. During investigations the technical stability in the Bogusz sense and the theory introduced in Section 2 is applied.

Zones of initial and admissible solutions have the forms

$$\omega \equiv \left\{ x_1^2 < r_1^2, x_2^2 < r_2^2, x_3^2 < r_3^2, x_4^2 < r_4^2 \right\},$$

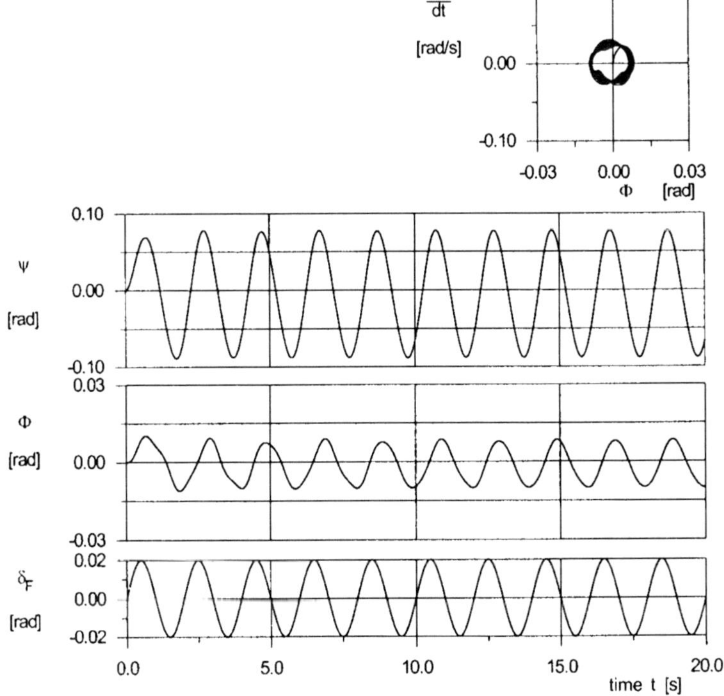

Figure 7.21. Time histories of the vehicle kinematics quantities during its curvilinear run ($K_y = 75 \cdot 10^3 N/rad$).

$$\Omega \equiv \left\{x_1^2 \leqslant R_1^2, x_2^2 \leqslant R_2^2, x_3^2 \leqslant R_3^2, x_4^2 \leqslant R_4^2\right\},$$

respectively.

The scalar Bogusz function $V_B(x_1, x_2, x_3, x_4)$ has the form

$$V_B(x_1, x_2, x_3, x_4) = \frac{1}{2}\left(AA \cdot x_1^2 + BB \cdot x_2^2 + CC \cdot x_3^2 + DD \cdot x_4^2\right),$$

and the coefficients

$$AA > 0, \quad BB > 0, \quad CC > 0, \quad DD > 0.$$

The stability conditions with respect to the introduced ω and Ω follow.

If there exists such a number C_0 satisfying the inequality $C_0 \geq V_B(x_1, x_2, x_3, x_4)$ for $(x_1, x_2, x_3, x)4) \in \omega$, $C_0 > 0$, i.e.

$$C_0 = \sup_{x_i \in \omega} V_B(x_1, x_2, x_3, x_4) = \frac{1}{2} \cdot \left(AA \cdot r_1^2 + BB \cdot r_2^2 + CC \cdot r_3^2 + DD \cdot r_4^2\right), \tag{7.28}$$

and the number C_1 satisfying the inequality $C_1 \leq V_B(x_1, x_2, x_3, x_4)$ for $(x_1, x_2, x_3, x_4) \in \Omega$, $C_1 > 0$, i.e.

$$C_1 = \inf_{x_i \notin \Omega} V_B(x_1, x_2, x_3, x_4)$$

$$= \frac{1}{2} \cdot \left(AA \cdot R_1^2 + BB \cdot R_2^2 + CC \cdot R_3^2 + DD \cdot R_4^2\right), \tag{7.29}$$

then for the derivative of V_B traced along the solutions (7.27) of the form

$$\frac{dV_B}{dt} = AA \cdot cc_1 \cdot x_1^2 + CC \cdot bb_3 \cdot x_3^2 + DD \cdot aa_4 \cdot x_4^2 + AA \cdot cc_5 \cdot x_1$$
$$+CC \cdot bb_5 \cdot x_3 + DD \cdot aa_5 \cdot x_4 + AA \cdot cc_2 \cdot x_1 \cdot x_2$$
$$+ (AA \cdot cc_3 + CC \cdot bb_1) \cdot x_1 \cdot x_3 + (AA \cdot cc_4 + DD \cdot aa_1) \cdot x_1 \cdot x_4$$
$$+ (BB + C \cdot bb_2) \cdot x_2 \cdot x_3 + DD \cdot aa_2 \cdot x_2 \cdot x_4 + (CC \cdot bb_4 + DD \cdot aa_3) \cdot x_3 \cdot x_4, \tag{7.30}$$

the Bogusz stability condition reads

$$\sup_{x_i \in \Omega/\omega \; t_1 \leqslant t \leqslant t_1 + T} \left\{ \frac{dV_B}{dt} \right\} < \frac{C1 - C0}{T}. \tag{7.31}$$

A satisfaction of condition (7.31) means that the system (7.27) is technically stable with respect to zones ω and Ω. The condition (7.31) has been verified numerically. In fact, it is reduced to numerical analysis of $\dot{V}_B$ in zone $\{x_1, x_2, x_3, x_4\} \in \Omega/\omega$, where:

$$\Omega/\omega \equiv \left\{r_1^2 < x_1^2 \leqslant R_1^2, r_2^2 < x_2^2 \leqslant R_2^2, r_3^2 < x_3^2 \leqslant R_3^2, r_4^2 < x_4^2 \leqslant R_4^2\right\}.$$

The technical stability factor is defined via the condition (7.31) through the relation

$$FT = \frac{C_1 - C_0}{T} - \frac{dV_B}{dt}.$$

Owing to relations (7.29), (7.30) and (7.31) the stability factor is defined in the following way:

$$FT = \frac{C_1 - C_0}{T} - AA \cdot cc_1 \cdot x_1^2 - CC \cdot bb_3 \cdot x_3^2 - DD \cdot aa_4 \cdot x_4^2$$
$$-AA \cdot cc_5 \cdot x_1 - CC \cdot bb_5 \cdot x_3 - DD \cdot aa_5 \cdot x_4 - AA \cdot cc_2 \cdot x_1 x_2$$
$$- (AA \cdot cc + CC \cdot bb_1) \cdot x_1 \cdot x_3 - (AA \cdot cc_4 + DD \cdot aa_1) \cdot x_1 \cdot x_4$$
$$- (BB + CC \cdot bb_2) \cdot x_2 \cdot x_3 - DD \cdot aa_2 \cdot x_2 \cdot x_4 - (CC \cdot bb_4 + DD \cdot aa_3) \cdot x_3 \cdot x_4.$$

If $FT > 0$ then the system is technically stable. Otherwise, it is unstable.

Table 7.1. Zones of ω and Ω

	ω			Ω	
$r_1(\beta)$	0.01	rad	$R_1(\beta)$	0.50	rad
$r_2(\beta)$	0.01	rad	$R_2(\varphi)$	0.05	rad
$r_3(\dot{\varphi})$	0.02	rad/s	$R_3(\dot{\varphi})$	0.25	rad/s
$r_4(\dot{\psi})$	0.01	rad/s	$R_4(\dot{\Psi})$	0.05	rad/s

EXAMPLE 7.1 *Display the Bogusz stability of the system governed by the equations* (7.27) *for the following data:* $m = 1200\,kg$, $v = 20\,m/s$, $a = 1.1\,m$, $b = 1.4\,m$, $h = 0.5\,m$, $K_{F1} = 15000\,N/rad$, $K_{F2} = 15000\,N/rad$, $K_{R1} = 15000\,N/rad$, $K_y = 49100\,N/rad$, $K_{R2} = 15000\,N/rad$, $B_{\gamma F1} = -1$, $B_{\gamma F2} = -1$, $B_{\gamma R1} = 1$, $B_{\gamma R2} = 1$, $A_{kF} = 2000\,N/rad$, $A_{kR} = 2400\,N/rad$, $A_{yF1} = 0.05\,m/rad$, $A_{yF2} = 0.05\,m/rad$, $A_{yR1} = 0.05\,m/rad$, $A_{yR2} = 0.05\,m/rad$, $C_y = 90\,N \cdot m \cdot s/rad$, $J_y = 1200\,kg \cdot m^2$ *and the observation time* $T = 2\,s$. *The initial and admissible zones are reported in Table 7.1.*

The analysis includes the following steps.

Input data procedure

1. *A vehicle parameters:* m, v, a, b, h, K_{F1}, K_{F2}, K_{R1}, K_y, K_{R2}, A_{kF}, A_{kR}, A_{yF1}, A_{yF2}, A_{yR1}, A_{yR2}, $B_{\gamma F1}$, $B_{\gamma F2}$, $B_{\gamma R1}$, $B_{\gamma R2}$, C_y, K_y, J_y, J.

2. *Data defining technical stability*

   2.1. *Zone of initial values.*

   2.2. *Zone of admissible values.*

   2.3. *Observation time* T *and the constants* AA, BB, CC *and* DD.

Maximal value $\left.\dfrac{dV_B}{dt}\right|_{x_i \in \omega}$ search procedure

1. *Calculation of* $\dfrac{dV_B}{dt}$ *for the starting values* $x_i = -r_i$.

2. *Starting value.*

3. *Search for the maximal value in the interval.*

Stability investigation procedure and calculation of FT factor

1 *Calculation of* C_0 *and* C_1.

2 *Verification of the Bogusz stability condition.*

3 Calculation of the technical stability factor

$$FT = \frac{C_1 - C_0}{T} - \left[\frac{dV_B}{dt}\right]_{\max} .$$

The numerical investigation results are shown in Figures 7.22 and 7.23, where stability factors versus the analysed quantities are shown.

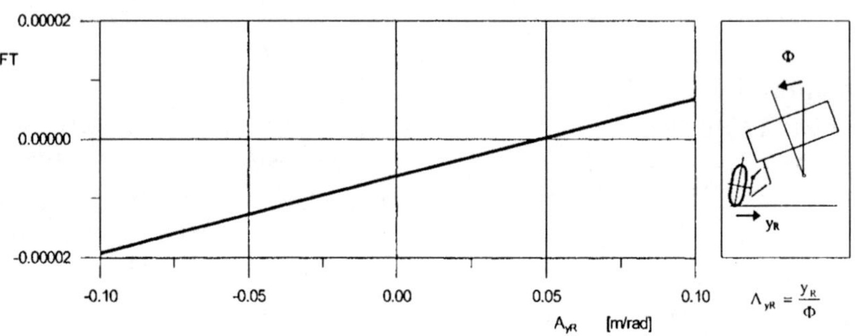

Figure 7.22. Stability factor versus the A_{yR} parameter.

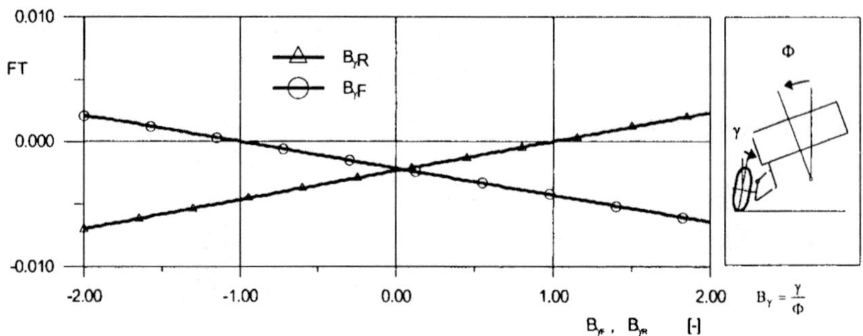

Figure 7.23. Stability factor versus $B_{\gamma F}$ and $B_{\gamma R}$ parameters.

The stability factor increases with an increase of the ratio A_{yR} of the tilting body motion and the transversal motion of the rear road wheels. For $A_{yR} < 0.05\, m/rad$ the analysed vehicle model is unstable with respect to the assumed zones of admissible solutions.

Depending on the tire characteristics the slope of the road wheel plane can either decrease or increase the wheel "resistance" to a transversal drift. The stability depends on the suspension choice and the relation between front and end suspension properties. In our case for the considered data the analysed vehicle motion is stable, when $B_{\gamma F} \leq -1$ and $B_{\gamma R} \geq +1$. It means that the

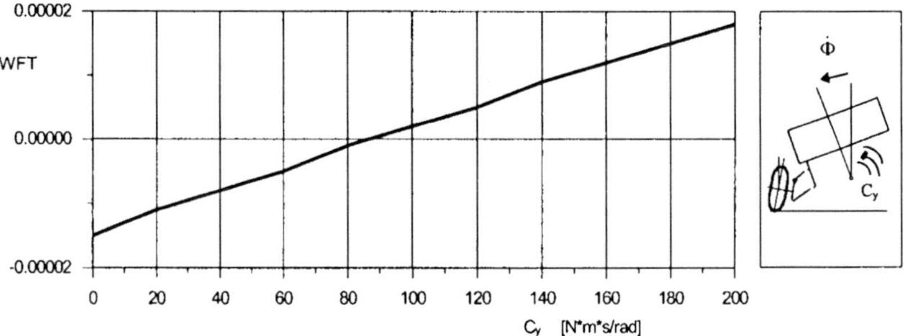

Figure 7.24. Stability factor versus damping coefficient C_y.

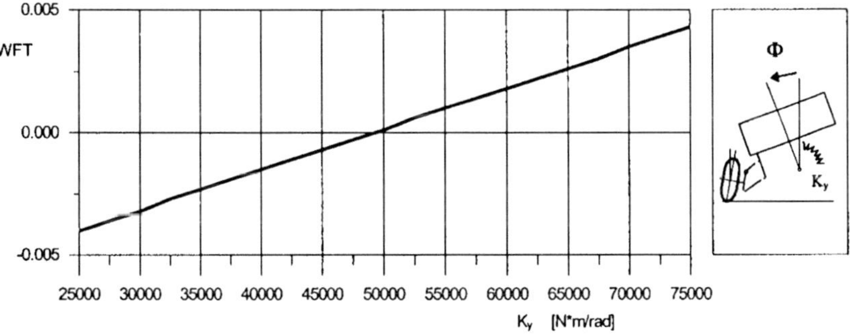

Figure 7.25. Stability factor versus damping coefficient K_y.

considered vehicle should have such front suspension that during the vehicle body tilt (on a turn) the front road wheels are tilted in the same direction as the car body, and the rear road wheels are tilted in the opposite direction to the car body movement.

Influence of the C_y coefficient on the stability factor is shown in Figure 7.24. For the considered data the stability factor decreases with a decrease of the damping value. For $C_y < 90\, N \cdot m \cdot s/rad$ instability appears.

Influence of the K_y coefficient on the vehicle stability factor is shown in Figure 7.25. For the introduced data the vehicle stability is preserved for $K_y > 5 \cdot 10^4\, N \cdot m/rad$.

Chapter 8

LONGITUDINAL TILT DYNAMICS

A longitudinal vehicle tilt occurs either when the vehicle runs on a road surface with transverse roughness like "waves" or "humps" or during the force excitation of the starting or braking processes. This phenomenon is particularly important when inside the vehicle there is a certain body which can shift with respect to the vehicle body. In this chapter the longitudinal vehicle tilt dynamics is studied for two vehicle systems: (i) vehicle body - shifting body; (ii) vehicle with a trailer - a shifting body is inside the trailer.

1. Dynamics of the System: Vehicle Body – Movable Body

The road vehicle model including the movable body with mass m_p and the associated constraints is shown in Figure 8.1. The body with mass m represents the "suspended mass" of the vehicle. The front vehicle suspension is modelled via the stiffness K_V and damping C_V.

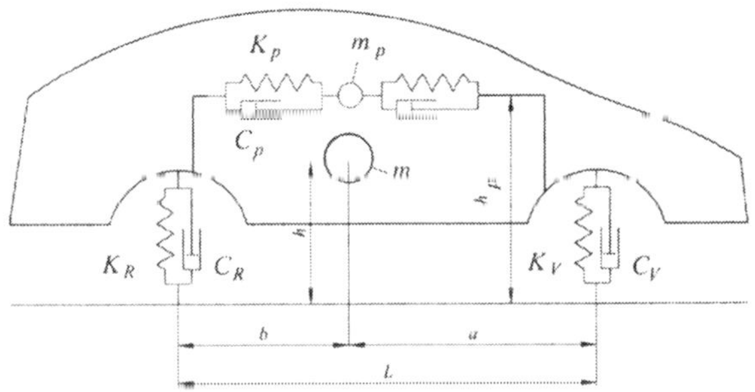

Figure 8.1. Vehicle diameters and the bodies m, m_p positions.

The rear vehicle suspension is modelled via the stiffness K_R and damping C_R. The one degree-of-freedom body with mass m_p moving along the vehicle longitudinal axle is situated at the height h_p. The mass m_p is coupled with the vehicle body through the elastic (K_p) and damping (C_p) elements.

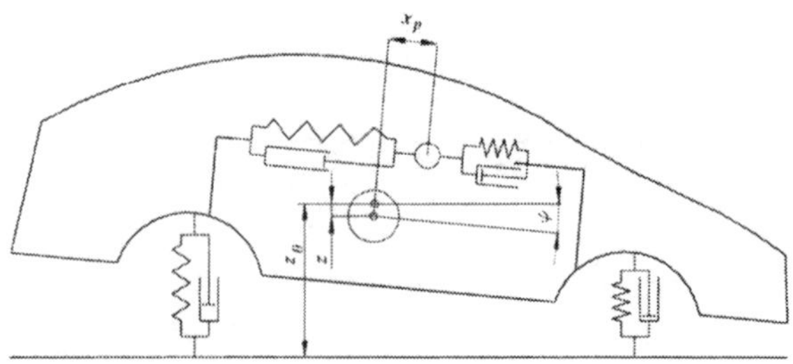

Figure 8.2. The vehicle model generalized coordinates x_p, z i φ.

In Figure 8.2 the introduced general coordinates are shown: x_p - horizontal displacement od the movable body with mass m_p; z - vertical displacement of mass m in its centre; φ - angular mass "m" displacement - pitch angle. The static mass centre m position is denoted by z_0.

The forces action on the road vehicle and the movable body are shown in Figure 8.3.

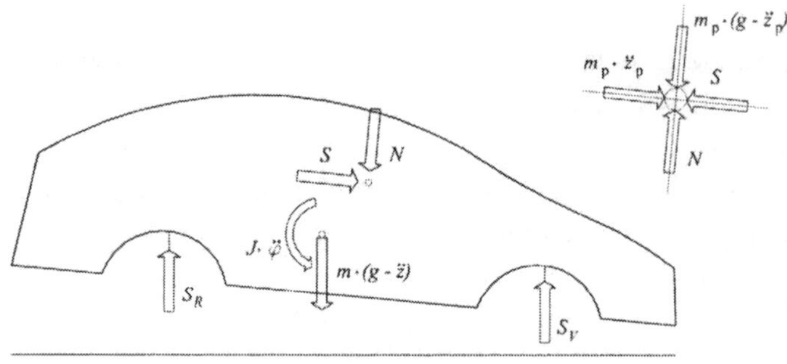

Figure 8.3. Forces action on the road vehicle and the movable mass m_p.

The static vertical vehicle suspension displacements are defined by the relations

$$z_{V0} = (m_p + m) \cdot g \cdot \frac{b}{L \cdot K_V}, \tag{8.1}$$

$$z_{R0} = (m_p + m) \cdot g \cdot \frac{a}{L \cdot K_R}. \tag{8.2}$$

Dynamical vertical vehicle displacements z_V and z_R are coupled with the displacement of the vehicle centre z and the tilting angle - pitch angle φ via dependencies

$$z_R = z - b \cdot \varphi, \tag{8.3}$$

$$z_V = z + a \cdot \varphi. \tag{8.4}$$

Accounting for the linear forces S_V and S_R occurring in suspension, the dynamics of our system is governed by the following differential equation:

$$z_V \cdot K_V + z_{V0} \cdot K_V + z_R \cdot K_R + z_{R0} \cdot K_R + \dot{z}_V \cdot C_V + \dot{z}_R \cdot C_R$$

$$-m \cdot g + m \cdot \ddot{z} - N \cdot \cos(\varphi) - S \cdot \sin(\varphi) = 0, \tag{8.5}$$

$$z_{R0} \cdot K_R \cdot L + \dot{z}_R \cdot C_R \cdot L - m \cdot g \cdot a + m \cdot \ddot{z} \cdot a - J \cdot \ddot{\varphi} - N \cdot \sin(\varphi) \cdot (h_p - z_p)$$

$$+z_R \cdot K_R \cdot L + S \cdot \cos(\varphi) \cdot (h_p - z_p) - S \cdot \sin(\varphi) \cdot (a - x_p) - N \cdot \cos(\varphi) \cdot (a - x_p) = 0, \tag{8.6}$$

$$z_p = z_R + (b + x_p) \cdot \varphi, \tag{8.7}$$

$$N = m_p \cdot (g - \ddot{z}_p) \cdot \cos(\varphi), \tag{8.8}$$

$$S = x_p \cdot K_p + \dot{x}_p \cdot C_p, \tag{8.9}$$

$$m_p \cdot \ddot{x}_p + S - m_p \cdot (g - \ddot{z}_p) \cdot \sin(\varphi) = 0. \tag{8.10}$$

After some transformations the following three second order ODEs are obtained

$$\ddot{z} = \frac{[-K_V - K_R]}{m} \cdot z + \frac{[-a \cdot K_V + b \cdot K_R + S]}{m} \cdot \varphi + \frac{[-C_V - C_R]}{m} \cdot \dot{z}$$

$$+ \frac{[-a \cdot C_V + b \cdot C_R]}{m} \cdot \dot{\varphi} + \frac{N}{m} - \frac{m_p}{m} \cdot g, \tag{8.11}$$

$$\ddot{\varphi} = \frac{[K_R \cdot b - K_V \cdot a]}{J} \cdot z + \frac{[-b^2 \cdot K_R - a^2 \cdot K_V - N \cdot (h_p - z_p) + S \cdot x_p]}{J} \cdot \varphi$$

$$+ \frac{[C_R \cdot b - C_V \cdot a]}{J} \cdot \dot{z} + \frac{[-b^2 \cdot C_R - a^2 \cdot C_V]}{J} \cdot \dot{\varphi} + \frac{S \cdot (h_p - z_p) + N \cdot x_p}{J}, \tag{8.12}$$

$$\ddot{x}_p = -S \cdot \frac{1}{m_p} + g \cdot \varphi - \ddot{z} \cdot \varphi - 2\dot{x}_p \cdot \varphi \cdot \dot{\varphi} - x_p \cdot \varphi \cdot \ddot{\varphi} - \ddot{x}_p \cdot \varphi^2, \tag{8.13}$$

where: z_p, N and S are defined via the following relations:

$$\ddot{z}_p = \ddot{z} + 2\dot{x}_p \cdot \dot{\varphi} + x_p \cdot \ddot{\varphi} + \ddot{x}_p \cdot \varphi, \quad N = m_p \cdot (g - \ddot{z}_p), \quad S = x_p \cdot K_p + \dot{x}_p \cdot C_p. \tag{8.14}$$

EXAMPLE 8.1 *The equation of motion governing the longitudinal vehicle tilt with the movable body are derived in accordance with the chosen model (see Figures 8.1–8.3). The numerical results are reported in Figures 8.4–8.6 for the earlier given data of a passenger car.*

The jump type moving body m_p displacement of the value $x_p = -0.5m$ serves as excitation. Three different constraints of the pair: movable mass – car body are studied.

Case 1. *A load moving freely on a (luggage boot) surface and elastically reflected on the transversal barriers.*
In this case the simulation results for the nominal car parameters are shown in Figure 8.4. The sudden shift of the body m_p yields an occurrence of undamped periodic oscillations of the road wheels pressure in the interval of $(+1000; -1000)N$. The moving body collisions (impacts) on the transversal (luggage boot) walls yield steep increase of vertical forces in road wheel suspension. This case can be treated as dangerous, since the road wheels are periodically unloaded with relatively large amplitude $(1000N)$.

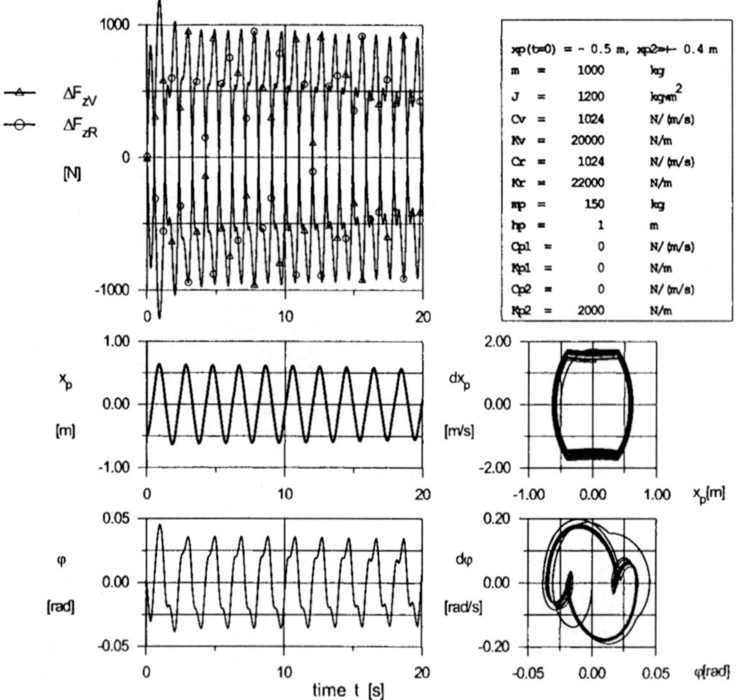

Figure 8.4. Time histories and phase projections of the system dynamics associated with Case 1 ($K_{p2} = 2000\,N/m$).

Case 2. *A load moving freely on a (luggage boot) surface and elastically reflected from the transverse barriers with energy dissipation.*
The numerical results are shown in Figure 8.5 ($C_{p2} = 2000\,N \cdot s/m$; $K_{p2} = 2000\,N/m$). A sudden shift of the mass m_p yields to occurrence

of instantaneous large amplitudes ΔF_{zV} and ΔF_{zR}. After a transitional vibrational process the mass m_p stops. The system undergoes not required dynamics in the first motion phase accompanied with the essential changes of the suspension forces. After duration of 2 s the vibrational process is damped.

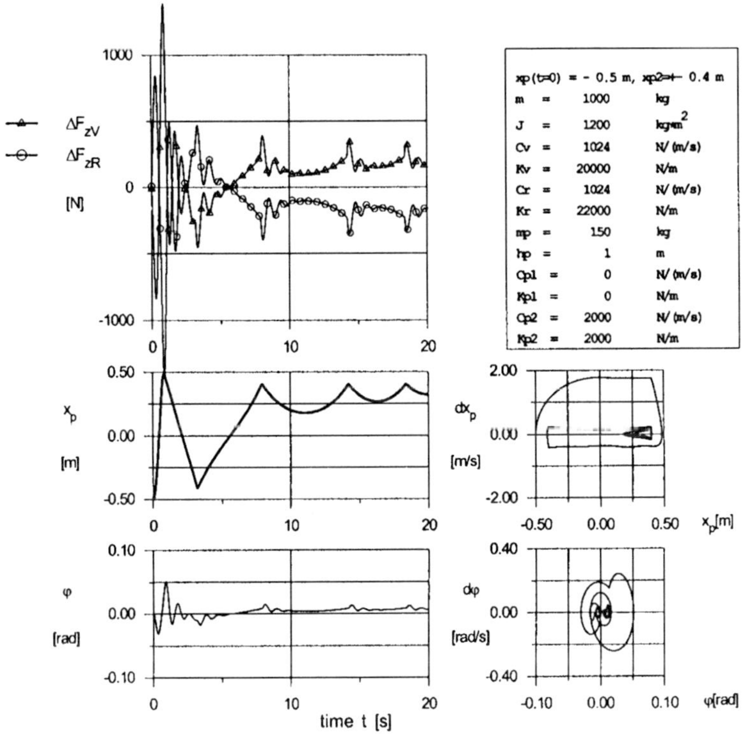

Figure 8.5. Time histories and phase projections of the system dynamics associated with Case 2 ($C_p = 2000\, N \cdot s/m$, $K_{p2} = 2000\, N/m$).

Case 3. *A load moves along longitudinal constraints (C_{p1}, K_{p1}) on a luggage boot surface with undefined length.*

The associated numerical results are shown in Figure 8.6. The sudden shift of the mass leads to instantaneous increase and decrease of the forces value in the road wheels suspension of about $\pm 500\, N$. Further dynamical behaviour can be recognized as stable.

In Figure 8.7 an influence of the mass m_p values on stability is shown. The investigations concern the constraints of car - movable body associated with Case 1. One may conclude, that for $m_p = 150\, kg$, the system exhibits damped vibrations around the equilibrium position $(0, 0)$. For $m_p = 360\, kg$

the motion is quasi-periodic. For $m_p > 370\,kg$ the solutions tend to infinity. Although the phase curve displays in the beginning, a tendency to achieve the equilibrium point $(0, -0.03)$, *it suddenly leaves the equilibrium neighbourhood and tends to* $-\infty$.

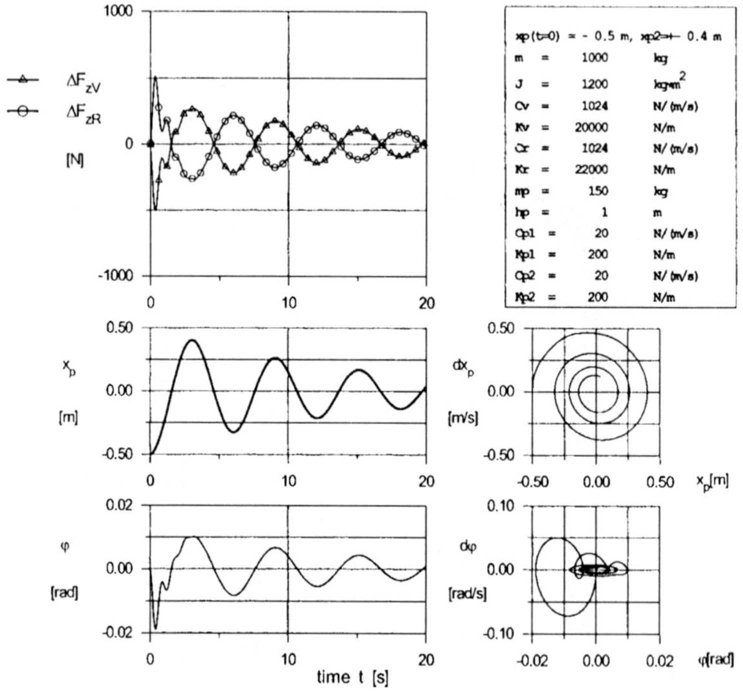

Figure 8.6. Time histories and phase projections of the system dynamics associated with Case 3 ($C_{p1} = 20\,N \cdot s/m$, $K_{p1} = 200\,N/m$).

2. Dynamics of a Longitudinal Tilt of a Road Vehicle with a Semi-trailer (trailer)

The phenomenon of the longitudinal tilt "pitch" of a vehicle with a trailer (for instance of a passenger vehicle with a house trailer) may cause a serious road safety problem. As it will be shown later, dynamics of such a system can exhibit many interesting behaviors. Therefore, in this section the model of a car with a semi-trailer, inside of which a body with mass m_p is situated (see Figure 8.8), is studied.

The mentioned body (for instance, a piece of luggage) may unexpectedly move loosening the additional constraints.

In Figure 8.8 the body with mass m models a *suspended mass of the pulling car*. The front car suspension is modelled via the elastic element with stiffness

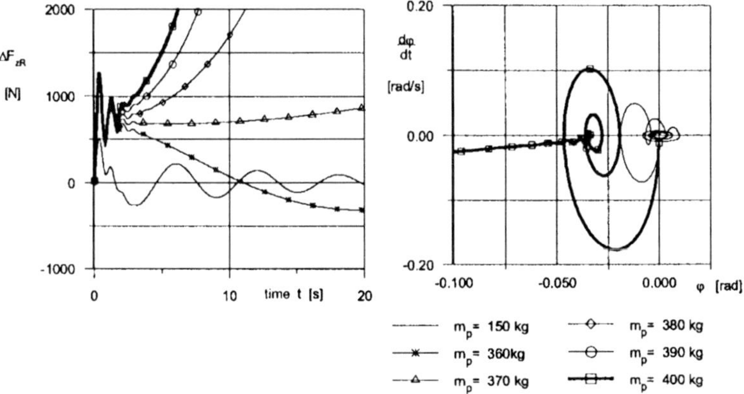

Figure 8.7. Time history ΔF_{zR} and phase projections $\dot{\varphi}(\varphi)$ for different values of m_p ($C_{p1} = 20\,N \cdot s/m$, $K_{p1} = 200\,N/m$).

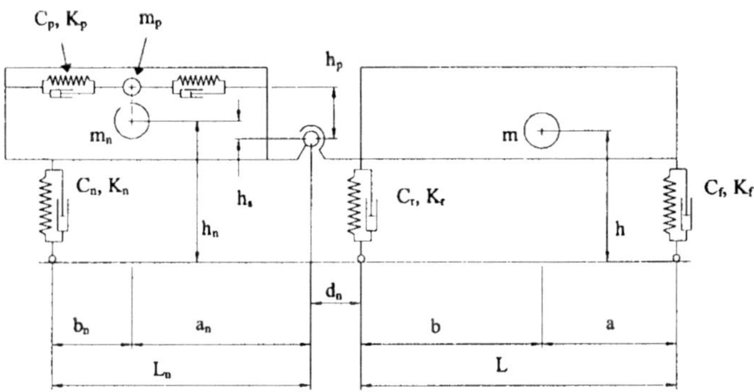

Figure 8.8. The mechanical scheme of a car with a semi-trailer including positions of the masses m, m_p and m_n.

K_f and via the damping element with damping coefficient C_f. The rear car suspension is modelled via stiffness K_r and via damping C_r. The body with mass m_u models the ***suspended semi-trailer mass***. Its suspension is represented by stiffness K_n and damping C_n. The body with mass m_p is situated on the height h_p and it has one degree-of-freedom along the longitudinal car axle. The massless elements K_p and C_p represent stiffness and damping of the constraints associated with the semi-trailer body.

In Figure 8.9 the introduced general coordinates are shown, whereas in Figure 8.10 the scheme of the forces and moments acting on the system are displayed.

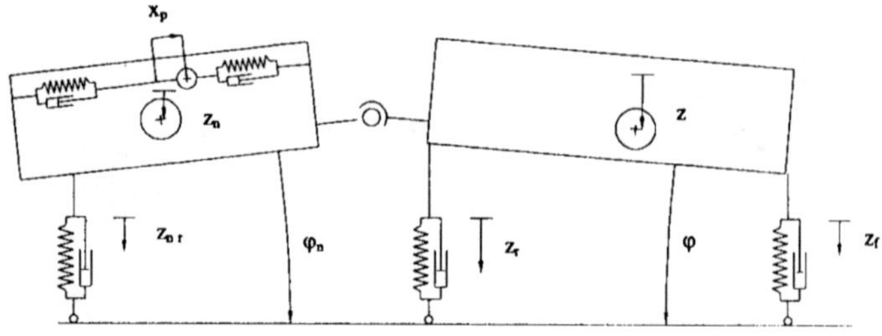

Figure 8.9. The generalized coordinates x_p, z, z_n and φ associated with the model (x_p - the body m_p displacement; z - displacement of the car centre; z_n - displacement of the semi-trailer centre; φ - angle of rotation of the suspended mass of the car - pitch angle).

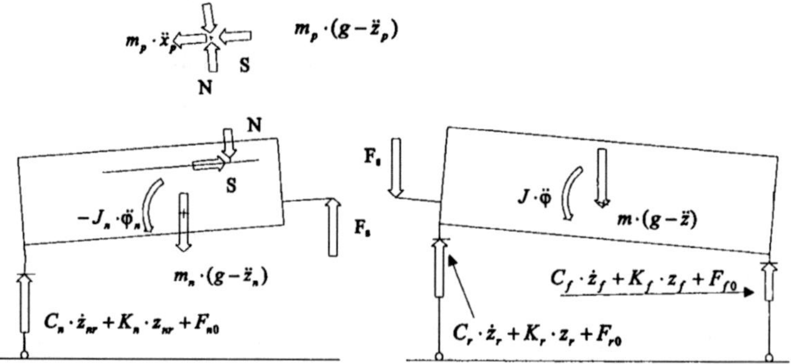

Figure 8.10. Forces and moments acting on the pulling car, semi-trailer and movable body with the mass m_p (F_s - vertical force on a hook (a joint); N - force perpendicular to the direction x_p acting on the stick of the mass m_p and the semi-trailer; S - force acting in direction x_p and generated in the constraints coupling the body m_p and the semi-trailer).

From Figures 8.8 and 8.9 the rotation semi-trailer angle is derived

$$\varphi_n = z_n \cdot \frac{1}{a_n} - z \cdot \frac{1}{a_n} + \frac{d_n + b}{a_n} \cdot \varphi, \tag{8.15}$$

whereas analysis of Figure 8.8 yields the following static forces:

(i) in the front car suspension

$$F_{f0} = m \cdot g \cdot \frac{b}{L} - (m_n + m_p) \cdot g \cdot \frac{b_n \cdot d_n}{L_n \cdot L}; \tag{8.16}$$

(ii) in the rear car suspension

$$F_{r0} = m \cdot g \cdot \frac{a}{L} + (m_n + m_p) \cdot g \cdot \frac{b_n \cdot (L + d_n)}{L_n \cdot L}; \tag{8.17}$$

(iii) in the semi-trailer suspension

$$F_{n0} = (m_n + m_p) \cdot g \cdot \frac{a_n}{L_n}. \tag{8.18}$$

Applying the d'Alembert rule the following differential equations governing: the car dynamics

$$\sum F_z = F_s - C_r \cdot \dot{z}_r - K_r \cdot z_r - F_{r0} + m \cdot (g - \ddot{z}) - C_f \cdot \dot{z}_f - K_f \cdot z_f - F_{f0} = 0, \tag{8.19}$$

$$\sum M_s = J \cdot \ddot{\varphi} - m \cdot (g - \ddot{z}) \cdot (b + d_n) + C_r \cdot \dot{z}_r \cdot d_n + K_r \cdot z_r \cdot d_n + F_{r0} \cdot d_n$$
$$+ C_f \cdot \dot{z}_f \cdot (L + d_n) + K_f \cdot z_f \cdot (L + d_n) + F_{f0} \cdot (L + d_n) = 0, \tag{8.20}$$

the semi-trailer dynamics

$$\sum F_z = -F_s + N + m_n \cdot (g - \ddot{z}_n) - C_n \cdot \dot{z}_{nr} - K_{rn} \cdot z_{nr} - F_{n0} = 0, \tag{8.21}$$

$$\sum M_s = J_n \cdot \ddot{\varphi}_n + m_n \cdot (g - \ddot{z}_n) \cdot a_n - C_n \cdot \dot{z}_{nr} \cdot L_n$$
$$- K_n \cdot z_{nr} \cdot L_n - F_{n0} \cdot L_n + N \cdot (a_n - x_p) - S \cdot h_p = 0, \tag{8.22}$$

and the movable body dynamics

$$N = m_p \cdot (g - \ddot{z}_p), \tag{8.23}$$

$$S = x_p \cdot K_p + \dot{x}_p \cdot C_p, \tag{8.24}$$

$$m_p \cdot \ddot{x}_p + S - m_p \cdot (g - \ddot{z}_p) \cdot \varphi_n = 0, \tag{8.25}$$

are obtained, where:

$$z_r = z - b \cdot \varphi, \tag{8.26}$$

$$z_f = z + a \cdot \varphi, \tag{8.27}$$

$$z_{nr} = z_n + b_n \cdot \varphi_n. \tag{8.28}$$

Owing to relations (8.19), (8.21), (8.26), (8.27) and (8.28), (8.15) and their transformations the following three second order differential equations are obtained:

$$\ddot{z} = a_1 \cdot \dot{z}_n + a_2 \cdot z_n + a_3 \cdot \dot{z} + a_4 \cdot z + a_5 \cdot \dot{\varphi} + a_6 \cdot \varphi + a_7 \cdot \ddot{z}_n + a_8, \tag{8.29}$$

$$\ddot{\varphi} = b_1 \cdot \dot{\varphi} + b_2 \cdot \varphi + b_3 \cdot \dot{z} + b_4 \cdot z + b_5 \cdot \ddot{z} + b_6, \tag{8.30}$$

$$\ddot{z}_n = c_1 \cdot \dot{z}_n + c_2 \cdot z_n + c_3 \cdot \dot{z} + c_4 \cdot z + c_5 \cdot \dot{\varphi} + c_6 \cdot \varphi + c_7 \cdot \ddot{z} + c_8, \tag{8.31}$$

where:

$$a_1 = \left[\frac{C_n \cdot b_n - C_f - C_r}{m \cdot a_n}\right]; \quad a_2 = \left[\frac{K_n \cdot b_n - K_f - K_r}{m \cdot a_n}\right];$$

$$a_3 = \left[\frac{C_r \cdot b - C_f \cdot a - C_n \cdot b_n \cdot (d_n + b)}{m \cdot a_n}\right];$$

$$a_4 = \left[\frac{K_r \cdot b - K_f \cdot a - K_n \cdot b_n \cdot (d_n + b)}{m \cdot a_n}\right];$$

$$a_5 = -\left[\frac{C_n + C_n \cdot b_n}{m \cdot a_n}\right]; \quad a_6 = -\left[\frac{K_n + K_n \cdot b_n}{m \cdot a_n}\right];$$

$$a_7 = -\frac{m_n}{m}; \quad a_8 = \frac{[-F_{n0} - F_{r0} - F_{f0} + m \cdot g + N + m_n \cdot g]}{m};$$

$$b_1 = \left[-\frac{1}{J} \cdot [C_f \cdot (L + d_n) \cdot a - C_r \cdot d_n \cdot b]\right];$$

$$b_2 = \left[-\frac{1}{J} \cdot [K_f \cdot (L + d_n) \cdot a - K_r \cdot d_n \cdot b]\right];$$

$$b_3 = \left[-\frac{1}{J} \cdot [C_r \cdot d_n + C_f \cdot (L + d_n)]\right]; \quad b_4 = \left[-\frac{1}{J} \cdot [K_r \cdot d_n + K_f \cdot (L + d_n)]\right];$$

$$b_5 = \left[-\frac{1}{J} \cdot m \cdot (b + d_n)\right]; \quad b_6 = \left[-\frac{1}{J} \cdot [F_{f0} \cdot (L + d_n) + F_{r0} \cdot d_n - m \cdot g \cdot (b + d_n)]\right];$$

$$c_1 = \left[\frac{-C_n \cdot L_n \cdot a_n - C_n \cdot L_n \cdot b_n}{J_n + m_n \cdot (a_n)^2}\right]; \quad c_2 = \left[\frac{-K_n \cdot L_n \cdot a_n - K_n \cdot L_n \cdot b_n}{J_n + m_n \cdot (a_n)^2}\right];$$

$$c_3 = \left[\frac{C_n \cdot L_n \cdot b_n}{J_n + m_n \cdot (a_n)^2} + \frac{J_n \cdot (d_n + b)}{J \cdot (J_n + m_n \cdot (a_n)^2)} \cdot [C_r \cdot d_n + C_f \cdot (L + d_n)]\right];$$

$$c_4 = \left[\frac{K_n \cdot L_n \cdot b_n}{J_n + m_n \cdot (a_n)^2} + \frac{J_n \cdot (d_n + b)}{J \cdot (J_n + m_n \cdot (a_n)^2)} \cdot [K_r \cdot d_n + K_f \cdot (L + d_n)]\right];$$

$$c_5 = \left[-\frac{C_n \cdot L_n \cdot b_n \cdot (d_n + b)}{J_n + m_n \cdot (a_n)^2} + \frac{J_n \cdot (d_n + b)}{J \cdot (J_n + m_n \cdot (a_n)^2)} \cdot [C_f \cdot (L + d_n) \cdot a - C_r \cdot d_n \cdot b]\right];$$

$$c_6 = \left[-\frac{K_n \cdot L_n \cdot b_n \cdot (d_n + b)}{J_n + m_n \cdot (a_n)^2} + \frac{J_n \cdot (d_n + b)}{J \cdot (J_n + m_n \cdot (a_n)^2)} \cdot [K_f \cdot (L + d_n) \cdot a - K_r \cdot d_n \cdot b]\right];$$

$$c_7 = \left[\frac{J \cdot J_n + J_n \cdot m \cdot (d_n + b)^2}{J_n + m_n \cdot (a_n)^2}\right];$$

$$c_8 = \left[-\frac{1}{J_n + m_n \cdot (a_n)^2} [F_{n0} \cdot L_n \cdot a_n + N \cdot (a_n - x_p) \cdot a_n - S \cdot h_p \cdot a_n + m_n \cdot a_n \cdot a_n \cdot g] + \right.$$
$$\left. \frac{J_n \cdot (d_n + b)}{J \cdot (J_n + m_n \cdot (a_n)^2)} \cdot [F_{f0} \cdot (L + d_n) + F_{r0} \cdot d_n - m \cdot g \cdot (b + d_n)]\right].$$

Introducing the following further notation

$$a_{11} = \frac{a_3 + a_7 \cdot c_3}{1 - a_7 \cdot c_7}; \quad a_{12} = \frac{a_5 + a_7 \cdot c_5}{1 - a_7 \cdot c_7}; \quad a_{13} = \frac{a_1 + a_7 \cdot c_1}{1 - a_7 \cdot c_7};$$

$$b_{11} = \frac{a_7 \cdot c_4 + a_4}{1 - a_7 \cdot c_7}; \quad b_{12} = \frac{a_7 \cdot c_6 + a_6}{1 - a_7 \cdot c_7}; \quad b_{13} = \frac{a_7 \cdot c_2 + a_2}{1 - a_7 \cdot c_7};$$

$$e_1 = \frac{a_8 + a_7 \cdot c_8}{1 - a_7 \cdot c_7};$$

$$a_{21} = \left(b_3 + b_5 \cdot \frac{a_3 + a_7 \cdot c_3}{1 - a_7 \cdot c_7} \right); \quad a_{22} = \left(b_1 + b_5 \cdot \frac{a_5 + a_7 \cdot c_5}{1 - a_7 \cdot c_7} \right);$$

$$a_{23} = \left(b_5 \cdot \frac{a_1 + a_7 \cdot c_1}{1 - a_7 \cdot c_7} \right); \quad b_{21} = \left(b_5 \cdot \frac{a_7 \cdot c_4 + a_4}{1 - a_7 \cdot c_7} + b_4 \right);$$

$$b_{22} = \left(b_5 \cdot \frac{a_7 \cdot c_6 + a_6}{1 - a_7 \cdot c_7} + b_2 \right); \quad b_{23} = \left(b_5 \cdot \frac{a_7 \cdot c_2 + a_2}{1 - a_7 \cdot c7} \right);$$

$$e_2 = \left(b_5 \cdot \frac{a_8 + a_7 \cdot c_8}{1 - a_7 \cdot c_7} + b_6 \right);$$

$$a_{31} = \left(\frac{c_7 \cdot a_3 + c_3}{1 - c_7 \cdot a_7} \right); \quad a_{32} = \left(\frac{c_7 \cdot a_5 + c_5}{1 - c_7 \cdot a_7} \right); \quad a_{33} = \left(\frac{c_7 \cdot a_1 + c_1}{1 - c_7 \cdot a_7} \right);$$

$$b_{31} = \left(\frac{c_4 + c_7 \cdot a_4}{1 - c_7 \cdot a_7} \right); \quad b_{32} = \left(\frac{c_6 + c_7 \cdot a_6}{1 - c_7 \cdot a_7} \right); \quad b_{33} = \left(\frac{c_7 \cdot a_2 + c_2}{1 - c_7 \cdot a_7} \right);$$

$$e_3 = \left(\frac{c_7 \cdot a_8 + c_8}{1 - c_7 \cdot a_7} \right);$$

the governing differential equations are transformed to the following form

$$\ddot{z} = a_{11} \cdot \dot{z} + a_{12} \cdot \dot{\varphi} + a_{13} \cdot \dot{z}_n + b_{11} \cdot z + b_{12} \cdot \varphi + b_{13} \cdot z_n + e_1, \quad (8.32)$$

$$\ddot{\varphi} = a_{21} \cdot \dot{z} + a_{22} \cdot \dot{\varphi} + a_{23} \cdot \dot{z}_n + b_{21} \cdot z + b_{22} \cdot \varphi + b_{23} \cdot z_n + e_2, \quad (8.33)$$

$$\ddot{z}_n = a_{31} \cdot \dot{z} + a_{32} \cdot \dot{\varphi} + a_{33} \cdot \dot{z}_n + b_{31} \cdot z + b_{32} \cdot z_n + b_{33} \cdot \varphi + e_3. \quad (8.34)$$

Assuming for small displacements $z_p = z$ and owing to relation (8.25), (8.24) and (8.15), the following equation is obtained

$$\ddot{x}_p = -\frac{C_p}{m_p} \cdot \dot{x}_p - \frac{K_p}{m_p} \cdot x_p - \frac{g}{a_n} \cdot z + \frac{g \cdot (d_n + b)}{a_n} \cdot \varphi +$$

$$\frac{g}{a_n} \cdot z_n + \frac{1}{a_n} \cdot z \cdot \ddot{z} - \frac{(d_n + b)}{a_n} \cdot \varphi \cdot \ddot{z} - \frac{1}{a_n} \cdot z_n \cdot \ddot{z}. \quad (8.35)$$

To conclude, the differential equations (8.32)–(8.35) govern dynamics of the car pulling the semi trailer with the movable body attached to the semi-trailer via constraints being the lumped stiffness and damping elements.

EXAMPLE 8.2 *The numerical simulations of the differential equations* (8.32)–(8.35) *governing dynamics of the pulling car with the semi-trailer and movable mass model are carried out. The initial conditions are taken with zero values, and the jump type displacement of the movable body on amount of* $x_p = 0.75\,m$ *serves as excitation.*

The following data are taken: mass of the pulling vehicle $m = 1000\,kg$; *mass of the semi-trailer* $m_n = 600\,kg$; *mass of the movable body* $m_p =$

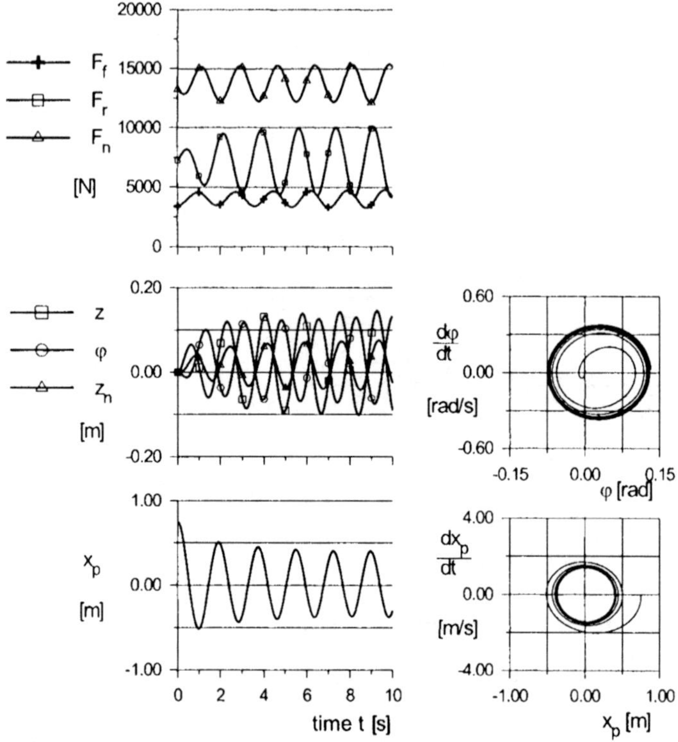

Figure 8.11. The simulation results of the car-semi-trailer-movable body dynamics ($K_r = 12 \cdot 10^3\, N/m$ - stiffness of the end car suspension; $d_n = 0.5\, m$ - the distance of the car and semi-trailer linking joint to the end car axle; F_f - vertical force in the front suspension; F_r - vertical force in the rear suspension; z - vertical displacement of the car centre; z_n - vertical displacement of the semi-trailer mass centre; φ - angular car displacement or the longitudinal tilt, pitch angle; x_p - displacement of the movable body).

$100\, kg$; *inertial mass moment of the car* $J_x = 1250\, kg \cdot m^2$; *inertial mass moment of the trailer* $J_n = 1000\, kg \cdot m^2$; *damping of the front suspension* $C_f = 1500\, N \cdot s/m$; *damping of the rear suspension* $C_r = 1500 N \cdot s/m$; *damping of the semi-trailer suspension* $C_n = 1400\, N \cdot s/m$; *stiffness of the front suspension* $K_f = 10000\, N/m$; *stiffness of the end suspension* $K_r = 12 \cdot 10^3 \div 20 \cdot 10^3\, N/m$; *stiffness of the semi-trailer suspension* $K_n = 10 \cdot 10^3\, N/m$; *damping of the constraints of the movable body* $C_p = 100\, N \cdot s/m$ *and semi-trailer stiffness of the constraints of the movable body* $K_p = 1000\, N/m$; *parameters characterizing positions of vehicle and semi-trailer mass centres:* $a = 1.5\, m$, $b = 1\, m$, $a_n = 1.2\, m$, $b_n = 0.8\, m$, $h = 0.6\, m$, $h_n = 0.7\, m$, $h_p = 0.6\, m$; *parameters of the joint (linking car and semi-trailer) position:* $d_n = 0.5 \div 0.1\, m$, $h_2 = 0.4\, m$ *(see Figures 8.8–8.10).*

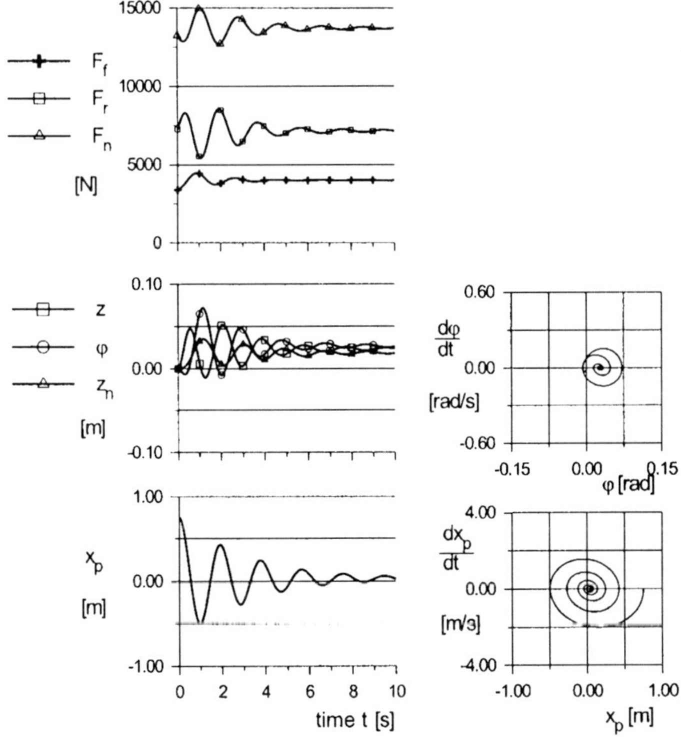

Figure 8.12. The simulation results of the car-semi-trailer-movable body dynamics ($K_r = 20 \cdot 10^3\ N/m$, other parameters and notation are the same as in Figure 8.11).

Case 1. *(See Figure 8.11) In this case a small stiffness of the end road wheels suspension and a standard distance of the joint linking the second vehicle to the rear wheels axle of the pulling car are assumed.*
One may conclude from Figure 8.11 that the vertical vibrations of the vehicle and semi-trailer as well as the longitudinal tilt (φ) of the pulling car have increasing amplitudes. Also oscillations of the vertical forces in the front wheels suspension increase. This case can be treated as a dangerous one.

Case 2. *In this case a large stiffness of the rear road wheels suspension and a standard distance of the joint linking two vehicles to the rear wheels axle of the pulling car are assumed.*
Since the parameters of the first case led to improper dynamics, the suspension stiffness has been changed from $K_r = 12 \cdot 10^3\ N/m$ to $K_r = 20 \cdot 10^3\ N/m$. It occurred that the dynamics is significantly improved (see Figure 8.12), and system stability is preserved.

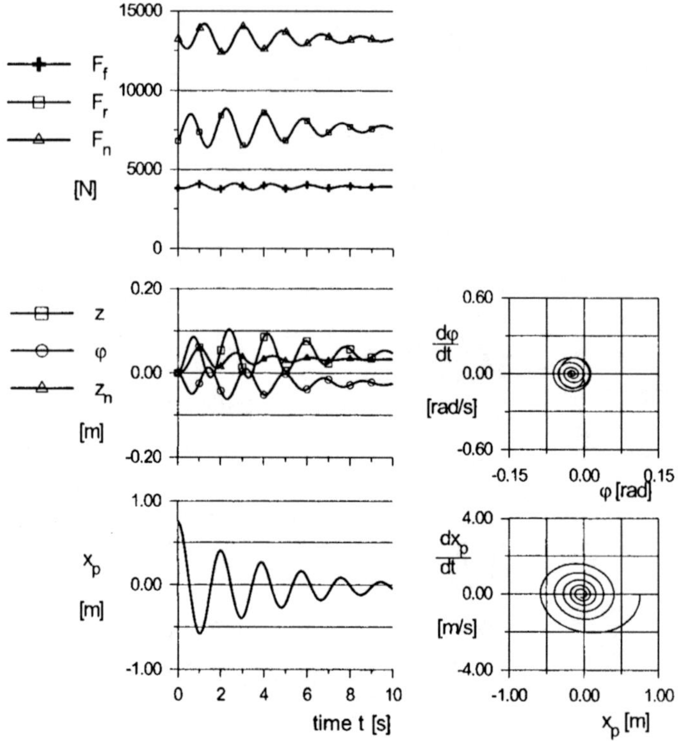

Figure 8.13. The simulation results of the car-semi-trailer-movable body dynamics ($K_r = 12 \cdot 10^3\, N/m$, $d_n = 0.1\, m$, other parameters and notation are the same as in Figure 8.11).

Case 3. *In this case small stiffness of the rear road wheels suspension and a small distance of the joint linking two vehicles of the rear wheels axle of the pulling car* d_n *are assumed.*

In particular, now the influence of d_n *is investigated. The distance* d_n *is decreased in comparison to Case 1 to the value of* $d_n = 0.1\, m$. *As it is seen in the Figures 8.12–8.13 the decrease of* d_n *does not lead to as good results as those achieved in Case 2, but the dynamics has been sufficiently improved in comparison to that of Case 1. The system dynamics is stable and the vibrational process is damped achieving the equilibrium state* $(0, 0)$.

Chapter 9

ROAD WHEEL ROTATIONAL DYNAMICS

In this chapter two principal cases are studied. Namely, dynamics of the road wheel rotational motion during braking and driving processes is investigated. In particular, stability of the system composed of the road wheel and braking process controller (including the adhesion tire characteristics) is analysed. In addition, the road wheel driven process during start of the road vehicle is quantified.

1. Dynamics of the Road Wheel Rotational Motion During Braking Process [8, 95]

Dynamics of the road wheel rotational motion during the braking process is analysed. The rotational moment $M_R(t)$ is applied to the wheel axle, and its value is controlled via a controller (for instance ABS or ASR). An action of the rotational moment yields the circumferential friction force F_x tangent to the road surface. The road wheel kinematics is defined via the vectors $\bar{v}_{kx}$ and $\overline{\dot{\varphi}_k}$.

The rotational motion equation (Figure 9.1) reads

$$F_x \cdot R_k - J_k \cdot \ddot{\varphi}_k - M_R(t) = 0, \tag{9.1}$$

where: J_k is the inertial wheel mass moment including the attached rotational elements.

The relative circumferential (longitudinal) slip (see Section 3.1) during the braking of the road wheel is defined via the relation

$$s_x = 1 - \frac{\dot{\varphi}_k \cdot R_k}{v_{kx}}. \tag{9.2}$$

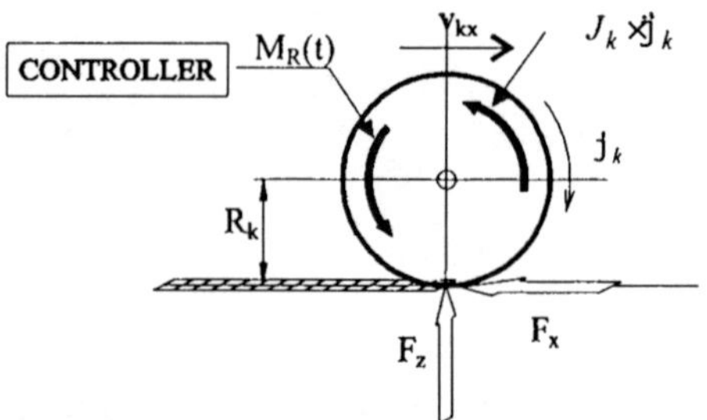

Figure 9.1. The road wheel rotational motion.

Assuming that the process is monitored for in a short time, i.e. $v_{kx} = const.$, and differentiating (9.2), one gets

$$\frac{ds_x}{dt} \cdot v_{kx} = -\ddot{\varphi}_k \cdot R_k. \tag{9.3}$$

Substituting (9.3) and $F_x = F_z \cdot \mu_x$ into (9.1) the following equation is obtained

$$F_z \cdot \mu_x \cdot R_k + J_k \cdot \frac{v_{kx}}{R_k} \cdot \frac{ds_x}{dt} - M_R(t) = 0\,. \tag{9.4}$$

Although the real tire characteristics $\mu_x(s_x)$ are non-, in our further considerations we use their linear approximation in the form

$$\mu_x = A \cdot s_x + B\,. \tag{9.5}$$

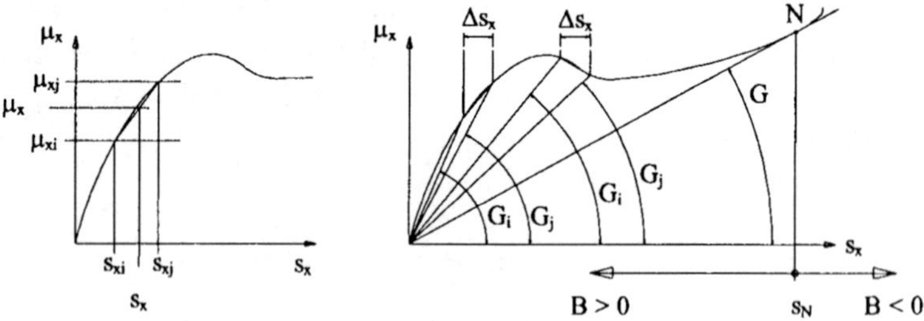

Figure 9.2. Linearization of the $\mu_x(s_x)$ function.

If the linearization step is taken as $\Delta s_x = s_{xj} - s_{xi}$ (see Figure 9.2) then the constants A, B are defined in the following way

$$A = \frac{\mu_{xj} - \mu_{xi}}{s_{xj} - s_{xi}}; \quad B = \mu_{xj} - s_{xj} \cdot \frac{\mu_{xj} - \mu_{xi}}{s_{xj} - s_{xi}}. \tag{9.6}$$

Observe that owing to notations used in Figure 9.2 $s_{xj} - s_{xi} = \Delta s_x > 0$, $A > 0$ when $\mu_{xj} > \mu_{xi}$, i.e. when the function $\mu_x(s_x)$ is increased.

The second equation of (9.6) is transformed to the relation

$$B = \frac{1}{\Delta s_x \cdot s_{xi} \cdot s_{xj}} \cdot \left(\frac{\mu_{xi}}{s_{xi}} - \frac{\mu_{xj}}{s_{xj}} \right),$$

where: $G_i = \mu_{xi}/s_{xi}$, $G_j = \mu_{xj}/s_{xj}$.

Note that $B > 0$ if $G_j < G_i$ for $s_x > 0$. Taking into account (9.6) and (9.5) one gets

$$\left(J_k \cdot \frac{v_{kx}}{R_k} \right) \cdot \frac{ds_x}{dt} + (A \cdot F_z \cdot R_k) \cdot s_x = M_R(t) - (B \cdot F_z \cdot R_k) \; . \tag{9.7}$$

Owing to the Lyapunov stability, the system governed by a first order differential equation is stable if and only if all coefficients of the associated homogeneous equation are positive (or negative). The stability condition applied to equation (9.7), for $v_{xk} > 0$ and $F_z > 0$, reads

$$A > 0 \; . \tag{9.8}$$

This condition obviously holds for $s_{xj} - s_{xi} > 0$ and takes place (see (9.6)) when the following inequality is satisfied

$$\mu_{xj} > \mu_{xi}. \tag{9.9}$$

The condition (9.9) is satisfied for s_x when $\mu_x(s_x)$ is the growing function. The exemplary tire characteristics with the marked stability zones are shown in Figure 9.3.

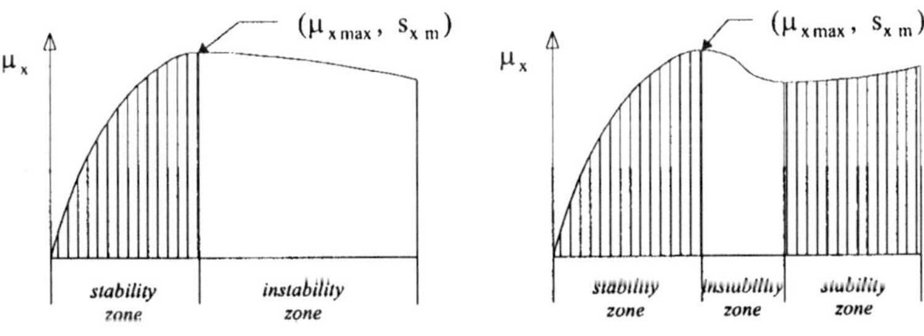

Figure 9.3. Tires characteristics $\mu_x(s_x)$ during the road wheel rotational motion with marked stability zones.

Let us now proceed to analysis of the equation (9.7) investigating influence of its coefficients and permanently acting disturbances on the technical stability of our investigated process. The equation (9.7) is transformed to the form

$$\frac{ds_x}{dt} = -K \cdot s_x + R(t), \tag{9.10}$$

where:

$$K = \frac{F_z \cdot R_k^2}{J_k \cdot v_{kx}} \cdot A; \quad R(t) = \frac{R_k}{J_k \cdot v_{kx}} \cdot M_R(t) - \frac{F_z \cdot R_k^2}{J_k \cdot v_{kx}} \cdot B\,.$$

The technical stability of the system governed by the equation (9.10) is studied owing to Theorem 2 of Section 2.1.5. The initial ω and admissible Ω zones associated with the circumferential slip s_x are assumed in the following form

$$\omega = \left\{ s_x^2 < s_{x0}^2 \right\}, \quad \Omega = \left\{ s_x^2 < s_{xd}^2 \right\},$$

where: s_{x0} is the possible distribution of the initial conditions of the circumferential slip; s_{xd} is the admissible distribution of the circumferential slip values.

The Bogusz function is taken in the form

$$V_B(s_x) = \frac{1}{2} s_x^2.$$

Its maximum in zone ω achieves

$$C_0 = \frac{1}{2} s_{x0}^2,$$

whereas its minimum in the space E_n/Ω is

$$C_1 = \frac{1}{2} s_{xd}^2.$$

The derivative of the function V_B along the solutions of (9.10) reads

$$\frac{dV_B}{dt} = -K \cdot s_x^2 + R(t) \cdot s_x. \tag{9.11}$$

The system is technically stable if the following inequality is satisfied

$$\left. \frac{dV_B}{dt} \right|_{s_x \in \Omega/\omega} < \frac{C_1 - C_0}{T}, \tag{9.12}$$

where: T is the finite monitoring time.

In order to find a maximum of the function $\dot{V}_B$ in the space Ω/ω, the sign of the coefficient K and the perturbation $R(t)$ should be taken into account.

Case 1. The function $\mu_x(s_x)$ is increased: $A > 0$, $K > 0$, $R(t) > 0$.

The maximum of $\dot{V}_B$ in the space Ω/ω reads

$$\inf_{s_x \in \Omega/\omega} \left[\frac{dV_B}{dt} \right] = -K \cdot s_{x0}^2 + R(t) \cdot s_{x0},$$

and the condition (9.12) takes the following form

$$-K \cdot s_{x0}^2 + R(t) \cdot s_{x0} < \frac{s_{xd}^2 - s_{x0}^2}{2T}.$$

Owing to (9.10) one gets

$$-\frac{F_z \cdot R_k^2}{J_k \cdot v_{kx}} \cdot A \cdot s_{x0}^2 + \frac{R_k}{J_k \cdot v_{kx}} \cdot M_R(t) \cdot s_{x0} - \frac{F_z \cdot R_k^2}{J_k \cdot v_{kx}} \cdot B \cdot s_{x0} < \frac{s_{xd}^2 - s_{x0}^2}{2T},$$

and hence

$$M_R(t) < \frac{s_{xd}^2 - s_{x0}^2}{2T \cdot s_{x0}} \cdot \frac{J_k \cdot v_{kx}}{R_k} + F_z \cdot R_k \cdot (A \cdot s_{x0} + B)\,.$$

Furthermore, two subspaces are considered.
1) $B > 0\,(K > 0 \Rightarrow A > 0)$
In this case the technical stability condition reads

$$M_R(t) < \frac{s_{xd}^2 - s_{x0}^2}{2T \cdot s_{x0}} \cdot \frac{J_k \cdot v_{kx}}{R_k} + F_z \cdot R_k \cdot (|A| \cdot s_{x0} + |B|)$$

for

$$M_R(t) > +F_z \cdot R_k \cdot |B|\,. \tag{9.13}$$

2) $B < 0\,(K > 0 \Rightarrow A > 0)$
In this case the technical stability condition reads

$$M_R(t) < \frac{s_{xd}^2 - s_{x0}^2}{2T \cdot s_{x0}} \cdot \frac{J_k \cdot v_{kx}}{R_k} + F_z \cdot R_k \cdot (|A| \cdot s_{x0} - |B|)$$

for

$$M_R(t) > -F_z \cdot R_k \cdot |B|\,. \tag{9.14}$$

Case 2. The function $\mu_x(s_x)$ is increased: $A > 0$, $K > 0$, $R(t) < 0$.
The maximum of V_B in the space Ω/ω reads

$$\inf_{s_x \in \Omega/\omega} \left| \frac{dV_B}{dt} \right| = -K \cdot s_{x0}^2 + R(t) \cdot s_{x0},$$

and the condition (9.12) takes the following form

$$-K \cdot s_{x0}^2 + R(t) \cdot s_{x0} < \frac{s_{xd}^2 - s_{x0}^2}{2T}.$$

Owing to (9.10) one gets

$$-\frac{F_z \cdot R_k^2}{J_k \cdot v_{kx}} \cdot A \cdot s_{x0}^2 + \frac{R_k}{J_k \cdot v_{kx}} \cdot M_R(t) \cdot s_{x0} - \frac{F_z \cdot R_k^2}{J_k \cdot v_{kx}} \cdot B \cdot s_{x0} < \frac{s_{xd}^2 - s_{x0}^2}{2T},$$

and hence

$$M_R(t) < \frac{s_{xd}^2 - s_{x0}^2}{2T \cdot s_{x0}} \cdot \frac{J_k \cdot v_{kx}}{R_k} + F_z \cdot R_k \cdot (A \cdot s_{x0} + B) .$$

Two subcases are further studied.
1) $B > 0\,(K > 0 \;\Rightarrow\; A > 0)$
In this case the technical stability conditions reads

$$M_R(t) < \frac{s_{xd}^2 - s_{x0}^2}{2T \cdot s_{x0}} \cdot \frac{J_k \cdot v_{kx}}{R_k} + F_z \cdot R_k \cdot (|A| \cdot s_{x0} + |B|)$$

for

$$M_R(t) < +F_z \cdot R_k \cdot |B| .$$

However, only one inequality is sufficient, namely

$$M_R(t) < +F_z \cdot R_k \cdot |B| . \tag{9.15}$$

2) $B < 0\,(K > 0 \;\Rightarrow\; A > 0)$
In this case the technical stability conditions read

$$M_R(t) < \frac{s_{xd}^2 - s_{x0}^2}{2T \cdot s_{x0}} \cdot \frac{J_k \cdot v_{kx}}{R_k} + F_z \cdot R_k \cdot (|A| \cdot s_{x0} - |B|)$$

for

$$M_R(t) < -F_z \cdot R_k \cdot |B| ,$$

but only the second inequality can be taken, namely

$$M_R(t) < -F_z \cdot R_k \cdot |B| . \tag{9.16}$$

Case 3. The function $\mu_x(s_x)$ is decreased: $A < 0$, $K < 0$, $R(t) > 0$.
The maximum of the function $\dot{V}_B$ in the space Ω/ω is

$$\inf_{s_x \in \Omega/\omega} \left[\frac{dV_B}{dt} \right] = -K \cdot s_{xd}^2 + R(t) \cdot s_{xd},$$

and the condition (9.12) reads

$$-K \cdot s_{xd}^2 + R(t) \cdot s_{xd} < \frac{s_{xd}^2 - s_{x0}^2}{2T} .$$

Owing to (9.10) one gets

$$-\frac{F_z \cdot R_k^2}{J_k \cdot v_{kx}} \cdot A \cdot s_{xd}^2 + \frac{R_k}{J_k \cdot v_{kx}} \cdot M_R(t) \cdot s_{xd} - \frac{F_z \cdot R_k^2}{J_k \cdot v_{kx}} \cdot B \cdot s_{xd} < \frac{s_{xd}^2 - s_{x0}^2}{2T} ,$$

and hence

$$M_R(t) < \frac{s_{xd}^2 - s_{x0}^2}{2T \cdot s_{xd}} \cdot \frac{J_k \cdot v_{kx}}{R_k} + F_z \cdot R_k \cdot (A \cdot s_{xd} + B).$$

Two subcases are further considered.

1) $B > 0\,(K > 0 \Rightarrow A > 0)$

In this case the technical stability conditions read

$$M_R(t) < \frac{s_{xd}^2 - s_{x0}^2}{2T \cdot s_{xd}} \cdot \frac{J_k \cdot v_{kx}}{R_k} + F_z \cdot R_k \cdot (-|A| \cdot s_{xd} + |B|)$$

for

$$M_R(t) > +|B|. \tag{9.17}$$

2) $B < 0\,(K > 0 \Rightarrow A > 0)$

In this case the technical stability condition has the following form

$$M_R(t) < \frac{s_{xd}^2 - s_{x0}^2}{2T \cdot s_{xd}} \cdot \frac{J_k \cdot v_{kx}}{R_k} + F_z \cdot R_k \cdot (-|A| \cdot s_{xd} - |B|)$$

for

$$M_R(t) > -|B|. \tag{9.18}$$

Case 4. The function $\mu_x(s_x)$ is decreased: $A < 0$, $K < 0$, $R(t) < 0$.

The maximum of the function $\dot{V}_B$ in the space Ω/ω is

$$\inf_{s_x \in \Omega/\omega} \left[\frac{dV_B}{dt} \right] = -K \cdot s_{xd}^2 + R(t) \cdot s_{xd},$$

and the condition (9.12) has the form

$$-K \cdot s_{xd}^2 + R(t) \cdot s_{xd} < \frac{s_{xd}^2 - s_{x0}^2}{2T}.$$

Owing to (9.10) one obtains

$$-\frac{F_z \cdot R_k^2}{J_k \cdot v_{kx}} \cdot A \cdot s_{xd}^2 + \frac{R_k}{J_k \cdot v_{kx}} \cdot M_R(t) \cdot s_{xd} - \frac{F_z \cdot R_k^2}{J_k \cdot v_{kx}} \cdot B \cdot s_{xd} < \frac{s_{xd}^2 - s_{x0}^2}{2T}.$$

and hence

$$M_R(t) < \frac{s_{xd}^2 - s_{x0}^2}{2T \cdot s_{xd}} \cdot \frac{J_k \cdot v_{kx}}{R_k} + F_z \cdot R_k \cdot (A \cdot s_{xd} + B).$$

Two subcases are further distinguished.

1) $B > 0\,(K > 0 \Rightarrow A > 0)$

In this case the technical stability condition

$$M_R(t) < \frac{s_{xd}^2 - s_{x0}^2}{2T \cdot s_{xd}} \cdot \frac{J_k \cdot v_{kx}}{R_k} + F_z \cdot R_k \cdot (-|A| \cdot s_{xd} + |B|)$$

for

$$M_R(t) < +|B|\,. \tag{9.19}$$

2) $B < 0\,(K > 0 \;\Rightarrow\; A > 0)$
In this case the technical stability conditions has the form

$$M_R(t) < \frac{s_{xd}^2 - s_{x0}^2}{2T \cdot s_{xd}} \cdot \frac{J_k \cdot v_{kx}}{R_k} + F_z \cdot R_k \cdot (-|A| \cdot s_{xd} - |B|)$$

for

$$M_R(t) < -|B|\,. \tag{9.20}$$

EXAMPLE 9.1 *For a tire characteristics $\mu_x(s_x)$ presented in Figure 9.3 a linearization is introduced by dividing the circumferential slip s_x into 20 equal intervals. The linearizing coefficients A and B are found from the relation* (9.6).

A stability investigation is carried out for the following data: the dynamical radius assumed to be equal with the kinematical radius of the road wheel $R_k = 0.283\,m$; the inertial mass moment $K_k = 2\,kg{\cdot}m^2$; the linear road wheel velocity component $v_{xk} = 20\,m/s$; the passive road pressure $F_z = 2500\,N$.

The admissible (owing to stability conditions) road wheel rotational moment $M_R(t)$, which can be generated by the controller *(see Figure 9.1), is defined via the earlier derived relations for different values of the coefficients A, B and K.*

The computational results displaying the function $M_{Rm}(s_x)$ are shown in Figure 9.4. Owing to the obtained results one can choose an appropriate value of the rotational moment acting on the braking mechanism wile the road wheel stability motion is preserved.

2. The Driven Road Wheel Rotational Motion

2.1 Experimental Rig

The following driven system is analysed: a road vehicle-tired road wheel-road surface. This system is reduced to analysis of a single road wheel, where the vehicle dynamics is modelled via the inertial drum shown in Figure 9.5.

It is assumed that the road wheel is driven via the driven engine through the moment $M_n(t)$. The driven process is controlled via a controller, i.e. a device of the ASR type, where: $M_{REG}(t) = M_n(t) - M_H(t)$, and $M_H(t)$ is the rotational moment generated by ASR with a use of, for instance, the road wheel braking mechanism. The road wheel tire is assumed to be the Magnum model (see Section 3). In addition, a generator simulating various vehicle motion resistance is attached to the inertial drum.

The chosen action algorithm of ASR accounts for a delay (occurring in real systems) between the generated controlling moment $M_H(t)$ and controlling

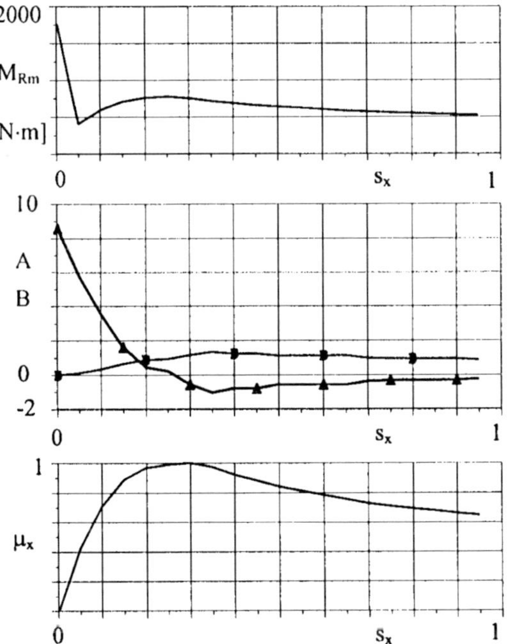

Figure 9.4. The wheel tire characteristics. The admissible value of the road wheel braking rotational moment $M_R = M_{Rm}(s_x)$.

signal. Two controlling quantities are used in the algorithm, i.e. the relative slip s_x and its derivative $\dot{s}_x$.

Algorithm of ASR

$$\text{if } s_x(t-\tau) > 0.32 \text{ and } \frac{ds_x(t-\tau)}{dt} > 0 \text{ then } \frac{dM_H}{dt} = \frac{M_{h\,max} - M_h}{S_{T\,w}}$$

$$\text{if } s_x(t-\tau) < 0.28 \text{ and } \frac{ds_x(t-\tau)}{dt} \le 0 \text{ then } \frac{dM_H}{dt} = -\frac{M_h}{S_{T\,s}}$$

$$j = \mathrm{INT}\left(\frac{\Delta t}{\tau}\right); \quad s_x(j) = s_x; \quad \text{if } v_k > 0 \text{ then } s_x = 1 - \frac{v}{v_k};$$

$$\text{for } i = 1 \text{ to } j-1 \quad s_x(i) = s_x(i+1) \quad \text{next } i;$$

$$s_x(t-\tau) = s_x(2) \quad \frac{ds_x(t-\tau)}{dt} = \frac{s_x(2) - s_x(1)}{\Delta t}$$

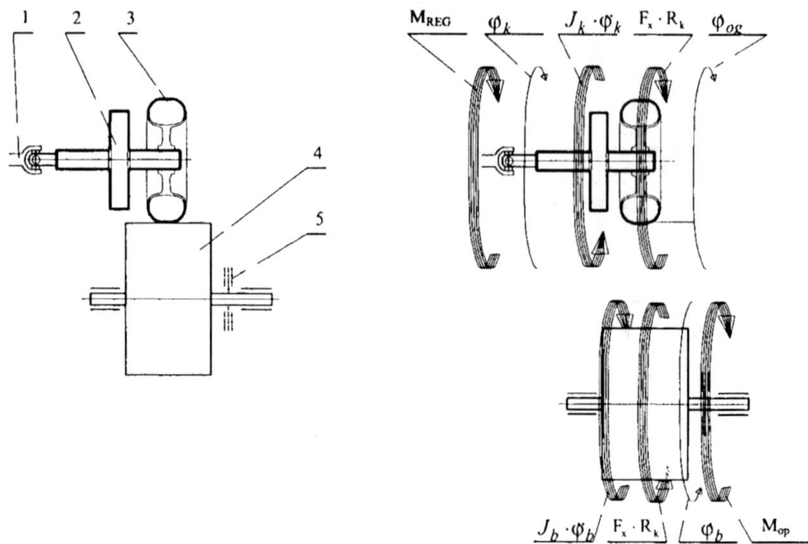

Figure 9.5. Model of the system: road wheel-road (1 - drive with the rotational moment controller $M_{REG} = M_n - M_H$; 2 - reduced body of the elements attached to the rotational axle of the road wheel having the inertial mass moment J_k; 3 - driven road wheel; 4 - road; 5 - simulator of the vehicle motion resistance; J_b - inertial mass moment of the drum of the experimental rig).

The following notation is used in the algorithm: τ - time delay; Δt - discretization step; j - number of delayed operations depending on the time step and the assumed delay value.

The simulation results of the driving system (Figure 9.5) are shown in Figures 9.6–9.8 for the data: $M_{Hm} = 1500\, N \cdot m$, $J_{k1} = 35\, N \cdot m^2$, $J_{k2} = 10\, M \cdot m^2$, $C_{og} = 16000\, N \cdot m \cdot s/rad$, $K_{og} = 100\, N \cdot m/rad$, $R_k = 0.5\, m$, $F_h = 1800\, N$, $F_c = 1200\, N$, $bet = 5$, $A_v = 20\, N \cdot s/m$, $m = 3140\, kg$, $f_t = 0.02\, N/N$.

The driving process and the starting vehicle process in condition of the excessive driving moment simulations are reported in Figure 9.6. Since the vehicle driving system does not have an ASR controller, the road wheel starts to rotate owing to the applied excessive driving moment. After some time of this process a circumferential friction force F_x decrease occurs, and the traced dynamics is improper one.

The starting vehicle process with the drive controller ASR is shown in Figure 9.7. The driven moment with a controlled value is applied to the road wheel. Note that now the wheel undergoes only a bounded circumferential slip. The applied ASR is characterized by a small time delay of $\tau = 0.02\, s$, and this dynamics is the proper one.

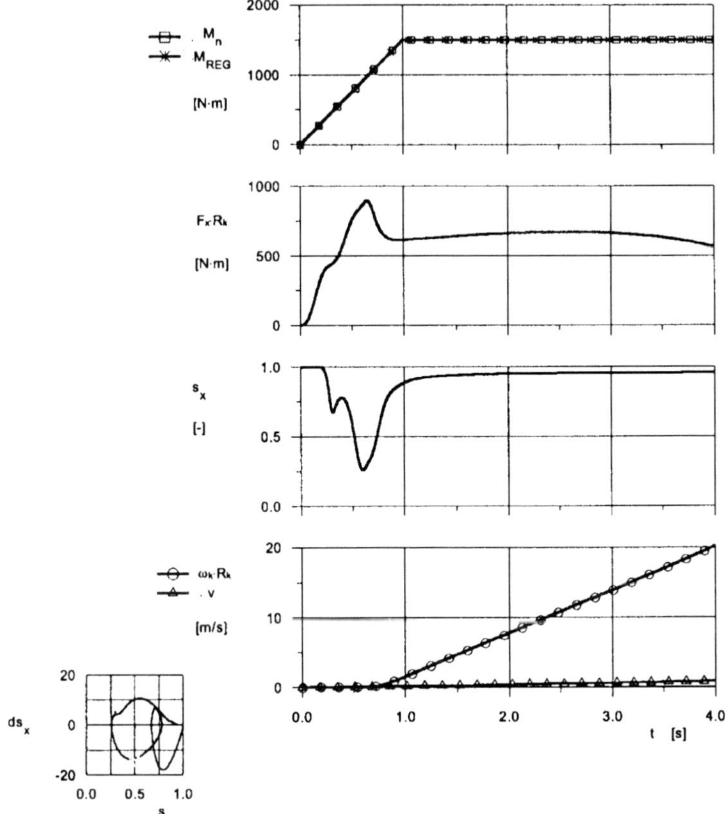

Figure 9.6. Simulation of the vehicle starting process ($M_n(t)$ - driving moment; $M_{REG}(t)$ - rotational torque generated by ASR; F_x - circumferential friction (adhesion) force; ω_k - road wheel rotational velocity; v - vehicle velocity; s_x - circumferential relative slip).

Finally, a similar simulation as in the latter case is displayed in Figure 9.8, but now the relatively large delay is applied ($\tau = 0.10\,s$). One may observe that a large delay is not suitable and the road wheel moves with a large circumferential slip s_x.

2.2 The Driven Road Wheel Rotational Motion (Two Axle Vehicle)

Following the steps of the previous section, the driven road wheel rotational motion in the system road wheel-two axle vehicle is studied.

The external forces acting on the vehicle body and the road wheels are shown in Figure 9.9. Applying the d'Alembert rule the following differential

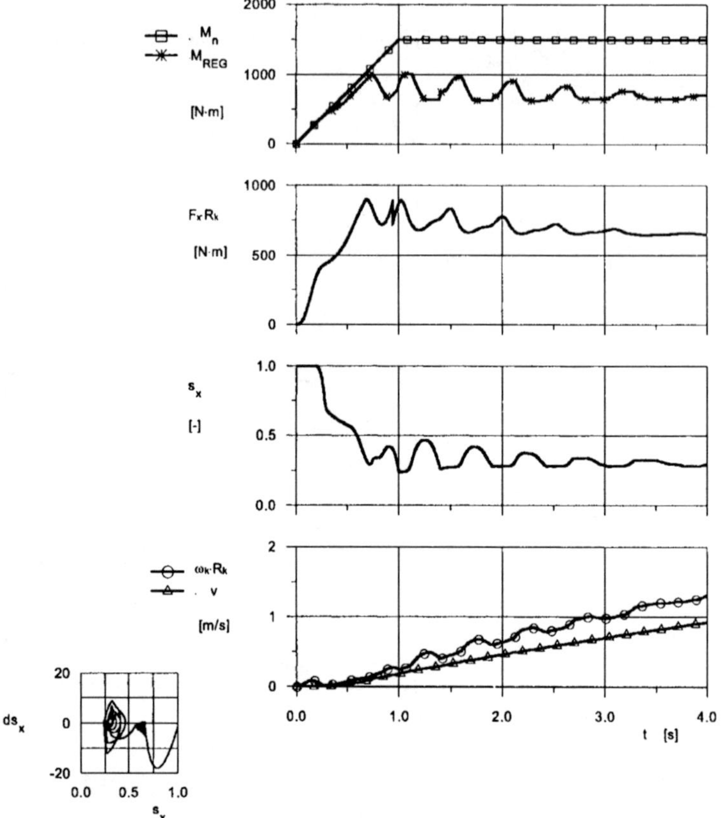

Figure 9.7. Same as in Figure 9.6, but ASR is not applied, and $\tau = 0.02\,s$.

equations are obtained

$$m \cdot \frac{dv}{dt} = F_{xf} + F_{xr}, \tag{9.21}$$

$$m \cdot g = F_{zf} + F_{zr}, \tag{9.22}$$

$$m \cdot g \cdot a + m \cdot \frac{dv}{dt} \cdot a + J_{kf} \cdot \ddot{\varphi}_{kf} + J_{kr} \cdot \ddot{\varphi}_{kr} = F_{zf} \cdot L, \tag{9.23}$$

where: $F_{xf,r}$ is the circumferential force generated on the tire-road stick; $J_{kf,r}$ - inertial mass moment of the front (end) road wheels; $\ddot{\varphi}_{kf,r}$ - rotational velocity of the front (end) wheel.

The differential equations governing road wheels dynamics read

$$J_{kf} \cdot \ddot{\varphi}_{kf} = M_{nf} + F_{xf} \cdot R_k, \tag{9.24}$$

$$J_{kr} \cdot \ddot{\varphi}_{kr} = M_{nr} + F_{xr} \cdot R_k, \tag{9.25}$$

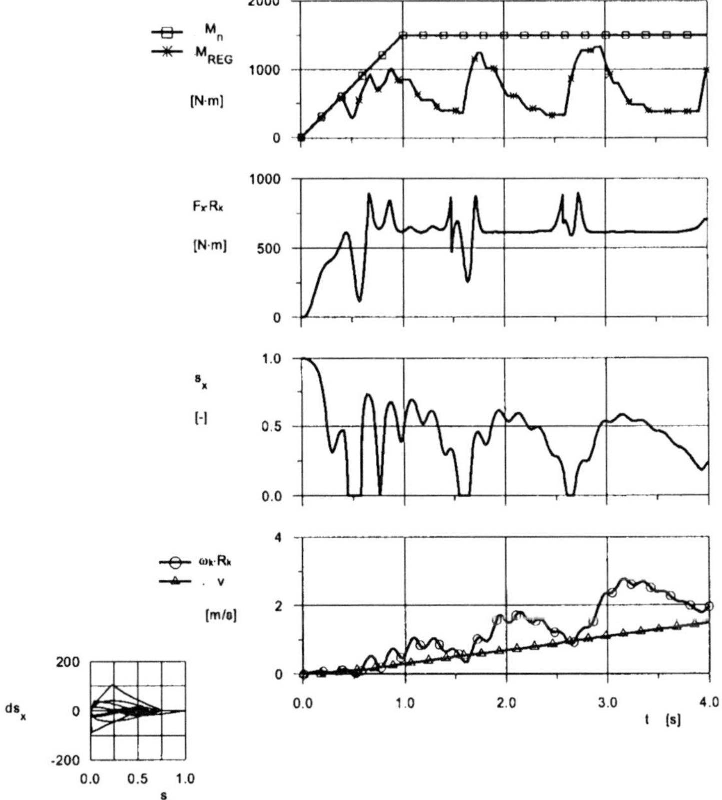

Figure 9.8. Same as in Figure 9.7 ($\tau = 0.10\ s$).

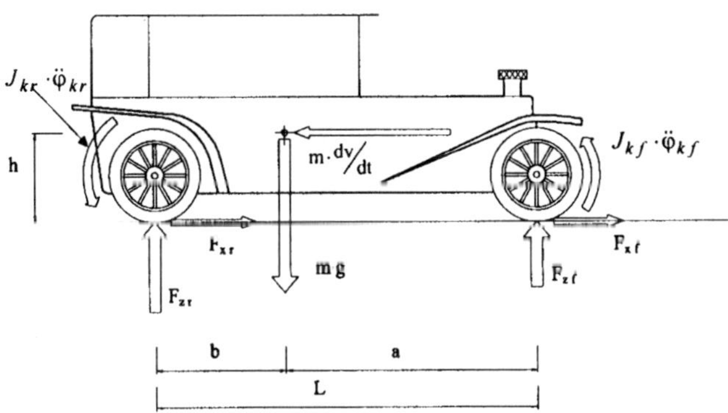

Figure 9.9. External forces acting on the driven vehicle.

whereas the relative circumferential slip of the driven road wheels is defined via the relations

$$s_{x_f} = 1 - \frac{v_{kxf}}{\omega_{kf} \cdot R_k}, \tag{9.26}$$

$$s_{x_r} = 1 - \frac{v_{kxr}}{\omega_{kr} \cdot R_k}. \tag{9.27}$$

The relations (9.21)–(9.27) govern the vehicle dynamics with the driven road wheels.

EXAMPLE 9.2 *The model introduced in this subsection is analysed with the following additional assumptions:*

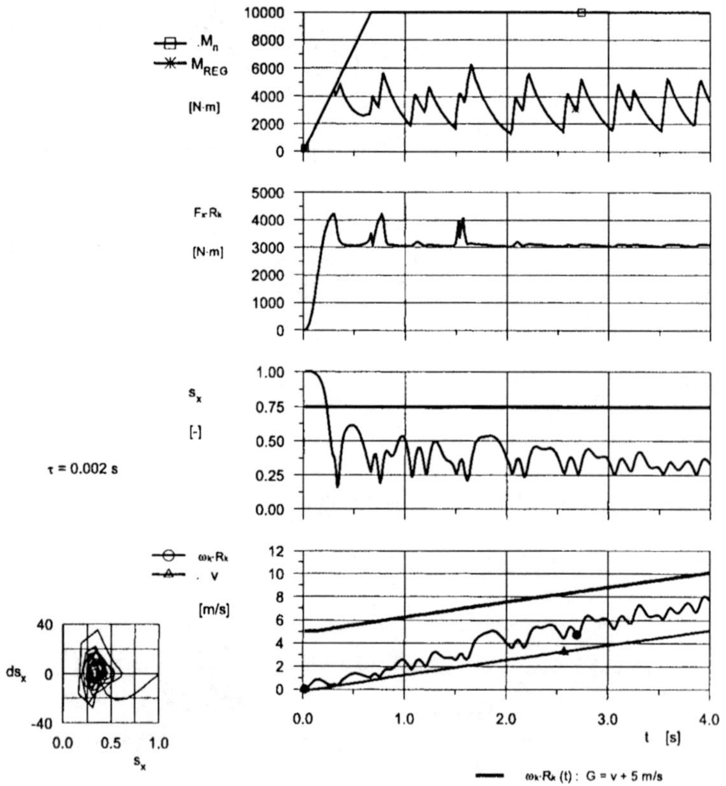

Figure 9.10. Simulation of the vehicle drive during start ($\tau = 0.002\ s$).

- *only end road wheels are driven;*
- *vehicle motion is driven via the drive torque $M_n(t)$ applied to the driven wheels;*

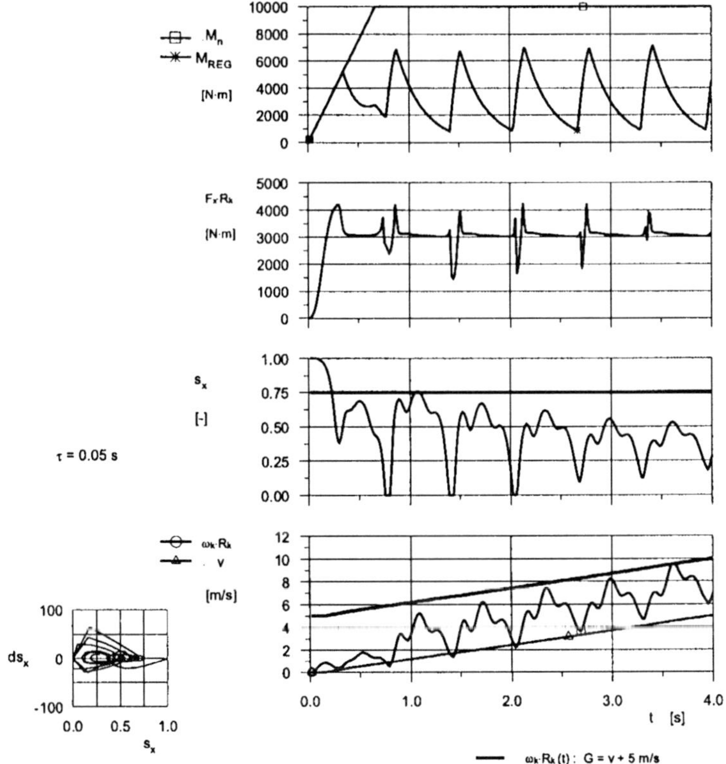

Figure 9.11. Simulation of the vehicle drive during start ($\tau = 0.05\, s$).

- *ASR controller with algorithm described in Section 2.1 is used and τ is delay;*
- *'Magnum' tire model is applied (see Section 3.2);*
- *circumferential tire stiffness $C_{og} = 16000\, N \cdot m \cdot s/rad$;*
- *circumferential tire damping $K_{og} = 100\, N \cdot m/rad$;*
- *tire energy dissipation coefficient $V_e = 0$;*
- *tire adhesion coefficient $\beta = 5$;*
- *tire adhesion coefficient $A_\eta = 20\, N \cdot s/m$;*
- *mass inertial moment of the road wheel and attached rotating elements $J_{k1} = 35\, N \cdot m^2$;*
- *mass inertial moment of the tire tread belt $J_{k2} = 10\, N \cdot m^2$;*

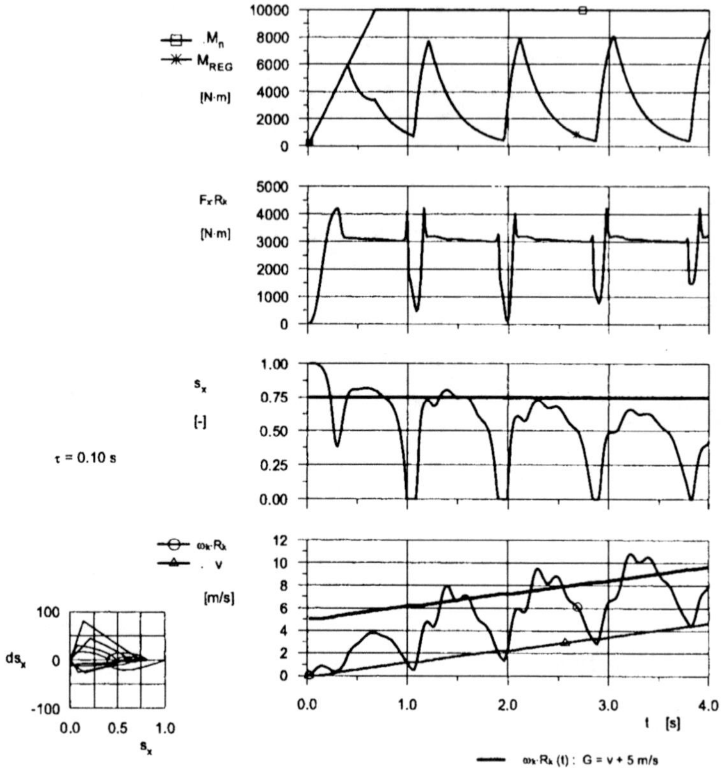

Figure 9.12. Simulation of the vehicle drive during start ($\tau = 0.1\,s$).

- *Coulomb friction coefficient* $\mu_{xC} = 0.18$;
- *adhesion friction coefficient* $\mu_{xH} = 0.24$;
- *friction forces:* $Fx_C = \mu x_C \cdot F_z$, $Fx_H = \mu_{xH} \cdot F_z$;
- *maximal friction torque of the braking mechanism ASR* $M_{Hm} = 10000\,N \cdot m$;
- *physical and geometrical vehicle parameters:* $R_k = 0.5\,m$, $a = 1.5\,m$, $b = 2.5\,m$, $h = 1.2\,m$, $m = 8373\,kg$.

The simulation results of the starting process of the vehicles with the assumptions for different delays mentioned in the above are reported in Figures 9.10–9.12 for the different delays of $\tau = 0.002\,s; 0.05\,s; 0.1\,s$, *respectively.*

All reported results indicate that all solutions are stable in the Lyapunov sense, and that they are similar both qualitatively and quantitatively. This phenomenon can be explained by the use of the brake as the execution unit of the controller.

Estimating the technical stability and defining the admissible solutions zone, it is assumed that road wheel slip velocity should not exceed the value of $5\,m/s$. *In Figures 9.10–9.12 the mentioned value is denoted by* $G = v + 5$. *Only the simulation results shown in Figures 9.10, 9.11 satisfy the introduced technical stability conditions. For* $\tau \leq 0.1s$ *(Figure 9.12) the obtained runs do not preserve the assumed technical stability conditions, since the ASR control introduces relatively large values of the wheel rotational velocity amplitudes.*

Chapter 10

MODELING OF A PISTON – CONNECTING ROD – CRANKSHAFT SYSTEM

In this chapter the inverted triple pendulum is proposed to model a real piston-connecting rod-crankshaft system of a mono-cylinder combination engine. After a short introduction, the model of rigid multi-body mechanical system with unilateral frictionless constraints is proposed. Then the generalized impact law is described. Sliding states along some obstacles are discussed and the computational model is introduced. Dynamics of a triple physical pendulum with barriers is illustrated and discussed in section 10.6. The piston-connecting rod-crankshaft system is next analysed, and the given numerical examples show a good agreement with a real mono-cylinder combustion engine behaviour.

1. Introduction

As known, even a single harmonically or parametrically excited pendulum may exhibit a rich spectrum of non-linear phenomena including various local and global bifurcations, attractors and repellers, stable and unstable manifolds, scenarios leading to chaos and out of chaos, symmetry breaking and crisis bifurcations, steady-state and transitional chaos, oscillatory-rotational attractors, etc (see references [26, 33, 105, 112]).

On the other hand, many real processes, asfor instance earth-quake caused vibrations of high buildings, can be modeled via coupled pendulums. It is clear that a coupled system of pendulums may exhibit more complex non-linear dynamics, and hence it attracts the attention of mathematicians, physicists and engineers, who show a particular interest in the examination or control of models of such systems. In addition, it may be expected that many theoretical unsolved problems of non-linear dynamics can be explained using a model of rigid multi-body coupled pendulums.

It turns out that although many technological and design-oriented details have been neglected, the inverted triple pendulum can be used to model a real piston – connecting rod – crankshaft system of a mono-cylinder combustion engine.

Modeling of multi-body systems with activity states of constraints varying during dynamics is addressed in references [30, 87, 112]. On the other hand, the problem of stiff impacts in multi-body systems is illustrated and discussed in the monographs [39, 27].

2. The Model of a Rigid Multi–Body Mechanical System With Unilateral Frictionless Constraints

The concept of modeling of the rigid multi-body mechanical system with unilateral constraints, when the activity state of a particular constraint may change during the system evolution, used in the present study, follows the references [25, 30, 87, 112].

Let us consider a Lagrange rigid multi-body mechanical system of n-degrees-of-freedom represented by the vector of generalized coordinates

$$\mathbf{q}(t) = [q_1(t), ..., q_n(t)]^T$$

and subjected to the m unilateral constraints representing rigid frictionless obstacles imposed on the system position. If none of the unilateral constraints is active, the system dynamics is governed by the following set of second-order differential equations:

$$\mathbf{M}(\mathbf{q}, t)\ddot{\mathbf{q}}(t) = \mathbf{f}(\mathbf{q}, \dot{\mathbf{q}}, t) \ , \qquad (10.1)$$

where: $\mathbf{M}(\mathbf{q}, t)$ is the symmetric $n \times n$ inertia matrix, $\mathbf{f}(\mathbf{q}, \ddot{\mathbf{q}}, t)$ is the $n \times 1$ vector containing gyroscopic, damping, potential and exciting forces, $\dot{q}(t)$ and $\ddot{q}(t)$ are the $n \times 1$ vectors of generalized velocities and accelerations, respectively, and t is the independent time variable.

The rigid unilateral constraints are defined by the following set of algebraic inequalities:

$$\mathbf{h}(\mathbf{q}, t) \geq \mathbf{0} \ , \qquad (10.2)$$

where: $\mathbf{h}(\mathbf{q}, t) = [h_1(\mathbf{q}, t), ..., h_m(\mathbf{q}, t)]$ is the vector of m smooth scalar functions, and each of them should have a direct physical sense of the normal distance from the appropriate physical obstacle.

In order to describe the state of each potential constraint, the following index sets are introduced:

$$I = \{1, 2, \ldots, m\} \ ,$$

$$I_I = \left\{ i \in I; \left(h_i = 0 \wedge \dot{h}_i < 0 \right) \right\} ,$$

$$I_S = \left\{ i \in I; \left(h_i = 0 \wedge \dot{h}_i = 0 \wedge \lambda_i \geqslant 0 \right) \right\}. \tag{10.3}$$

The set I consists of m indexes of all potentially active constraints. The elements of the set I_I are represented by m_I indexes of the unilateral constraints with vanishing normal distance and negative relative velocity in the normal direction (describing the system just before an impact with respect to these constraints). In the set I_S there are m_S indexes of the constraints with vanishing normal distance and vanishing relative velocity in the normal direction. The system is in the state of sliding on these obstacles, acting on the system via non-negative normal reactions represented by non-negative Lagrange multipliers λ_i ($i \in I_S$) that assure a continuous contact between bodies and the active obstacles ($\ddot{h}_i = 0$, $i \in I_S$). The Lagrange multiplier λ_i represents a generalized normal force acting along the generalized coordinate h_i, where h_i is a real physical distance from the i-th obstacle.

3. Generalized Impact Law

Since the investigated system is a Lagrangian one, it can be represented by a point moving in its configuration space $\mathbf{q}$ (see reference [39] for more details). The unilateral constraints define domains of this space, and the point representing the system strikes the boundaries of these domains. At the time instants $t = t_k$ of these "generalized" collisions, the system velocities undergo jumps, i.e. the vector of generalized velocities just before an impact $\dot{\mathbf{q}}(t_k^-) = \dot{\mathbf{q}}^-$ is transformed to the vector $\dot{\mathbf{q}}(\mathbf{t}_k^+) = \dot{\mathbf{q}}^+$. It is assumed that at time $t = t_k$, the point that represents the configuration of the system strikes only one of the smooth surfaces $h_i(\mathbf{q}, t) = 0$. In other words, the set I_I consists of $m_I = 1$ element i in the instant $t = t_k^-$. Therefore, the problem of multiple impacts is omitted in our considerations.

Now the method of calculating the post-impact velocities for a system with many degrees-of-freedom in the case of a single impact will be outlined briefly in accordance with the description given in [39].

During an impact, the following shock dynamics equation is valid:

$$\mathbf{M}(\mathbf{q}, t) \left(\dot{\mathbf{q}}^+ - \dot{\mathbf{q}}^- \right) = \mathbf{p}_\mathbf{q}, \tag{10.4}$$

where: $\mathbf{p}_\mathbf{q}$ is the generalized percussion vector for coordinates $\mathbf{q}$ (the force impulse vector generated by the impact).

For frictionless constraints, $\mathbf{p}_\mathbf{q}$ occurs along $\nabla_\mathbf{q} h_i(\mathbf{q}, t)$ (because the interaction force due to the impact occurs along a Euclidean normal to the surface $h_i(\mathbf{q}, t) = 0$, which is yielded by the virtual work principle). Now, from (10.4) the following $n - 1$ algebraic equations versus n unknowns can be obtained:

$$\mathbf{t}_{\mathbf{q},i,j}^T \mathbf{M}(\mathbf{q}, t) \left(\dot{\mathbf{q}}^+ - \dot{\mathbf{q}}^- \right) = 0, \qquad j = 1, \ldots, n - 1\,. \tag{10.5}$$

In the above, $\mathbf{t}_{\mathbf{q},i,j}$ are vectors tangent to the surface $h_i(\mathbf{q},t)=0$, chosen to be mutually independent, i.e. $\mathbf{t}^T_{\mathbf{q},i,j}\nabla_{\mathbf{q}}h_i(\mathbf{q},t)=0$ $(j=1,...,n-1)$ and $\mathbf{t}^T_{\mathbf{q},i,j}\mathbf{t}_{\mathbf{q},i,k}=0$ $(j,k=1,...,n-1)$.

The last lacking equation represents the extended Newton (restitution coefficient) rule applied to the relative velocity in the normal direction to the constraint

$$\dot{h}_i^+=-e_i\dot{h}_i^-, \tag{10.6}$$

where: e_i is the restitution coefficient attached to the i-th obstacle.

From (10.6) we obtain

$$\frac{\partial h_i}{\partial \mathbf{q}^T}\dot{\mathbf{q}}^+ + \frac{\partial h_i}{\partial t} = -e_i\left(\frac{\partial h_i}{\partial \mathbf{q}^T}\dot{\mathbf{q}}^- + \frac{\partial h_i}{\partial t}\right). \tag{10.7}$$

Owing to (10.5), it is seen that there is (in general) a discontinuity in the tangential velocity due to the inertia coupling.

A change of the kinetic energy at impact is defined by the formula

$$\Delta T(t_k)=\frac{1}{2}\left(\dot{\mathbf{q}}^+\right)^T\mathbf{M}(\mathbf{q},t)\dot{\mathbf{q}}^+ - \frac{1}{2}\left(\dot{\mathbf{q}}^-\right)^T\mathbf{M}(\mathbf{q},t)\dot{\mathbf{q}}^-. \tag{10.8}$$

From (10.8), using (10.5) and (10.7) and due to of the assumption that the constraint is stationary ($\partial h_i/\partial t=0$), one arrives at

$$\Delta T(t_k)=\frac{1}{2}\left(e_i^2-1\right)\left(\frac{\mathbf{M}(\mathbf{q},t)^{-1}\nabla_{\mathbf{q}}h_i}{\sqrt{(\nabla_{\mathbf{q}}h_i)^T\mathbf{M}(\mathbf{q},t)^{-1}\nabla_{\mathbf{q}}h_i}}\mathbf{M}(\mathbf{q},t)\dot{\mathbf{q}}^-\right)^2. \tag{10.9}$$

Observe that for $0\le e_i\le 1$, the kinetic energy change due to impact is $\Delta T(t_k)\le 0$, for $e_i=1$, it is $\Delta T(t_k)=0$, and for $e_i=0$, $\Delta T(t_k)$ reaches its lowest possible value.

To summarize, when the system achieves a discontinuity point at the time instant $t=t_k$ indicated by $h_i(\mathbf{q}(t_k),t_k)=0$, the vector of velocities is transformed via the relation

$$\dot{\mathbf{q}}\left(t_k^+\right)=\mathbf{g}_i\left(\mathbf{q}(t_k),\dot{\mathbf{q}}\left(t_k^-\right),t_k\right), \tag{10.10}$$

where the vector function $\mathbf{g}_i(\mathbf{q},\dot{\mathbf{q}},t)$ is derived from (10.5) and (10.7), and has the form

$$\mathbf{g}_i(\mathbf{q},\dot{\mathbf{q}},t)=\begin{bmatrix}(\nabla_{\mathbf{q}}h_i(\mathbf{q},t))^T\\ \begin{bmatrix}\mathbf{t}^T_{\mathbf{q},i,1}\\ \cdots\\ \mathbf{t}^T_{\mathbf{q},i,n-1}\end{bmatrix}\cdot\mathbf{M}(\mathbf{q},t)\end{bmatrix}^{-1}$$

$$\cdot \left(\left[\begin{array}{c} -e_i \left(\nabla_{\mathbf{q}} h_i (\mathbf{q}, t)\right)^T \\ \left[\begin{array}{c} \mathbf{t}_{\mathbf{q},i,1}^T \\ \cdots \\ \mathbf{t}_{\mathbf{q},i,n-1}^T \end{array} \right] \cdot \mathbf{M}(\mathbf{q}, t) \end{array} \right] \dot{\mathbf{q}} + \left\{ \begin{array}{c} -(e_i + 1) \frac{\partial h_i(\mathbf{q},t)}{\partial t} \\ 0 \\ \cdots \\ 0 \end{array} \right\} \right). \quad (10.11)$$

4. Sliding States Along Some Obstacles

Let us assume that the set I_I is empty ($m_I = 0$) and the set of indexes of continuously active constraints has the form

$$I_S = \{i_1, i_2, ..., i_{m_S}\} . \quad (10.12)$$

In what follows, the $m_S \times 1$ vector corresponding to notation (10.12) is defined as

$$\mathbf{h}_S (\mathbf{q}, t) = \left\{ \begin{array}{c} h_{i_1} (\mathbf{q}, t) \\ h_{i_2} (\mathbf{q}, t) \\ \cdots \\ h_{i_{m_S}} (\mathbf{q}, t) \end{array} \right\}, \quad (10.13)$$

containing all functions $h_i(\mathbf{q}, t)$ that represent continuously active obstacles.

The vector of Lagrange multipliers

$$\lambda_S = \left\{ \begin{array}{c} \lambda_{i_1} \\ \lambda_{i_2} \\ \cdots \\ \lambda_{i_{m_S}} \end{array} \right\} \quad (10.14)$$

is attached to vector (10.13), where $\lambda_{i_k} \geq 0$ for $k = 1, 2, ..., m_S$.

In order to describe the system dynamics, normal reactions generated by active constraints acting on the system are introduced to eqs. (10.1) via Lagrange multipliers

$$\mathbf{M}\ddot{\mathbf{q}} = \mathbf{f} + \left(\frac{\partial \mathbf{h}_S}{\partial \mathbf{q}^T} \right)^T \lambda_S . \quad (10.15)$$

The continuous contact between the active constraint surfaces and the bodies of the system implies the condition

$$\ddot{\mathbf{h}}_S = \frac{\partial \mathbf{h}_S}{\partial \mathbf{q}^T} \ddot{\mathbf{q}} + \dot{\mathbf{q}}^T \frac{\partial^2 \mathbf{h}_S}{\partial \mathbf{q} \partial \mathbf{q}^T} \dot{\mathbf{q}} + 2 \frac{\partial^2 \mathbf{h}_S}{\partial t \partial \mathbf{q}^T} \dot{\mathbf{q}} + \frac{\partial^2 \mathbf{h}_S}{\partial t^2} = 0 . \quad (10.16)$$

Equations (10.15) and (10.16) create the following set of differential-algebraic equations governing the system [102]

$$\left[\begin{array}{cc} \mathbf{M} & \mathbf{A}^T \\ \mathbf{A} & \mathbf{0} \end{array} \right] \left\{ \begin{array}{c} \ddot{\mathbf{q}} \\ \lambda_S \end{array} \right\} = \left\{ \begin{array}{c} \mathbf{f} \\ \mathbf{d} \end{array} \right\}, \quad (10.17)$$

where:

$$\mathbf{A} = -\frac{\partial \mathbf{h}_S}{\partial \mathbf{q}^T},$$

$$\mathbf{d} = \dot{\mathbf{q}}^T \frac{\partial^2 \mathbf{h}_S}{\partial \mathbf{q} \partial \mathbf{q}^T} \dot{\mathbf{q}} + 2 \frac{\partial^2 \mathbf{h}_S}{\partial t \partial \mathbf{q}^T} \dot{\mathbf{q}} + \frac{\partial^2 \mathbf{h}_S}{\partial t^2} . \tag{10.18}$$

Introducing the vector

$$\mathbf{U} = \begin{bmatrix} \mathbf{U}_{11} & \mathbf{U}_{12} \\ \mathbf{U}_{21} & \mathbf{U}_{22} \end{bmatrix} = \begin{bmatrix} \mathbf{M} & \mathbf{A}^T \\ \mathbf{A} & \mathbf{0} \end{bmatrix}^{-1}, \tag{10.19}$$

the following set of differential-algebraic equations is obtained:

$$\ddot{\mathbf{q}} = \mathbf{U}_{11}\mathbf{f} + \mathbf{U}_{12}\mathbf{d},$$

$$\lambda_S = \mathbf{U}_{21}\mathbf{f} + \mathbf{U}_{22}\mathbf{d}. \tag{10.20}$$

5. Computational Model

A numerical scheme for calculation of the system response is presented in Figure 10.1. The scheme applies the Runge-Kutta method of fourth order with a nominal time step h_{RK4} in time intervals (t_j, t_{j+1}), where the set I_I is empty (there is no impact) and the set I_S does not change. The discontinuity points are detected with a finite precision by halving the integration step. At those points, appropriate changes of the set I_S members are introduced as well as possible transformations of the velocity vector generated by the impact. The symbols ε_h, $\varepsilon_{\dot{h}}$ and ε_λ denote the obstacle detection accuracy, the detection accuracy of the zero of the normal component of the relative velocity, and the detection accuracy of the zero of the normal force with respect to the barrier surface, respectively.

Note that a multiple impact [30, 39] with more than one obstacle $h_i = 0$ $(i \in I_{I0})$ at the same time instant t_k is recognized as a sequence of single impacts with an arbitrary succession. The latter description is motivated by the observation that the set I_{I0} consists of the indices of the constraints without continuous contact with the bodies of the system at t_k; contact with them is reached in this time or it is lost. These points are detected with a certain finite precision, and it can be assumed that they create a sequence of points with an arbitrarily chosen succession. In the case of an impact with more than one obstacle $h_i = 0$ $(i \in I_{IS})$ we have a strict multiple impact since the constraints I_{IS} are continuously active at the time moment t_k.

6. Triple Physical Pendulum With Barriers

A sketch of the general model of the triple physical pendulum moving in the plane of a global co-ordinate system $\bar{x}, \bar{y}$ (with origin at point O_I) and being

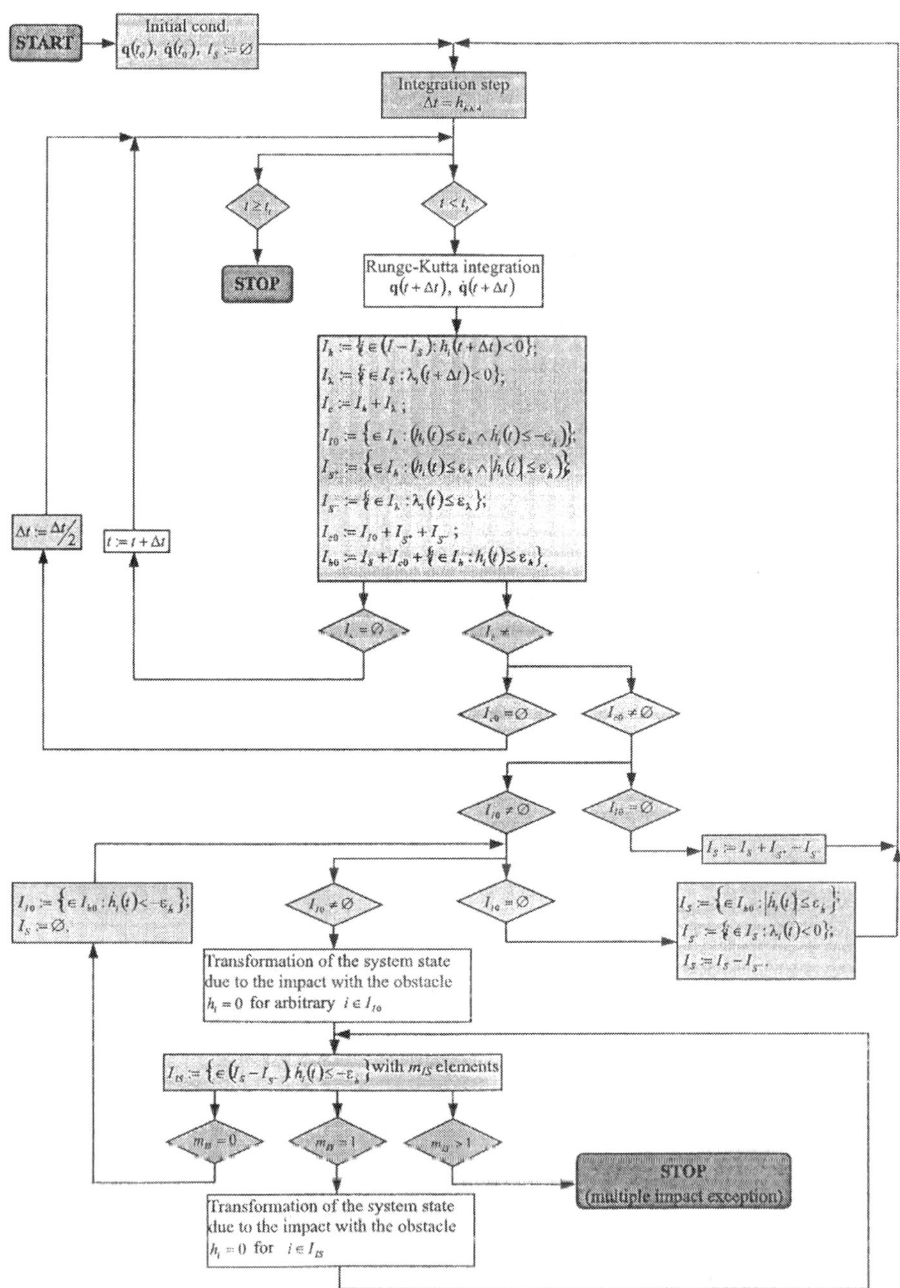

Figure 10.1. Scheme of numerical simulation.

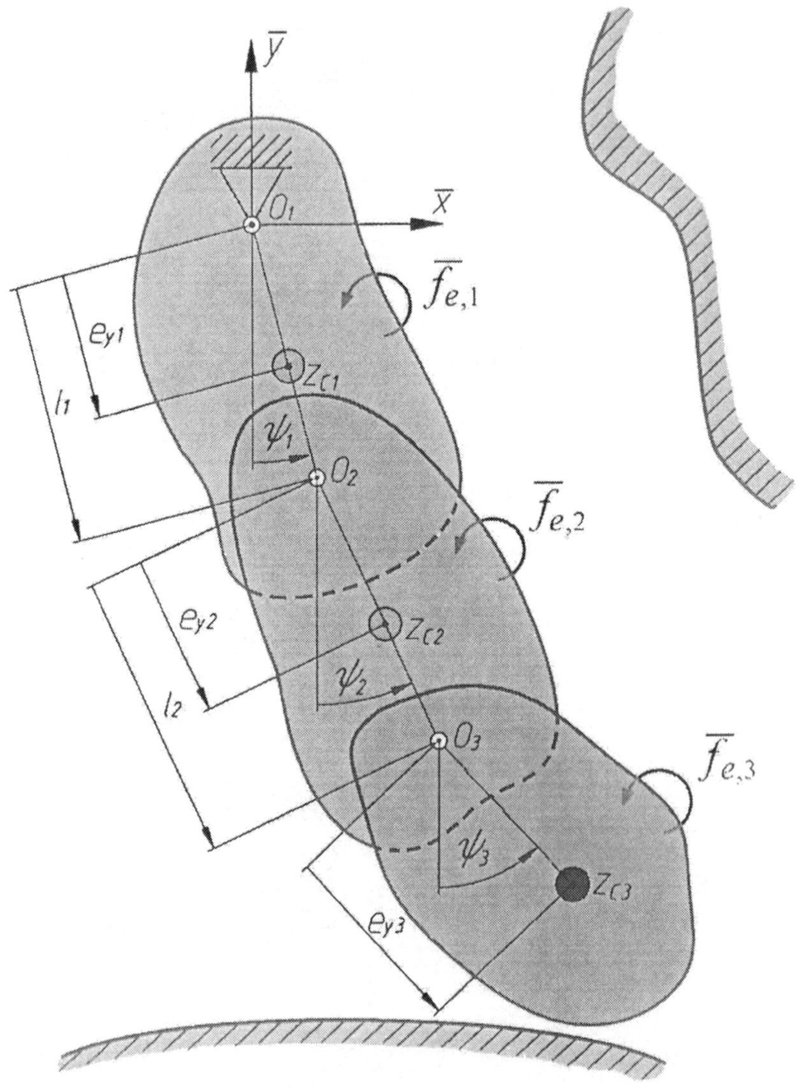

Figure 10.2. Triple physical pendulum with obstacles.

a special case of the model introduced in the previous sections is presented in Fig. 10.2. It is assumed that the links are absolute stiff bodies moving in a vacuum and coupled by viscous damping with coefficients $\bar{c}_i$ $(i = 1, 2, 3)$. Each of the links can be externally forced by $\bar{f}_{e,i}$ $(i = 1, 2, 3)$. It is also assumed that: (i) the mass centers of the links lie on the lines including joints O_i; (ii) one of the principal central inertia axle of each link (z_{Ci}) is perpendicular to a movement plane $\bar{x} - \bar{y}$. The position of the system is described by three angles ψ_i $(i = 1, 2, 3)$.

On the position of the system, the set of arbitrarily situated stiff and frictionless barriers can be imposed. The system is governed by a non-dimensional set of ordinary differential equations without obstacles plus a set of algebraic inequalities (unilateral constraints) representing barriers with the corresponding restitution coefficients (e_i) of the form:

$$\mathbf{M}(\psi)\,\ddot{\psi} + \mathbf{N}(\psi)\,\dot{\psi}^{\mathbf{2}} + \mathbf{C}\dot{\psi} + \mathbf{p}(\psi) = \mathbf{f}_e\left(\psi, \dot{\psi}, t\right), \tag{10.21}$$

$$h_i(\psi) \geqslant 0, \quad i = 1, ..., m, \tag{10.22}$$

where:

$$\psi = \left\{ \begin{array}{c} \psi_1 \\ \psi_2 \\ \psi_3 \end{array} \right\}, \; \ddot{\psi} = \left\{ \begin{array}{c} \ddot{\psi}_1 \\ \ddot{\psi}_2 \\ \ddot{\psi}_3 \end{array} \right\}, \; \dot{\psi}^{\mathbf{2}} = \left\{ \begin{array}{c} \dot{\psi}_1^2 \\ \dot{\psi}_2^2 \\ \dot{\psi}_3^2 \end{array} \right\}, \; \dot{\psi} = \left\{ \begin{array}{c} \dot{\psi}_1 \\ \dot{\psi}_2 \\ \dot{\psi}_3 \end{array} \right\}, \tag{10.23}$$

$$\mathbf{M}(\psi) = \left[\begin{array}{ccc} 1 & \nu_{12}\cos(\psi_1 - \psi_2) & \nu_{13}\cos(\psi_1 - \psi_3) \\ \nu_{12}\cos(\psi_1 - \psi_2) & \beta_2 & \nu_{23}\cos(\psi_2 - \psi_3) \\ \nu_{13}\cos(\psi_1 - \psi_3) & \nu_{23}\cos(\psi_2 - \psi_3) & \beta_3 \end{array} \right],$$

$$\mathbf{N}(\psi) = \left[\begin{array}{ccc} 0 & \nu_{12}\sin(\psi_1 - \psi_2) & \nu_{13}\sin(\psi_1 - \psi_3) \\ -\nu_{12}\sin(\psi_1 - \psi_2) & 0 & \nu_{23}\sin(\psi_2 - \psi_3) \\ -\nu_{13}\sin(\psi_1 - \psi_3) & -\nu_{23}\sin(\psi_2 - \psi_3) & 0 \end{array} \right],$$

$$\mathbf{C} = \left[\begin{array}{ccc} c_1 + c_2 & -c_2 & 0 \\ -c_2 & c_2 + c_3 & -c_3 \\ 0 & -c_3 & c_3 \end{array} \right], \quad \mathbf{p}(\psi) = \left\{ \begin{array}{c} \sin\psi_1 \\ \mu_2 \sin\psi_2 \\ \mu_3 \sin\psi_3 \end{array} \right\},$$

$$\mathbf{f}_e\left(\psi, \dot{\psi}, t\right) = \left\{ \begin{array}{c} f_{e,1}\left(\psi, \dot{\psi}, t\right) \\ f_{e,2}\left(\psi, \dot{\psi}, t\right) \\ f_{e,3}\left(\psi, \dot{\psi}, t\right) \end{array} \right\}. \tag{10.24}$$

The following relations hold between the non-dimensional quantities and the real ones

$$t = \alpha_1 \tau,$$

$$\dot{\psi}_j = \alpha_1^{-1} \overset{|}{\underset{j}{\psi}}, \quad \ddot{\psi}_j = \alpha_1^{-2} \overset{||}{\underset{j}{\psi}}, \quad j = 1, 2, 3, \tag{10.25}$$

$$\alpha_1 = \left(M_1 B_1^{-1}\right)^{\frac{1}{2}}, \tag{10.26}$$

$$\beta_2 = \frac{B_2}{B_1}, \quad \beta_3 = \frac{B_3}{B_1},$$

$$\nu_{12} = \frac{N_{12}}{B_1}, \quad \nu_{13} = \frac{N_{13}}{B_1}, \quad \nu_{23} = \frac{N_{23}}{B_1}, \tag{10.27}$$

$$\mu_2 = \frac{M_2}{M_1}, \quad \mu_3 = \frac{M_3}{M_1},$$

$$c_j = \frac{\bar{c}_j}{\sqrt{M_1 B_1}}, \quad f_{e,j} = \frac{\bar{f}_{e,j}}{M_1}, \quad j = 1, 2, 3\,, \tag{10.28}$$

where symbols $(.\,!\,.)$ and $(.\,\ddot{}\,.)$ denote derivatives with respect to real time t and non-dimensional time τ, respectively, and where the following notation is used:

$$B_1 = J_{z1} + e_{y1}^2 m_1 + l_1^2 (m_2 + m_3)\,,$$
$$B_2 = J_{z2} + e_{y2}^2 m_2 + l_2^2 m_3,$$
$$B_3 = J_{z3} + e_{y3}^2 m_3,$$
$$N_{12} = m_2 e_{y2} l_1 + m_3 l_1 l_2,$$
$$N_{13} = m_3 e_{y3} l_1,$$
$$N_{23} = m_3 e_{y3} l_2,$$
$$M_1 = m_1 g e_{y1} + (m_2 + m_3)\, g l_1,$$
$$M_2 = m_2 g e_{y2} + m_3 g l_2,$$
$$M_3 = m_3 g e_{y3}\,. \tag{10.29}$$

In the above, J_{zi} $(i = 1, 2, 3)$ denote appropriate principal central moments of inertia, m_i $(i = 1, 2, 3)$ denote masses of respective links and g is the gravitational acceleration.

The functions $h_i(\psi)$ represent distances of the system from an appropriate obstacle. More details of the triple physical pendulum model can be found in references [24, 26, 66].

7. Piston – Connecting Rod – Crankshaft System

The general model of the triple physical pendulum with barriers, can be used to build a model of the piston - connecting rod - crankshaft system of the mono-cylinder combustion engine shown in Fig. 10.3. The first link represents the crankshaft (1), the second one is the connecting rod (2) and the third one is the piston (3). The links are connected by rotational joints with viscous damping. The cylinder barrel imposes restrictions on the position of the piston that moves in the cylinder with backlash. It is assumed that in the contact of the surfaces between the piston and the cylinder a tangent force does not appear. Observe that the model of the piston - connecting rod - crankshaft system can

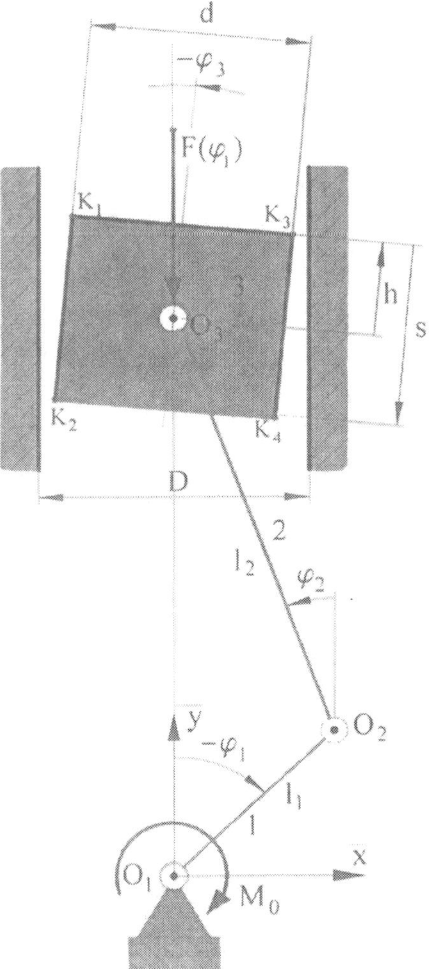

Figure 10.3. Piston – connecting rod – crankshaft system.

be treated as an inverted pendulum, so it is natural to introduce the following vector of angles describing the system position:

$$\phi = [\varphi_1, \varphi_2, \varphi_3]^T = [\psi_1 + \pi, \psi_2 + \pi, \psi_3 + \pi]^T, \tag{10.30}$$

where angles ψ_i $(i = 1, 2, 3)$ are the generalized co-ordinates used in section 2 for the description of the general triple physical pendulum.

The restriction on the position of the piston imposed by the cylinder barrel can be described using four conditions defined by four points K_i $(i = 1, 2, 3, 4)$:

$$\bar{x}_{K_1} = -l_1 \sin\varphi_1 - l_2 \sin\varphi_2 - h \sin\varphi_3 - \frac{d}{2}\cos\varphi_3 \geqslant -\frac{D}{2},$$

$$\bar{x}_{K_2} = -l_1 \sin\varphi_1 - l_2 \sin\varphi_2 - h \sin\varphi_3 - \frac{d}{2}\cos\varphi_3 + s\sin\varphi_3 \geqslant -\frac{D}{2},$$

$$\bar{x}_{K_3} = -l_1 \sin\varphi_1 - l_2 \sin\varphi_2 - h \sin\varphi_3 + \frac{d}{2}\cos\varphi_3 \leqslant \frac{D}{2},$$

$$\bar{x}_{K_4} = -l_1 \sin\varphi_1 - l_2 \sin\varphi_2 - h \sin\varphi_3 + \frac{d}{2}\cos\varphi_3 + s\sin\varphi_3 \leqslant \frac{D}{2}, \quad (10.31)$$

where: $\bar{x}_{K_1}, \bar{x}_{K_2}, \bar{x}_{K_3}$ and $\bar{x}_{K_4}$ are co-ordinates of four corners of the piston in the co-ordinate system $\bar{x} - \bar{y}$. Notice that the given description is valid only when the piston does not undergo full rotation (i.e. the backlash is sufficiently small).

The above inequalities are divided by l_1 and transformed to the form

$$h_1(\phi) = \frac{\bar{x}_{K_1} + \frac{D}{2}}{l_1} \geqslant 0, \quad h_2(\phi) = \frac{\bar{x}_{K_2} + \frac{D}{2}}{l_1} \geqslant 0,$$

$$h_3(\phi) = \frac{-\bar{x}_{K_3} + \frac{D}{2}}{l_1} \geqslant 0, \quad h_4(\phi) = \frac{-\bar{x}_{K_4} + \frac{D}{2}}{l_1} \geqslant 0, \quad (10.32)$$

yielding the final form of the non-dimensional set of inequalities (representing stiff unilateral constraints):

$$h_1(\phi) = \frac{\Delta}{2} - \sin\varphi_1 - \lambda_2 \sin\varphi_2 - \eta \sin\varphi_3 - \frac{\delta}{2}\cos\varphi_3 \geqslant 0,$$

$$h_2(\phi) = \frac{\Delta}{2} - \sin\varphi_1 - \lambda_2 \sin\varphi_2 + (\sigma - \eta)\sin\varphi_3 - \frac{\delta}{2}\cos\varphi_3 \geqslant 0,$$

$$h_3(\phi) = \frac{\Delta}{2} + \sin\varphi_1 + \lambda_2 \sin\varphi_2 + \eta \sin\varphi_3 - \frac{\delta}{2}\cos\varphi_3 \geqslant 0,$$

$$h_4(\phi) = \frac{\Delta}{2} + \sin\varphi_1 + \lambda_2 \sin\varphi_2 - (\sigma - \eta)\sin\varphi_3 - \frac{\delta}{2}\cos\varphi_3 \geqslant 0, \quad (10.33)$$

where the following non-dimensional parameters are introduced:

$$\lambda_2 = \frac{l_2}{l_1}, \quad \eta = \frac{h}{l_1}, \quad \sigma = \frac{s}{l_1}, \quad \delta = \frac{d}{l_1}, \quad \Delta = \frac{D}{l_1}. \quad (10.34)$$

The same restitution coefficient e is related to each of the unilateral constraints.

It is assumed that a gas pressure force can be reduced to the force acting along a line parallel to the axle of the cylinder and containing the piston pin axle

O_3. Moreover, this force is assumed to be a function of the angular position of the crankshaft φ_1,

$$\bar{F}(\varphi_1) = p_{\max}\frac{\pi d^2}{4} p(\varphi_1), \tag{10.35}$$

where: $p_{\max}$ is the maximal pressure over the piston, and $p(\varphi_1)$ is the non-dimensional pressure distribution such that its maximal value is 1. The pressure distribution function with period $2\pi N$ (where N is an integer) is developed into a Fourier series with a finite number of terms K,

$$p(\varphi_1) = a_0 + \sum_{i=1}^{K} a_i \cos\left(i\frac{\varphi_1}{N}\right) + \sum_{i=1}^{K} b_i \sin\left(i\frac{\varphi_1}{N}\right). \tag{10.36}$$

The crankshaft is externally driven by moment $\bar{M}_0$ originating from an external power receiver (brake) and acting contrary to the positive sense of the angle φ_1. It is also assumed that the rotational speed of the crankshaft is constant. The system defined in such a way is in fact an autonomous and self-excited system.

The external forces acting on the system can be reduced to the following generalized force vector:

$$\bar{\mathbf{f}}_e(\phi) = \left\{ \begin{array}{c} \bar{F}(\varphi_1) l_1 \sin\varphi_1 - \bar{M}_0 \\ \bar{F}(\varphi_1) l_2 \sin\varphi_2 \\ 0 \end{array} \right\}, \tag{10.37}$$

where: $\bar{f}_{e,1}$ is the moment of force acting on the crankshaft system, $\bar{f}_{e,2}$ is the moment acting on the connecting rod and $\bar{f}_{e,3}$ is equal to zero (the latter observation is yielded assuming that the gas pressure force acts along the line including point O_3).

The non-dimensional equations of motion, when none of the obstacles is active, are as follows:

$$\mathbf{M}(\phi)\ddot{\phi} + \mathbf{N}(\phi)\dot{\phi}^2 + \mathbf{C}\dot{\phi} + \mathbf{p}(\phi) = \mathbf{f}_e(\phi), \tag{10.38}$$

where:

$$\ddot{\phi} = \ddot{\psi}\Big|_{\psi=\phi}, \quad \dot{\phi}^2 = \dot{\psi}^2\Big|_{\psi=\phi}, \quad \dot{\phi} = \dot{\psi}\Big|_{\psi=\phi},$$
$$\mathbf{M}(\phi) = \mathbf{M}(\psi)|_{\psi=\phi}, \quad \mathbf{N}(\phi) = \mathbf{N}(\psi)|_{\psi=\phi}, \quad \mathbf{p}(\phi) = -\mathbf{p}(\psi)|_{\psi=\phi}. \tag{10.39}$$

In the above, the non-dimensional generalized force vector takes the following explicit form:

$$\mathbf{f}_e(\phi) = \frac{\bar{\mathbf{f}}_e(\phi)}{M_1} = \left\{ \begin{array}{c} F_0 p(\varphi_1) \sin\varphi_1 - M_0 \\ \lambda_2 F_0 p(\varphi_1) \sin\varphi_2 \\ 0 \end{array} \right\}, \tag{10.40}$$

where the following non-dimensional parameters are introduced:

$$F_0 = p_{\max}\frac{\pi d^2}{4}\frac{l_1}{M_1}, \quad M_0 = \frac{\bar{M}_0}{M_1}. \tag{10.41}$$

Predicting large rotational speeds of the crankshaft, it is reasonable to introduce a new non-dimensional time

$$t = \alpha_1 t\,, \tag{10.42}$$

where α_1 is the average non-dimensional angular velocity (measured in the non-dimensional time t) of the crankshaft per one period (corresponding to the period $2\pi N$ of the pressure function $p(\varphi_1)$) of the periodic steady state motion of the system without backlash between the piston and the cylinder ($d = D$).

In this special case ($d = D$) the analyzed system reduces to a one-degree-of-freedom system, and the generalized co-ordinates vector has independent components

$$\phi = \begin{Bmatrix} \varphi_1 \\ \varphi_2(\varphi_1) \\ 0 \end{Bmatrix}. \tag{10.43}$$

The geometric and kinematic relations between the co-ordinates of the given vector, and between their derivatives follow

$$\sin(\varphi_2) = -\lambda_2^{-1}\sin(\varphi_1)\,,$$

$$\cos(\varphi_2) = \sqrt{1-\lambda_2^{-2}\sin^2\varphi_1}, \quad \text{for } \lambda_2 > 1 \text{ and } \varphi_2 \in \left(-\frac{\pi}{2}, \frac{\pi}{2}\right),$$

$$\frac{\partial\varphi_2}{\partial\varphi_1} = -\lambda_2^{-1}\frac{\cos\varphi_1}{\cos\varphi_2} = -\lambda_2^{-1}\frac{\cos\varphi_1}{\sqrt{1-\lambda_2^{-2}\sin^2\varphi_1}},$$

$$\dot{\varphi}_2 = \frac{\partial\varphi_2}{\partial\varphi_1}\dot{\varphi}_1 = -\lambda_2^{-1}\frac{\cos\varphi_1}{\sqrt{1-\lambda_2^{-2}\sin^2\varphi_1}}\dot{\varphi}_1,$$

$$d\varphi_2 = \frac{\partial\varphi_2}{\partial\varphi_1}d\varphi_1 = -\lambda_2^{-1}\frac{\cos\varphi_1}{\sqrt{1-\lambda_2^{-2}\sin^2\varphi_1}}d\varphi_1\,. \tag{10.44}$$

The average angular velocity is defined by the formula

$$L_d + L_3 = 0, \tag{10.45}$$

where L_d and L_e denote correspondingly the viscous damping forces work and the external forces work during one period of the steady state motion of the system without backlash between the piston and the cylinder.

The viscous damping forces work can be written as

$$L_d = -\int_T \left(\mathbf{C}\dot{\phi}\right) d\phi^T$$

$$= -\int_0^{2\pi N} \left(\begin{bmatrix} c_1 + c_2 & -c_2 & 0 \\ -c_2 & c_2 + c_3 & -c_3 \\ 0 & -c_3 & c_3 \end{bmatrix} \begin{Bmatrix} \dot{\varphi}_1 \\ \frac{\partial \varphi_2}{\partial \varphi_1}\dot{\varphi}_1 \\ 0 \end{Bmatrix} \right) \begin{Bmatrix} d\varphi_1 \\ \frac{\partial \varphi_2}{\partial \varphi_1} d\varphi_1 \\ 0 \end{Bmatrix}^T . \quad (10.46)$$

Using relations (10.44) and replacing the angular velocity by the average velocity $\dot{\varphi}_1$, relation (10.46) can be expressed as follows:

$$L_d = -(c_1 + c_2)\, \alpha_1 \int_0^{2\pi N} d\varphi_1 - 2\lambda_2^{-1} c_2 \alpha_1 \int_0^{2\pi N} \frac{\cos\varphi_1}{\sqrt{1 - \lambda_2^{-2}\sin^2\varphi_1}} d\varphi_1$$

$$-\lambda_2^{-2}(c_2 + c_3)\, \alpha_1 \int_0^{2\pi N} \frac{\cos^2\varphi_1}{1 - \lambda_2^{-2}\sin^2\varphi_1} d\varphi_1 . \quad (10.47)$$

After integration, the following expression is obtained for the damping forces work as a function of the average angular velocity:

$$L_d = -2\pi N \left((c_1 + c_2) + \lambda_2^{-1}(c_2 + c_3)\left(1 - \sqrt{1 - \lambda_2^{-2}}\right)\right) \alpha_1 . \quad (10.48)$$

The non-restoring external forces work can be cast in the form

$$L_e = \int_T \mathbf{f}_e(\phi) d\phi^T = \int_0^{2\pi N} \begin{Bmatrix} F_0 p(\varphi_1)\sin\varphi_1 - M_0 \\ \lambda_2 F_0 p(\varphi_1)\sin\varphi_2 \\ 0 \end{Bmatrix} \begin{Bmatrix} d\varphi_1 \\ \frac{\partial \varphi_2}{\partial \varphi_1} d\varphi_1 \\ 0 \end{Bmatrix}^T . \quad (10.49)$$

From the given expression, and using (10.44) one gets

$$L_e = L_F + L_M, \quad (10.50)$$

where:

$$L_F = F_0 \int_0^{2\pi N} p(\varphi_1)\sin\varphi_1 \left(1 + \frac{\cos\varphi_1}{\lambda_2\sqrt{1 - \lambda_2^{-2}\sin^2\varphi_1}}\right) d\varphi_1,$$

$$L_M = -2\pi N M_0 \quad (10.51)$$

represent the work of the non-dimensional gas pressure forces and the work of the non-dimensional external moment loading the shaft, respectively.

Condition (10.45) and relations (10.48), (10.50), (10.51) yield the following form of the average angular velocity of the shaft

$$\not{\alpha}_1 = F_0 \frac{\int\limits_0^{2\pi N} p(\varphi_1)\sin\varphi_1 \left(1 + \frac{\cos\varphi_1}{\lambda_2\sqrt{1-\lambda_2^{-2}\sin^2\varphi_1}}\right) d\varphi_1 - 2\pi N M_0}{2\pi N\left((c_1+c_2) + \lambda_2^{-1}(c_2+c_3)\left(1-\sqrt{1-\lambda_2^{-2}}\right)\right)} . \tag{10.52}$$

Let $(.\mathring{.}.)$ denote the derivative with respect to time $\not{t}$, whereas relations between derivatives with respect to non-dimensional time t, real time τ and new non-dimensional time $\not{t}$ have the form

$$\frac{d(\ldots)}{dt} = \frac{d(\ldots)}{d\not{t}}\frac{d\not{t}(t)}{dt} = \not{\alpha}_1 \frac{d(\ldots)}{d\not{t}},$$

$$\frac{d(\ldots)}{d\tau} = \frac{d(\ldots)}{d\not{t}}\frac{d\not{t}(t)}{dt}\frac{dt(\tau)}{d\tau} = \not{\alpha}_1\alpha_1 \frac{d(\ldots)}{d\not{t}}, \tag{10.53}$$

and

$$\dot{\varphi}_j = \not{\alpha}_1 \varphi_j, \quad \ddot{\varphi}_j = \not{\alpha}_1^2 \varphi_j, \quad i = 1, 2, 3 . \tag{10.54}$$

In order to arrive at a more convenient setting of the system parameters, the following relation suitable for the mechanical efficiency of the engine estimation is introduced:

$$\eta_m = \frac{-L_M}{L_F}, \tag{10.55}$$

and on taking (10.51) into account, the following expression for the moment loading the engine is obtained

$$M_0 = \frac{\eta_m F_0 \int\limits_0^{2\pi N} p(\varphi_1)\sin\varphi_1 \left(1 + \frac{\cos\varphi_1}{\lambda_2\sqrt{1-\lambda_2^{-2}\sin^2\varphi_1}}\right) d\varphi_1}{2\pi N} . \tag{10.56}$$

Note that now the damping coefficients can be determined owing to the assumed rotational speed of the engine. Equations (10.45), (10.48), (10.50), (10.51), (10.55) yield

$$c_1 = \frac{F_0(1-\eta_m) \int\limits_0^{2\pi N} p(\varphi_1)\sin\varphi_1 \left(1 + \frac{\cos\varphi_1}{\lambda_2\sqrt{1-\lambda_2^{-2}\sin^2\varphi_1}}\right) d\varphi_1}{2\pi N\left((1+c_{21}) + \lambda_2^{-1}(c_{21}+c_{31})\left(1-\sqrt{1-\lambda_2^{-2}}\right)\right)\not{\alpha}_1}, \tag{10.57}$$

where:

$$c_{21} = \frac{c_2}{c_1}, \quad c_{31} = \frac{c_3}{c_1}, \tag{10.58}$$

and

$$\omega_1 = \alpha_1^{-1} \frac{\pi n}{30}. \tag{10.59}$$

In the above, n $[rot/min]$ represents the real average rotational speed of the crankshaft.

Finally, after introducing the non-dimensional time τ and when none of the obstacles is active, the equations of motion of the piston – connecting rod – crankshaft system are as follows:

$$\mathbf{M}(\phi)\,\phi + \mathbf{N}(\phi)\,\phi^2 + \mathbf{C}\phi + \mathbf{p}(\phi) = \mathbf{f}_e(\phi), \tag{10.60}$$

where:

$$\mathbf{C} = \omega_1^{-1} c_1 \begin{bmatrix} 1 + c_{21} & -c_{21} & 0 \\ -c_{21} & c_{21} + c_{31} & -c_{31} \\ 0 & -c_{31} & c_{31} \end{bmatrix},$$

$$\mathbf{p}(\phi) = -\omega_1^{-2} \begin{Bmatrix} \sin\varphi_1 \\ \mu_2 \sin\varphi_2 \\ \mu_3 \sin\varphi_3 \end{Bmatrix},$$

$$\mathbf{f}_e(\phi) = \omega_1^{-2} \begin{Bmatrix} F_0 p(\varphi_1) \sin\varphi_1 - M_0 \\ \lambda_2 F_0 p(\varphi_1) \sin\varphi_2 \\ 0 \end{Bmatrix}, \tag{10.61}$$

and M_0, c_1, ω_1 are determined from (10.56), (10.57), (10.59), respectively.

Observe that the proposed dynamical model of the piston – connecting rod – crankshaft system can be treated as a simplified model since some very important technological details are neglected. The most important simplifications are as follows: (i) tangent forces of interaction between the surfaces of the piston and the cylinder are neglected; (ii) interaction of the piston-cylinder introduced by the piston rings (by means of the friction forces in the ring grooves in the direction perpendicular to the cylinder surface) is neglected; (iii) a simplified friction model in every joint of the system (i.e. linear damping) is assumed.

In addition, the modelling of the impact between the piston and the cylinder, where an oil layer exists, requires an approach different from the generalized restitution coefficient rule.

In other words, a detailed modelling of the piston – connecting rod – crankshaft system with all essential technological details exceeds the scope of this work. However, we believe that the general model of the triple physical pendulum presented and investigated in some earlier works [24, 26, 66] can serve as a good starting point for a more advanced and closer to reality dynamical model of the piston - connecting rod - crankshaft system, taking into account the lateral motion and impacts between the piston and the cylinder barrel.

It should also be noticed that the presented model can govern steady state solutions of the system, and that the simulated transient motion does not correspond to the real piston – connecting rod – crankshaft system.

The dynamics of the piston - connecting rod - crankshaft system has been rigorously studied in the Habilitation Thesis [107]. In that monograph the following basic piston positions are assumed: (i) four piston positions in the cylinder barrel: two skew positions (with the contact between one corner of the piston and one side of the cylinder and between the opposite corner and the second side of the cylinder barrel), and two positions of the piston adjoining to one of the two sides of the cylinder surface; (ii) four displacements (turns) of the piston with one of the four corners being in contact with the cylinder.

In addition, in each of the four piston positions three equilibrium states of dynamic forces are distinguished, whereas in each of the four piston displacements two such states are distinguished. Therefore, the piston can be in one of the twenty equilibrium states of the dynamic forces. The piston movement from one side of the cylinder to the opposite side has been assumed to consist of two piston turns and one skew piston position. A direct piston movement with loss of the contact with cylinder has not been analysed.

Assuming the constant rotational speed of the crankshaft, the schedule of the forces acting on the piston and the connecting rod has been made, including tangent forces of interaction between the cylinder and the piston surfaces, forces of interaction between the piston and the cylinder via the piston rings, and assuming more real friction model in the bearings. In that way the system of six equations of equilibrium of the dynamic forces has been obtained for the piston with the connecting rod system.

The obtained equations can be solved for one of the possible piston states for each crankshaft position. The obtained values of both normal and friction forces verify the admissibility of the assumed piston state. If the piston state is not admissible, the next piston state is assumed and the calculations are repeated until the admissible piston state is found. In that way, by varying the crankshaft position with a small angle step, for each crankshaft position one admissible piston state can be found.

Summarizing, although the model presented in reference [107] satisfies the assumed role, it does not take into account the full piston dynamics including a lateral motion of the piston in the cylinder barrel. In contrast, the full dynamical model of the piston – connecting rod – crankshaft system presented in this work, although simplified, governs the full dynamics of the piston analysis including impacts between the piston and the cylinder.

8. Numerical Examples

The non-dimensional pressure distribution function $p(\varphi_1)$ used in this section and shown in Fig. 10.4 applies the data included in [107], and concerns the real pressure function obtained experimentally from the engine 1HC102. The period of the function is 4π ($N = 2$ for the four-stroke engine), maximal pressure $p_{\max} = 8\,MPa$ for the rotational crankshaft speed $n = 1200\ [rot/min]$ and the full engine loading.

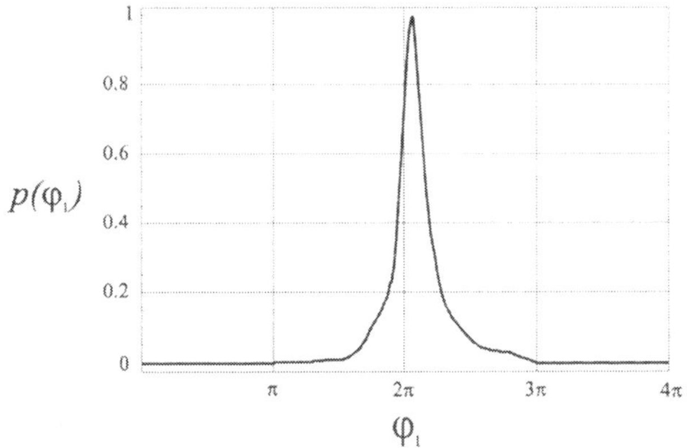

Figure 10.4. Gas pressure function used for calculations.

The function $p(\varphi_1)$ is developed into the Fourier series with $K = 25$ terms. The remaining parameters are as follows: $m_1 = 10\,kg$, $m_2 = 1\,kg$, $m_3 = 0.4\,kg$, $J_{z1} = 1\,kg \cdot m^2$, $J_{z2} = 0.0075\,kg \cdot m^2$, $J_{z3} = 0.001\,kg \cdot m^2$, $l_1 = 0.04\,m$, $l_2 = 0.15\,m$, $e_{y1} = 0\,m$, $e_{y2} = 0.12\,m$, $e_{y3} = 0.01\,m$, $d = 0.08\,m$, $s = 0.08\,m$, $h = 0.04\,m$, $\eta_m = 0.85$, $c_{21} = 0.2$, $c_{31} = 0.1$, $g = 9.81\,m\,s^{-2}$. The following real parameters are found owing to the introduced values: $\bar{M}_0 = 24.7\,Nm$ and $\bar{c}_1 = 0.0288\,N \cdot m^{-1}s$. The calculations are performed for different values of the restitution coefficient and the external diameter D.

The differential equations are integrated with the time step $k_{RK4} = 2\pi/400$, and the obstacle detection accuracy is $\varepsilon_h = 10^{-12}$, the detection accuracy of the zero of the normal component of the relative velocity is $\varepsilon_{\dot{h}} = 10^{-8}$ and the detection accuracy of the zero of the normal force to the barrier surface is $\varepsilon_\lambda = 10^{-12}$ (see Fig. 10.1).

The initial conditions at the time instant $t = 0$ are the same for all examples: $\varphi_{10} = \varphi_{20} = \varphi_{30} = 0$, $\dot{\varphi}_{10} = 1$, $\dot{\varphi}_{20} = \dot{\varphi}_{30}$. In Figs. 10.5–10.10 the steady state solution is shown within the time interval $t \in (5000, 5500)$.

In order to construct some diagrams, the following non-dimensional coordinates describing the position of the piston pin axle are used:

$$x_{O3} = \frac{\bar{x}_{O3}}{l_1} = -\sin\varphi_1 - \lambda_2 \sin\varphi_2\,,$$

$$y_{O3} = \frac{\bar{y}_{O3}}{l_1} = \cos\varphi_1 + \lambda_2 \cos\varphi_2\,. \qquad (10.62)$$

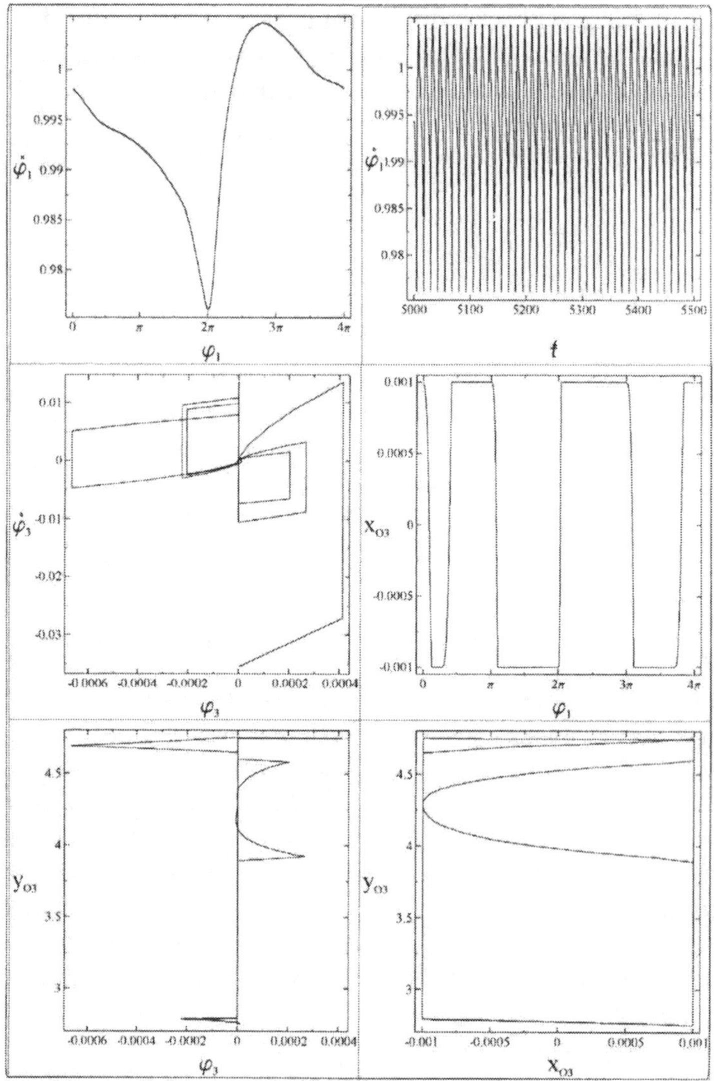

Figure 10.5. System response for $e = 0$ and $D = 0.08008\,m$.

The response of the system for the restitution coefficient $e = 0$ and the cylinder diameter $D = 0.08008\,m$ (the backlash of the piston in the barrel is $0.08\,mm$) is shown in Figure 10.5. It is seen from the figures that the piston moves six times from one side of the cylinder to the other side during one cycle of the engine work, and most of the time the piston adjoins to one or the other side of the cylinder surface. This result confirms the well-known fact and the results presented in [107]. However, the piston loses contact with the cylinder

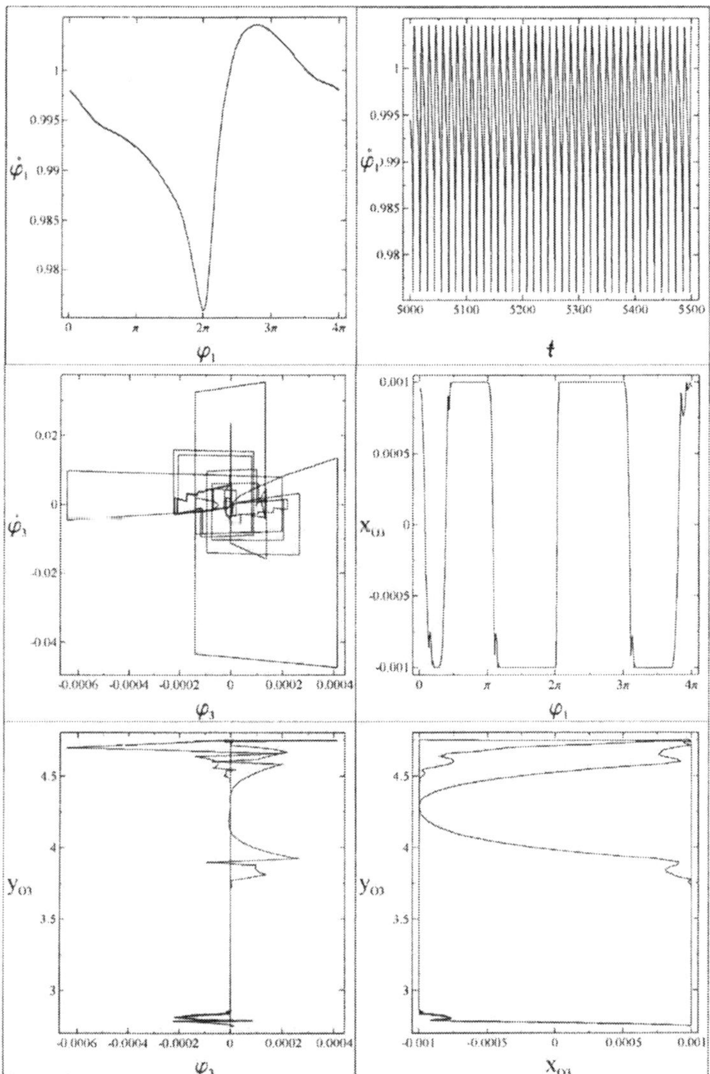

Figure 10.6. System response for $e = 0.5$ and $D = 0.08008\ m$.

while moving from one side of it to the other with a small rotation angle. This phenomenon differs from the results presented in [107], where it was assumed that the piston did not lose the contact with the cylinder. The crankshaft angular positions at the beginnings and the ends of the phases of the piston adjoining and sliding along the cylinder (see the example shown in Fig. 10.5) differ also from the results presented in [107] up to 35°. The exhibited differences follow

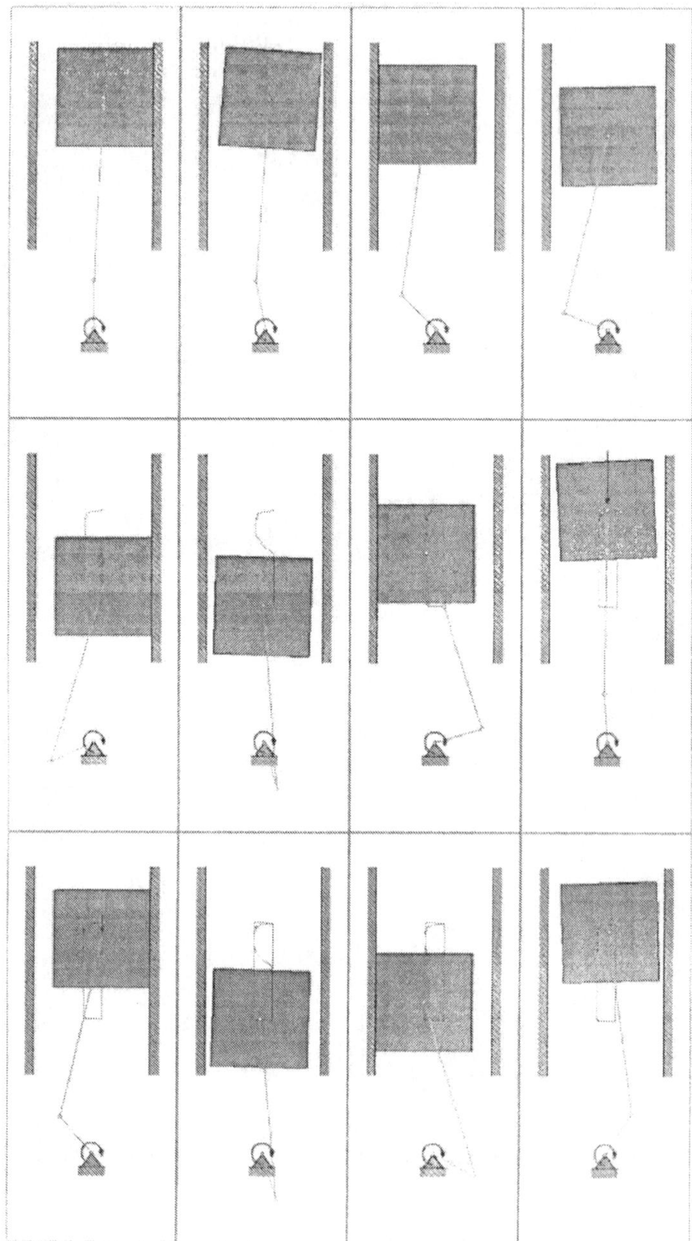

Figure 10.7. Successive positions of the piston in cylinder barrel for $e = 0.5$ and $D = 0.08008\,m$.

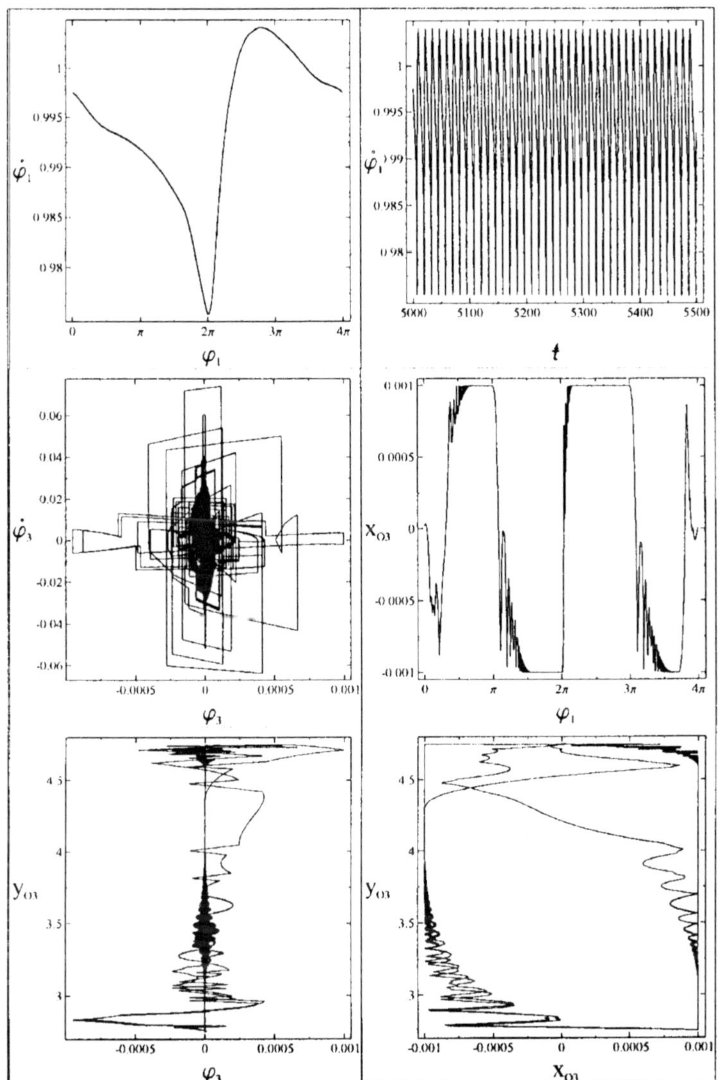

Figure 10.8. System response for $e = 0.9$ and $D = 0.08008\ m$.

straightforwardly from the neglecting of certain essential technological details in our study, as mentioned in the previous section.

In Fig. 10.6, the results for the larger restitution coefficient $e = 0.5$ are shown. It is seen that in general the states of the piston adjoining to the cylinder surface are the same as previously. Only the beginning of each of them is slightly delayed since the piston bounces against the cylinder a few times before the sliding occurs. Fig. 10.7 depicts successive piston positions

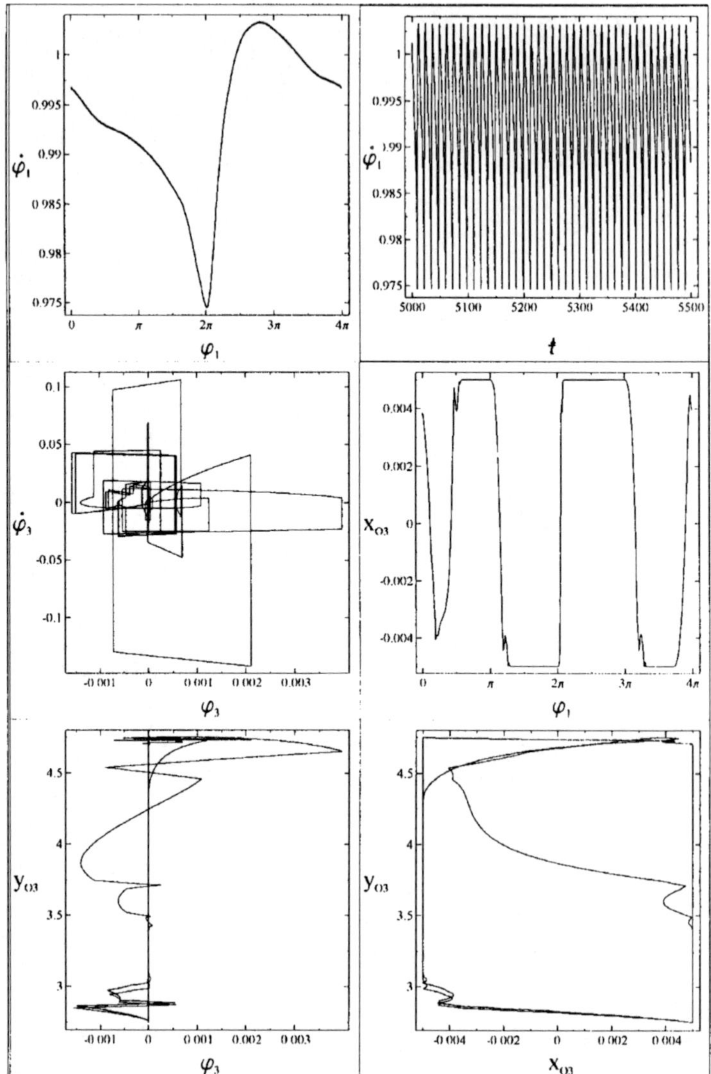

Figure 10.9. System response for $e = 0.5$ and $D = 0.08040\,m$.

yielded by this solution. The results for the restitution coefficient $e = 0.9$ (Fig. 10.8), for the five times larger backlash between the cylinder and the piston ($D = 0.08040\,m$), and for the restitution coefficient $e = 0.5$ (Fig. 10.9) and (Fig. 10.10) are also reported. It is worth noticing that the system is at least inclined to reach the same states of the piston adjoining and sliding along the cylinder, and lasting in the same crankshaft positions as previously. Since multiple impacts between the piston and the cylinder occur, it happens that

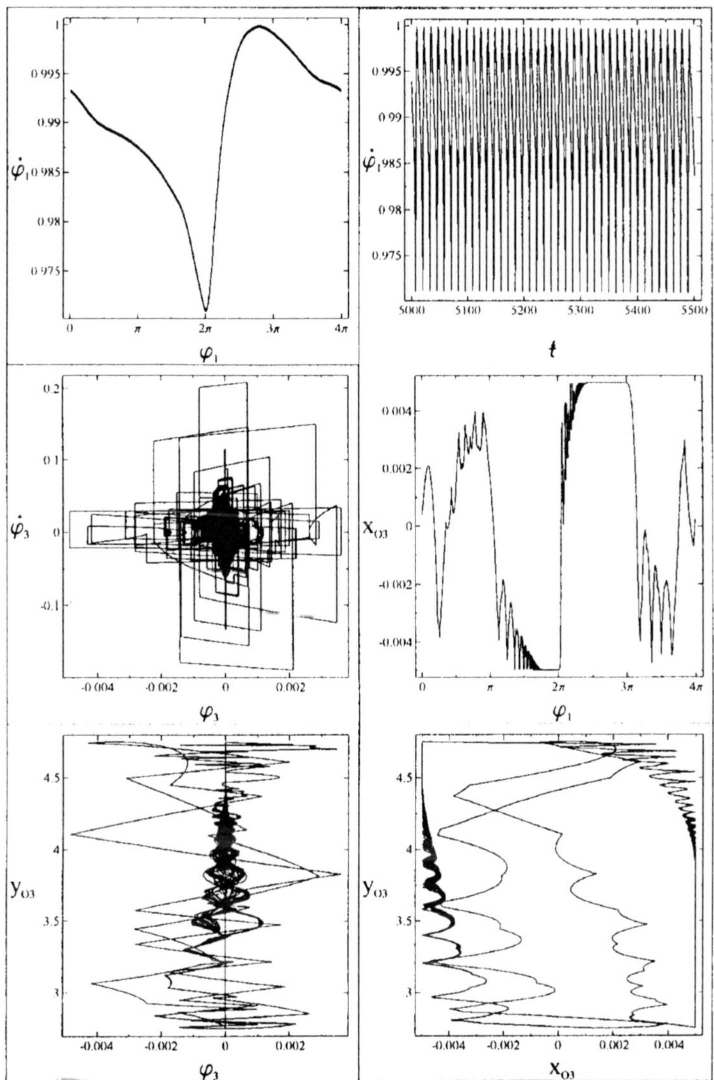

Figure 10.10. System response for $e = 0.9$ and $D = 0.08040\ m$.

before the piston gains the motion stabilisation at one cylinder side, it rapidly leaves the contact and transits into the other side of the cylinder.

Chapter 11

MODELING OF A DUO-SERVO BRAKE

In the previous chapter a combustion engine model and its dynamics have been considered. Without an engine a wheeled car (or a motor-cycle) can not run. On the other hand, it can not stop without a brake. Therefore, this section is devoted to modeling and analysis of one exemplary brake, namely, a duo-servo brake. Since friction during the braking process plays a key role a brief review of friction modeling is given in the Introduction. In section 11.2 the modeled system with friction is presented including the governing differential equations. Numerical analysis is carried out in section 11.3. Section 11.4 describes the experimental investigation of the introduced physical model. The last section 11.5 is devoted to conclusions.

1. Introduction

The nature of sliding components with an occurrence of intermittent stick and slip leads to unpredictable behaviors. These problems are exhibited in many industrial applications including bearings, disc brake systems, electric motor drives, rail mass transit systems, and machine tool/work piece systems [60]. A fuller understanding of the stick-slip phenomena, which in consequence might help in elimination of its effects, is of considerable importance for applications requiring high precision motion [18].

The relative sliding of two solid bodies is a non-equilibrium process where the kinetic energy of the motion is transferred into the energy of an irregular microscopic motion. This dissipative process is responsible for creation of a dry friction phenomenon. The phenomenological laws of dry friction, like Coulomb's laws, are well known and there is a well-established theory in the applied physics related to this subject.

The simplest models describe friction as a function of the difference in the velocity of sliding bodies. Models such as the Coulomb's friction one are called static models. In fact, Coulomb's dry friction laws simplify a highly complex behavior that involves mechanical, plastic, and chemical processes [96]. Application of Coulomb's law often brings results that bear experimentally observable differences [7]. Computer simulations of mechanical systems with friction are complicated not only due to the strongly non-linear behavior of the friction force near the zero velocity but also because of the lack of a universally considered friction model. For rigid bodies with dry friction, the classical Coulomb law of friction is usually applied in engineering contact problems exactly because of its simplicity. It can explain several phenomena associated with friction and it is commonly used for friction compensation [50].

A well-known velocity-limited friction model given by Oden and Martins [77] uses a smooth quadratic function when the sliding velocity is near zero. However, although the value of the limiting velocity is essential here, there is no standard method for its estimation.

Using a different approach, Antunes et al. [17] developed a spring-damper friction model that introduces both tangential spring and damper during sticking. The phase of motion is detected by a change in sign of the tangential velocity. The sticking spring force is expressed by a product of the adherence stiffness and contact distance from a zero tangential velocity. A viscous damper force is incorporated in parallel with the spring to damp out any residual numerical velocity.

Karnopp [62] developed a force-balance model for one-dimensional motion with a small-velocity window. Outside the velocity window, the friction force may be any function of the sliding velocity. Inside the window, the friction force is estimated so as to balance the other forces in the system, the velocity remains small constant until the loss of the contact value of force is reached. This model has been used to describe many practical actuator and mechanism problems. Basing on Karnopp's model, Tan and Rogers developed a two-dimensional friction model often used for simulations of multi degrees-of-freedom systems with friction acting on a surface.

The problem of friction force modeling is not solved because the physics and dynamical effects are not sufficiently understood. There are two main theoretical approaches to the modeling of dry friction interfaces: the macro-slip and micro-slip ones [96]. In the micro-slip approach, a relatively detailed analysis of the friction interfaces should be made. In this case investigations can provide accurate results only when the preload between interfaces is very high. In the macro-slip approach, the entire surface is assumed to be either sliding or sticking. The force necessary to keep sliding at a constant velocity depends on the sliding velocity of the contact surfaces. With this respect, smooth and non-smooth velocity-dependent friction laws have been cited in the references [86, 77, 70].

Generally speaking, the various modifications of the Coulomb model, and the Karnopp model are all static in the sense that the friction model is a function of velocity. They can be substituted when friction is recognized to be in fact a dynamic phenomenon that should be modeled as a dynamical system. Under this convention, some of the dynamical dry friction models need to be specified: Dahl's models [46]; Bliman-Sorine model [34]; LuGre model [41, 6].

There is an essential shortage of works that take into account modeling problems connected with the experimentally observable velocity-dependent friction force. The paper [38] deals with measurement of dry friction. A tribometer was developed to identify both sticking and sliding friction coefficients. The so-called Stribeck-curve was determined for any material in the contact zone. Similarly, a multi degrees-of-freedom model of friction was investigated in [35], where an experimentally observable friction characteristic expressed the kinetic friction force as a function of the relative velocity of motion. The experimental investigation of vibrations of a system composed of a steel-polyester pair confirmed that the friction static force increases with both the increasing adhesion time and the growing force. Additionally, the kinetic friction force depends also on the sign of acceleration.

Despite numerous papers concerned with the analysis of regular and chaotic dynamics of mechanical systems with friction, not all possible non-linear phenomena seem to have been properly understood or even detected and explained [19, 21, 22, 40, 51, 67, 73, 77]. Although this paper is devoted to numerical and experimental investigations, the problem is expected to be attacked also from an analytical point of view. The stick-slip chaos has been predicted analytically using the Melnikov technique by Awrejcewicz and Holicke [23], but such a prediction for a two degrees-of-freedom system is in general more complicated. Even if this problem has been solved, it will contain only a special type of non-linear terms and will be valid only for special systems. Therefore, in this paper we have focused on numerical simulations, which are free from the mentioned drawbacks.

A self-excited system with friction analysed in this work requires a suitable algorithm to avoid the problems that occur under integration of equations of motion with the sgn function and result in sudden jumps and undesired errors. In what follows the Hénon method [57, 28] is applied, which is particularly useful for obtaining of suitable uniform solutions of non-smooth systems (here with friction).

Estimation of Lyapunov exponents plays a crucial role in the analysis and identification of chaotic dynamics. The estimation belongs to the fundamental tools [111, 113] that allow learning about regularity of an attractor under analysis. At present, however, a novel approach, especially suitable for estimation of the Lyapunov exponents of non-smooth systems, i.e. including those with friction, is applied.

2. The Modeled System With Friction

The study and prevention of unstable vibration of systems with friction is of essential importance in industry and there is a need for the friction pair modeling that could provide a correct description of the kinetic and static friction forces change between two moveable surfaces.

The model can be further developed to govern also the dynamics of a duo-servo brake mechanism [29]. Our analysis is focused on modeling of friction phenomena exhibited, for instance, by a brake like the duo-servo system illustrated in Figure 11.1.

Figure 11.1. Duo-servo brake: 1 - hydraulic cylinder, 2 - brake shoes with friction linings, 3 - couple element, 4 - auxiliary long spring, 5 - auxiliary short spring, 6 - hand-brake mechanism, 7 - body.

The observed (in duo-servo brake) frictional mechanism can be modeled approximately by a two degrees-of-freedom mechanical system (see Figure 11.2). Therefore, the 2-DOF dynamical system illustrated schematically in Fig. 11.2 is analyzed numerically and investigated experimentally.

The self-excited system presented in Fig. 11.2 is equivalent to a real experimental rig in which *block* mass m is moving on the *belt* in x_1 direction, and where the *angle* body represented by moment mass of inertia J is rotating around point s with respect to angle direction ϕ. The analysed system consists of the following parts: two bodies are coupled by linear springs k_2 and k_3;

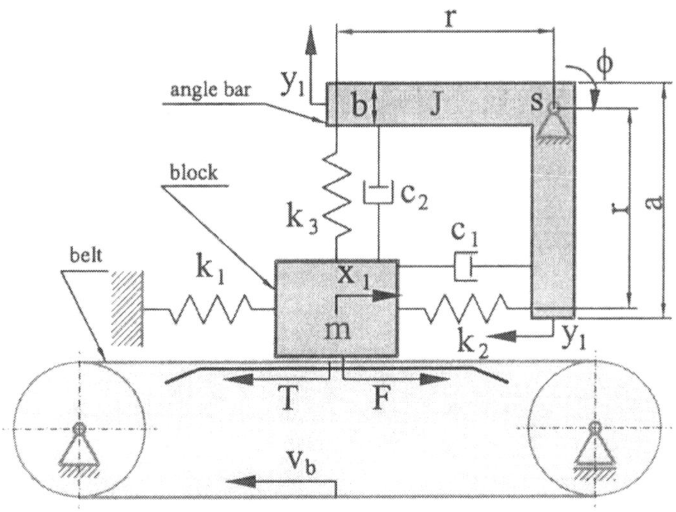

Figure 11.2. The 2-DOF system under analysis.

the block on the belt is additionally coupled to a fixed base by means of linear spring k_1; the angle body is excited only by spring forces; there are no extra mechanical actuators; rotational motion of the angle body is damped using virtual actuators characterizing air resistance and denoted by constants c_1 and c_2; damping of the block is neglected; it is assumed that the angle of rotation of the angle body is small and within the interval $[+5, -5]$ degrees (in this case rotation is equivalent to linear displacement y_1 of legs a of the angle body); the belt is moving with constant velocity v_b and there is no deformation of the belt in the contact zone.

Non-dimensional equations that govern the dynamics of the system under investigation have the following form:

$$\dot{x}_1 = x_2 \,,$$

$$x_2 = -x_1 - \alpha_1^{-1}\left[\eta_1\left(x_2 + y_2\right) - y_1 - T\right],$$

$$\dot{y}_1 = y_2 \,,$$

$$\dot{y}_2 = \alpha_2^{-1}\left(-\beta_3 y_1 - \eta_{12} y_2 - x_1 - \eta_1 x_2\right), \tag{11.1}$$

where: x_2, y_2 are velocities of the block and angle body, respectively; $v_{rel} = x_2 - v_b$ is a relative velocity between the bodies of the investigated system, $\alpha_1 = \dfrac{\omega^2 m}{k_2}, \alpha_2 = \dfrac{\omega^2 J}{k_2 r^2}, \beta_1 = \dfrac{k_1 + k_2}{k_2}, \beta_2 = \dfrac{\mu_0 k_3}{k_2}, \beta_3 = \dfrac{k_2 + k_3}{k_2}, \eta_1 = \frac{c_1 \omega}{k_2}$, $\eta_2 = \frac{c_2 \omega \mu_0}{k_2}, \eta_{12} = \frac{\omega(c_1 + c_2)}{k_2}$ are the remaining parameters; ω is a frequency of mass m. The friction force is described in the following manner:

$$\begin{cases} |T \leqslant (1 - \beta_2 y_1 - \eta_2 y_2)\,\mu_0| & \text{for} v_{rel} = 0, \\ T = \operatorname{sgn}(v_{rel})\,(1 - \beta_2 y_1 - \eta_2 y_2)\,\mu_i & \text{for} v_{rel} \neq 0, \end{cases} \tag{11.2}$$

where: μ_i $(i = 1, 2)$ are friction coefficients defined for two various cases of our numerical analysis, i.e. the negative slope characteristic

$$\mu_1(v_{rel}) = \frac{1}{1 + \gamma_1 |v_{rel}|}, \tag{11.3}$$

and the Stribeck curve

$$\mu_2(v_{rel}) = \frac{\mu_d}{\mu_0} + \left(1 - \frac{\mu_d}{\mu_0}\right) \exp\left(\frac{-\gamma_2 |v_{rel}|}{\mu_0 - \mu_d}\right), \tag{11.4}$$

where: μ_0, μ_d are the coefficients of sticking and sliding (for $v_{rel} \to \infty$) friction, respectively; γ_i $(i = 1, 2)$ are certain constant coefficients.

3. Numerical Analysis

In this paper the algorithm for numerical integration of the ODE, including the discontinuous term (see Eq. (11.2)) describing dry friction, is applied. It is worth noticing that our self-excited system with friction requires a special algorithm to avoid the problems occurring in the integration of Eq. (11.1). The algorithm used is based on the Hénon method [57, 28] that proves to be extremely useful for locating and tracking of the stick to slip and slip to stick transitions in non-smooth systems.

Although the Hénon algorithm uses the Runge-Kutta method of fourth order, separate procedures can be introduced to detect the points belonging either to stick or to slip phases for both increasing and decreasing velocities. Special transformations are applied to eliminate jumps. The procedure is carried out with an automatically chosen step of numerical integration.

3.1 Time Histories

It is obvious that the use of numerical methods can provide only approximate solutions of real trajectories of systems under analysis due to the finite step of numerical integrations and finite accuracy of the numbers used in the floating-point arithmetic. Nevertheless, if properly applied, numerical approximations prove sufficient for engineering purposes, which is important for our discontinuous system.

Duration of a transitional process depends on initial conditions and system parameters. In Fig. 11.3, one of the co-ordinates versus time is presented for $\tau_0 = 0$. In this case, the masses are in the equilibrium positions, whereas their initial velocities are equal to the belt velocity. In practice, from the

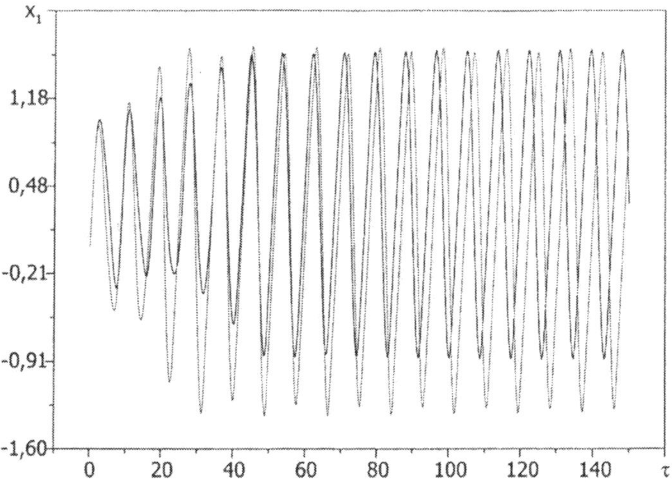

Figure 11.3. Quickly stabilized periodic motion for μ_1 (black curve) and μ_2 (gray curve) friction models for the parameters: $\alpha_1 = 3$, $\alpha_2 = 1.159$, $\beta_2 = 0.577$, $\beta_3 = 1.825$, $\eta_1 = \eta_2 = \eta_{12} = 0$, $\gamma_1 = 0.203$, $\gamma_2 = 0.813$, $v_b = 0.6$, $\mu_0 = 0.7$, $\mu_d = 0.18$ and initial conditions: $\tau_0 = 0$, $\tau_k = 150$, $x_1(0) = 0$, $x_2(0) = 0.6$, $y_1(0) = y_2(0) = 0$.

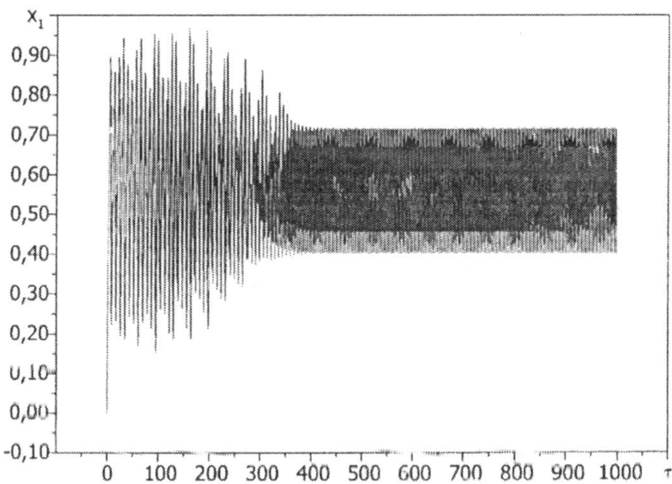

Figure 11.4. Slowly stabilized periodic motion for μ_1 (black curve) and μ_2 (red curve) friction models for the parameters: $\alpha_1 = 2.4$, $\alpha_2 = 0.928$, $\beta_2 = 0.84$, $\beta_3 = 2.2$, $\eta_1 = \eta_2 = \eta_{12} = 0$, $\gamma_1 = 0.203$, $\gamma_2 = 0.813$, $v_b = 0.2$, $\mu_0 = 0.7$, $\mu_d = 0.6$ and initial conditions: $\tau_0 = 0$, $\tau_k = 150$, $x_1(0) = 0$, $x_2(0) = 0.2$, $y_1(0) = y_2(0) = 0$.

very beginning the system starts to move on an attractor. When the initial conditions are changed, duration of the transitional process becomes equal to about $\tau = 400$ (see Fig. 11.4). Our numerical analysis shows that in some cases the transitional state can be many times longer, which may be of importance, especially in the engineering practice.

3.2 Phase Spaces

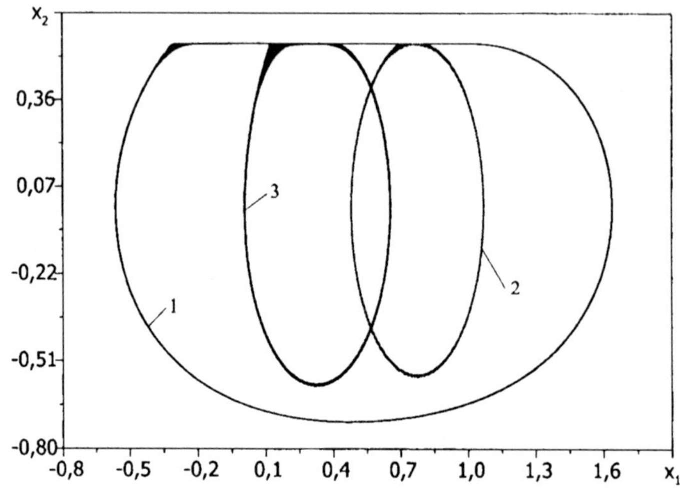

Figure 11.5. Phase space projection of periodic motion in (x_1, x_2) plane for μ_1 ((1) curve) and μ_2 ($\mu_d = 1.25$, (2) curve, $\mu_d = 0.3$, (3) curve) models for the parameters: $\alpha_1 = 2.4$, $\alpha_2 = 0.927$, $\beta_2 = 2.304$, $\beta_3 = 2.8$, $\eta_1 = \eta_2 = \eta_{12} = 0$, $\gamma_1 = 0.289$, $\gamma_2 = 1$, $v_b = 0.55$, $\mu_0 = 1.28$ and initial conditions: $\tau_0 = 3000$, $\tau_k = 3500$, $x_1(0) = 0$, $x_2(0) = -0.1$, $y_1(0) = y_2(0) = -0.1$.

The phase space of a dynamical system is a mathematical space in orthogonal coordinates that represents all of the variables which are necessary to determine a momentary state of the system. A typical projection of a trajectory associated with our system with friction is shown in Fig. 11.5. Two parts are easily distinguishable: a stick part (where $x_2 = v_b = 0.55$, which is represented by the horizontal line) and a slip part.

A more complicated motion is presented in Fig. 11.6. Time moments of an occurrence and duration of a stick are unpredictable. The mentioned phases appear with different velocities illustrated with small and large arcs in the phase plane. This means that the corresponding static friction force is smaller than an absolute value of the resulting horizontal forces.

3.3 Poincaré Sections

Construction of a Poincaré section can be performed to replace investigations of the properties of the n dimensional phase trajectory by an analysis of an n - 1

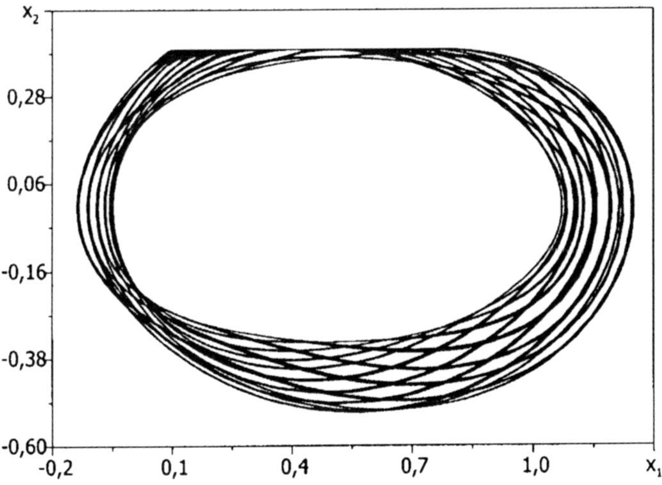

Figure 11.6. Phase space projection of chaotic motion in (x_1, x_2) plane for μ_1 model for the parameters: $\alpha_1 = 2.83$, $\alpha_2 = 1.093$, $\beta_2 = 1.873$, $\beta_3 = 2.441$, $\eta_1 = \eta_2 = \eta_{12} = 0$, $\gamma_1 = 0.132$, $\gamma_2 = 0.66$, $v_b = 0.4$, $\mu_0 = 1.3$ and initial conditions: $\tau_0 = 2000$, $\tau_k = 4000$, $x_1(0) = 0$, $x_2(0) = -0.1$, $y_1(0) = y_2(0) = -0.1$.

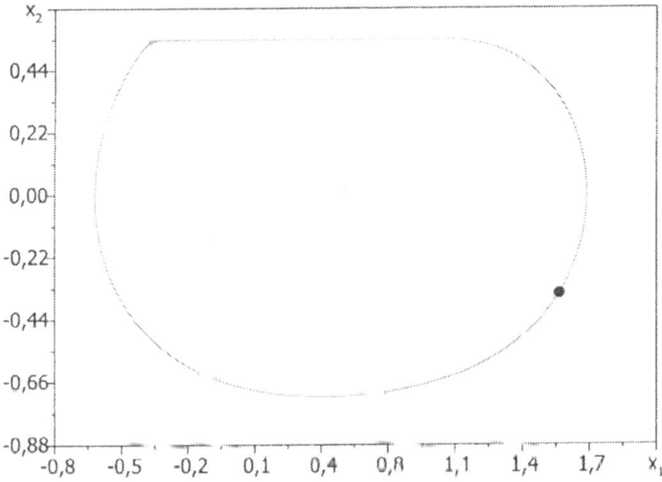

Figure 11.7. Phase space projection of 1-periodic motion of the block in (x_1, x_2) plane and its Poincaré section for μ_1 model for the parameters: $\alpha_1 = 2.4$, $\alpha_2 = 0.927$, $\beta_2 = 2.304$, $\beta_3 = 2.8$, $\eta_1 = \eta_2 = \eta_{12} = 0$, $\gamma_1 = 0.289$, $\gamma_2 = 0.999$, $v_b = 0.55$, $\mu_0 = 1.28$ and initial conditions: $\tau_0 = 4000$, $\tau_k = 4300$, $x_1(0) = 0$, $x_2(0) = -0.1$, $y_1(0) = 0$, $y_2(0) = -0.1$.

dimensional discrete system. Different definitions are assumed for autonomous and non-autonomous dynamical systems. For our autonomous system the map

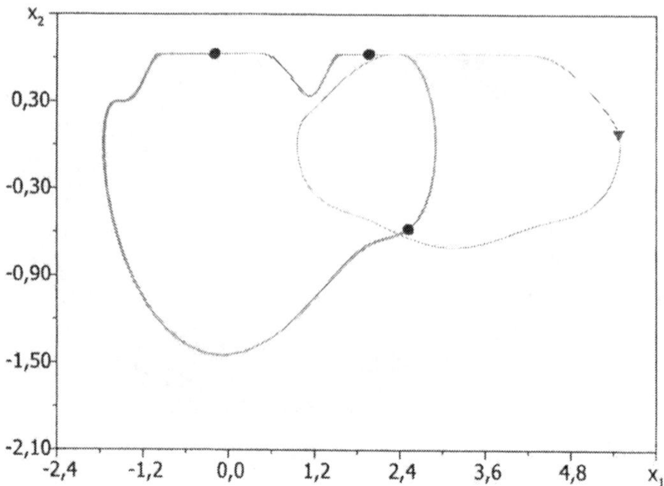

Figure 11.8. Phase space projection of periodic motions of the block in (x_1, x_2) plane and their Poincaré section for μ_1 (circle) and μ_2 (triangle) model for the parameters: $\alpha_1 = 2.5$, $\alpha_2 = 0.966$, $\beta_2 = 3.712$, $\beta_3 = 2.125$, $\eta_1 = \eta_2 = \eta_{12} = 0$, $\gamma_1 = 0.577$, $\gamma_2 = 2.047$, $v_b = 0.63$, $\mu_0 = 3.3$, $\mu_d = 0.3$ and initial conditions: $\tau_0 = 700$, $\tau_k = 1000$, $x_1(0) = 0$, $x_2(0) = -0.1$, $y_1(0) = 0$, $y_2(0) = -0.1$.

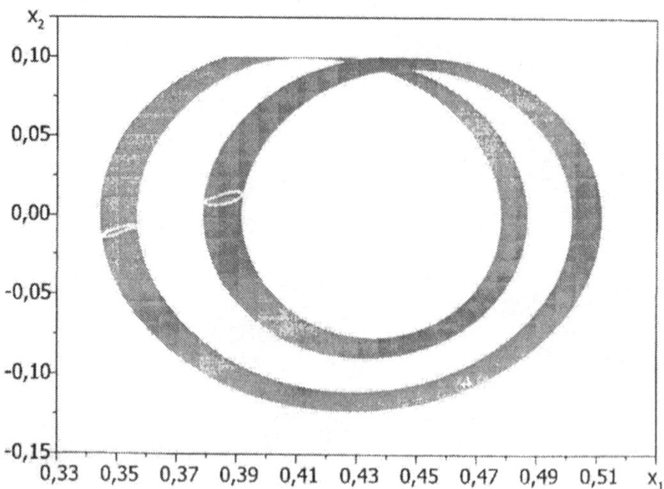

Figure 11.9. Phase space projection of quasi-periodic motion of the block in (x_1, x_2) plane and its Poincaré section for μ_1 model for the parameters: $\alpha_1 = 2.874$, $\alpha_2 = 1.111$, $\beta_2 = 0.412$, $\beta_3 = 2.472$, $\eta_1 = \eta_2 = \eta_{12} = 0$, $\gamma_1 = 0.035$, $\gamma_2 = 0.141$, $v_b = 0.1$, $\mu_0 = 0.28$ and initial conditions: $\tau_0 = 4000$, $\tau_k = 7000$, $x_1(0) = x_2(0) = 0$, $y_1(0) = y_2(0) = 0$.

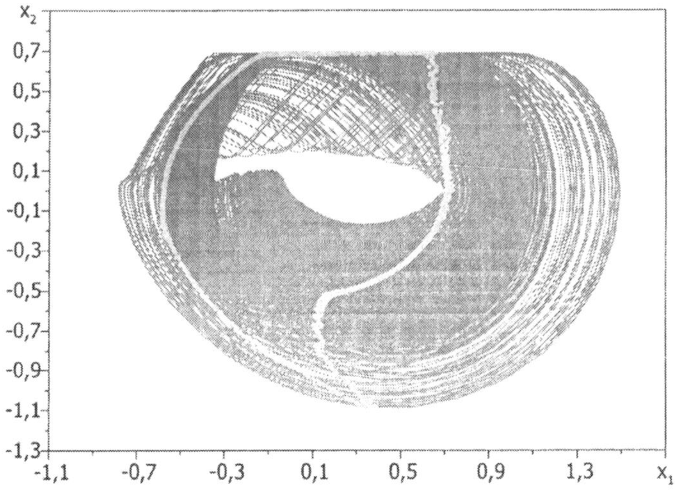

Figure 11.10. Phase space projection of chaotic motion of the block in (x_1, x_2) plane and its Poincaré section for μ_1 model for the parameters: $\alpha_1 = 4.5$, $\alpha_2 = 1.739$, $\beta_2 = 6.766$, $\beta_3 = 3.333$, $\eta_1 = \eta_2 = \eta_{12} = 0$, $\gamma_1 = 0.976$, $\gamma_2 = 3.551$, $v_b = 0.689$, $\mu_0 = 2.9$ and initial conditions: $\tau_0 = 2000$, $\tau_k = 20000$, $x_1(0) = 0$, $x_2(0) = -0.1$, $y_1(0) = 0$, $y_2(0) = -0.1$.

construction is defined in the following way:

$$\begin{array}{llll} \text{if} & x_{1,i-1} < x_{1,i} > x_{1,i+1} & \Rightarrow & y_{m,i} = [y_{1,i}, y_{2,i}], \\ \text{if} & y_{1,i-1} < y_{1,i} > y_{1,i+1} & \Rightarrow & x_{m,i} = [x_{1,i}, x_{2,i}], \end{array} \tag{11.5}$$

where: $y_{m,i}$, $x_{m,i}$ are Poincaré map points of the angle body and the block, respectively; i denotes the iteration number of the solution of Eq. (11.1).

In what follows, we are going to show that our autonomous system can exhibit stick-slip periodic (Figs. 11.7, 11.8), quasi-periodic (Fig. 11.9) as well as chaotic dynamics (Fig. 11.10).

Points of the Poincaré sections presented in Figs. 11.7–11.10 are matched by solid black circles against the background of their phase spaces. Moreover, Fig. 11.9 presents attractors in the shape of black closed curves confirming occurrence of a quasi-periodic motion. An unpredictable form of the chaotic attractor is presented in Fig. 11.10. Long-time solution provides location of randomized points.

3.4 Bifurcation Diagrams

Bifurcation diagrams are constructed by changing a parameter in the interval $[0.2, 0.7]$ with step 0.001, wherefrom 500 Poincaré maps are obtained. Then, one of the phase axle is selected and all results are presented versus the parameter. Another way is to increase parameter values by changing the initial

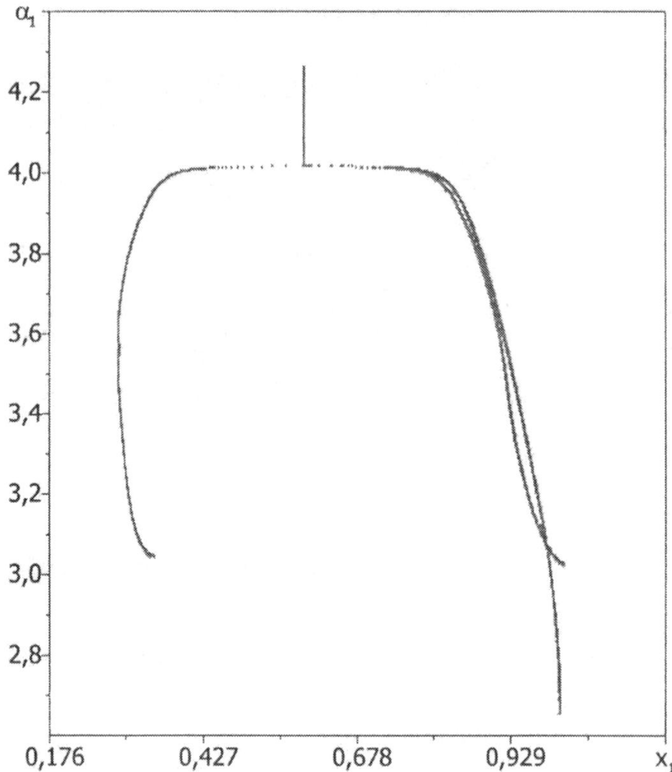

Figure 11.11. Bifurcation diagram of $\alpha_1 \in (2.72; 4.32)$ parameter versus x_1 displacement in the time interval from $\tau_0 = 1000$ to $\tau_k = 51000$ (remaining parameters are as in Fig. 11.3).

conditions. In this case we leave an attractor, contrary to the first case in which we stayed on an attractor all the time.

An example of the bifurcation diagram is shown in Fig. 11.11. Beginning from the smallest considered values of α_1 we observe the one-periodic motion, but for $\alpha_1 \approx 3.1$ a period tripled bifurcation with an increase of α_1 occurs. In the vicinity of bifurcation point $\alpha_1 \approx 4.0$, the period tripled bifurcation with a decrease of the bifurcation parameter is observed once again. It should be emphasized that for a large interval of changes of the bifurcation parameter only the periodic motion can be reached by our analysed system.

An interesting example of more complex bifurcations is shown in Fig. 11.12. It may be traced how the successive period doubling (accompanying the decrease of α_1 parameter) leads to a chaotic motion, which exists for $\alpha_1 \approx 2.9$. Additionally, period-n windows for $\alpha_1 \approx 3.15, 3.05, 2.95$ are reported.

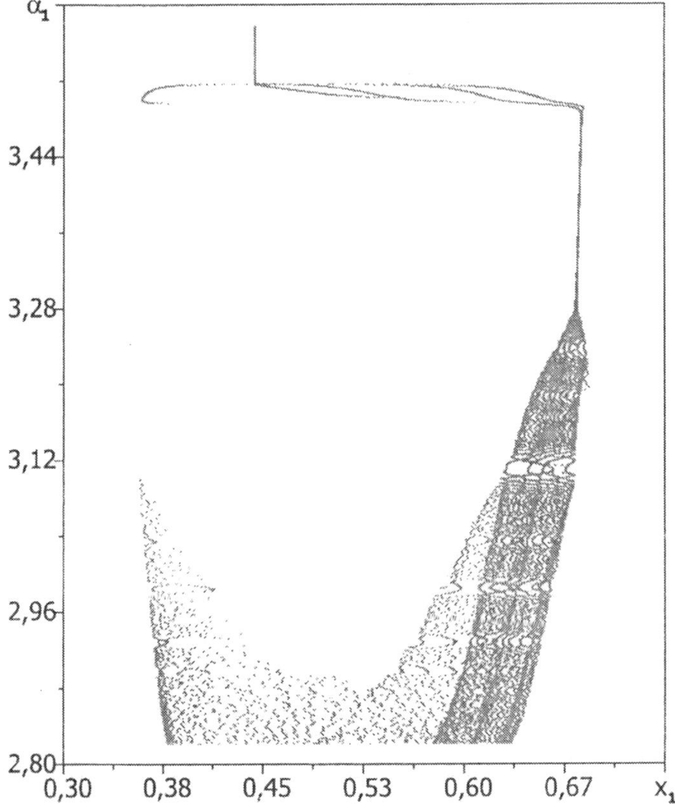

Figure 11.12. Bifurcation diagram of $\alpha_1 \in (2.83; 3.58)$ parameter versus x_1 displacement for μ_2 model for the parameters: $\alpha_2 = 1.093$, $\beta_2 = 1.729$, $\beta_3 = 2.441$, $\eta_1 = \eta_2 = \eta_{12} = 0$, $\gamma_1 = 0.152$, $\gamma_2 = 0.609$, $v_b = 0.1$, $\mu_0 = 1.2$, $\mu_d = 0.3$ and initial conditions: $\tau_0 = 1000$, $\tau_k = 51000$, $x_1(0) = 0$, $x_2(0) = 0.1$, $y_1(0) = y_2(0) = 0$.

3.5 Lagrange Interpolation and Lyapunov Exponents

The Lagrange polynomial scheme is applied to estimate Lyapunov exponents numerically from a stored time series in hard computer memory. The time series with a variable time step of the Hénon integration procedure is obtained by solving Eq. (11.1). The Lagrange interpolation is performed for each time history of the phase coordinates of the system. The obtained series is then interpolated through standard Lagrange interpolation scheme by a new one, where points are distributed in equal intervals of time. The latter ones are then used to estimate Lyapunov exponents. The convergence of Lyapunov's exponents versus iterations n for a chaotic attractor are illustrated in Fig. (11.13).

A good convergence is achieved after about $n = 500$ iterations of the computation procedure. The friction still affects the presented convergence

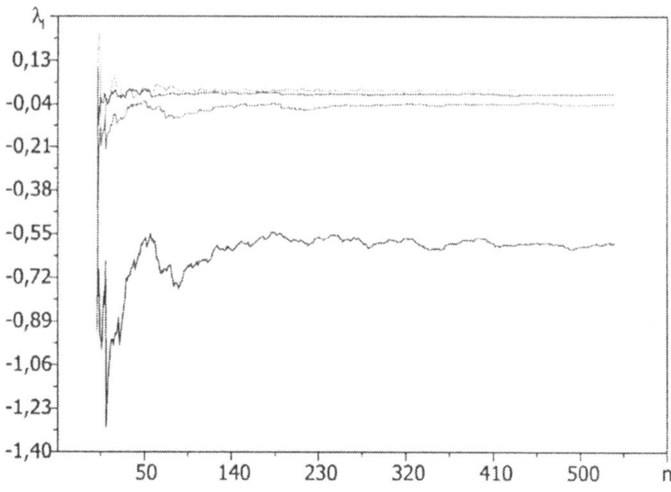

Figure 11.13. Lyapunov exponents λ_i convergence for the chaotic motion shown in Fig. 11.10.

Table 11.1. Lyapunov exponents spectrum for the motions illustrated in Figs. 11.7–11.10.

	Lyapunov exponents					
Fig.	λ_1	λ_2	λ_3	λ_4	R	τ $[\cdot 10^3]$
6	**-0.0047**	-0.0174	-0.0480	-0.4566	30	3-12
7	**-0.0031**	-0.0419	-0.1521	-0.3645	28	3-12
8	**+0.0001**	**-0.0002**	-0.0733	-0.8381	23	4-11
9	+0.0146	**-0.0030**	-0.0400	-0.5869	21	2-10

form in an essential way. Lyapunov exponents have been estimated for Figs. 11.7–11.10 and summarized in Table 11.1.

In Table 11.1, R denotes a relation between the number of points of the trajectory solved by means of the Hénon method and its Lagrange interpolated equivalent.

4. Experimental Investigations

In this section, the numerical analysis is supported by experimental investigations. For that purpose a laboratory rig designed for observations and experimental research of the friction effects including the friction force measurement is constructed. Photos of the rig are presented in Fig. 11.14.

The general view, component parts, and some connectors, like coil springs, correspond to the elements schematically indicated and presented in Fig. 11.2.

Displacement of the block and the angle of rotation are measured using a laser proximity switch and a Hall-effect device, which guaranties a non-sticking

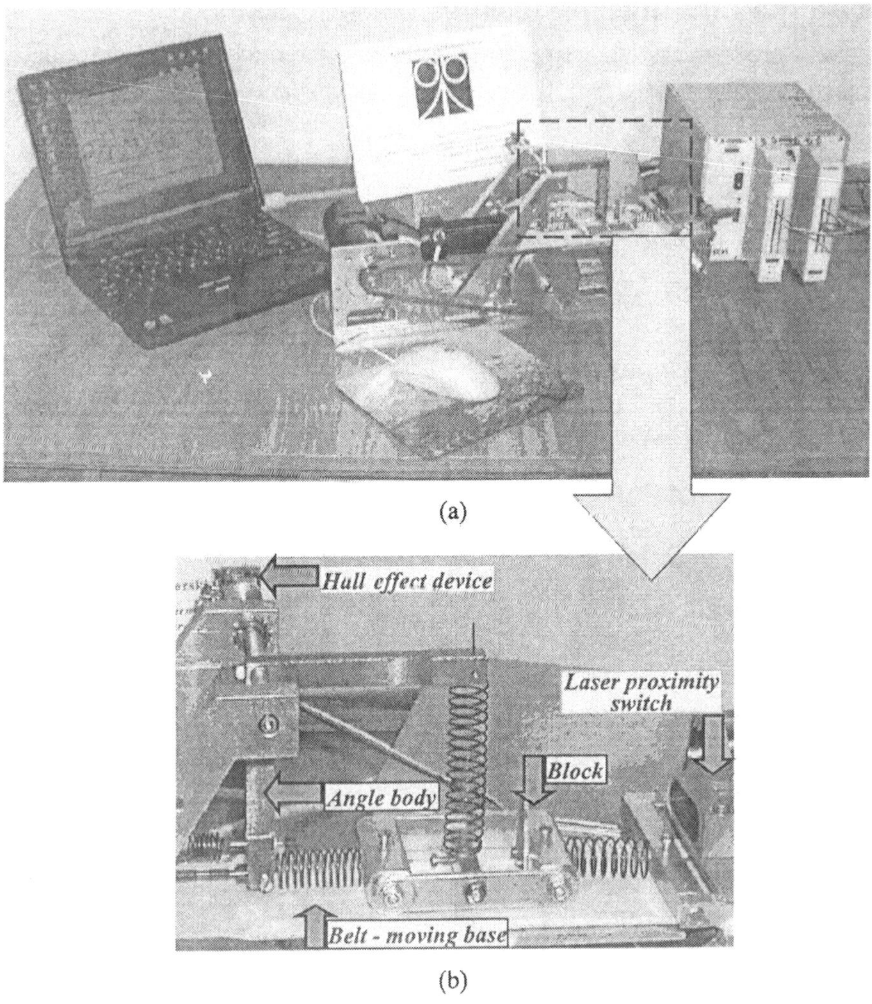

(a)

(b)

Figure 11.14. The laboratory rig (a) general view, (b) its vibratory subsystem.

method of the measurement. Both of them provide the linear dependency of the measured quantity versus the analogue voltage output. Measurement instruments connected through PCI computer card to LabView software enable the dynamic acquisition of the two measured signals. Disturbances of the entire construction, noise in electrical circuits, and other additional maintenances influence significantly the accuracy of any measured signals. Therefore, some signals are filtered digitally (elliptic topology) and a real differentiation preventing formation of high peaks is applied.

4.1 Results of Experimental Measurements

Measurement results are obtained following the methodology described in Sec. 11.3. The examples of time characteristics of state variables are shown in Fig. 11.15.

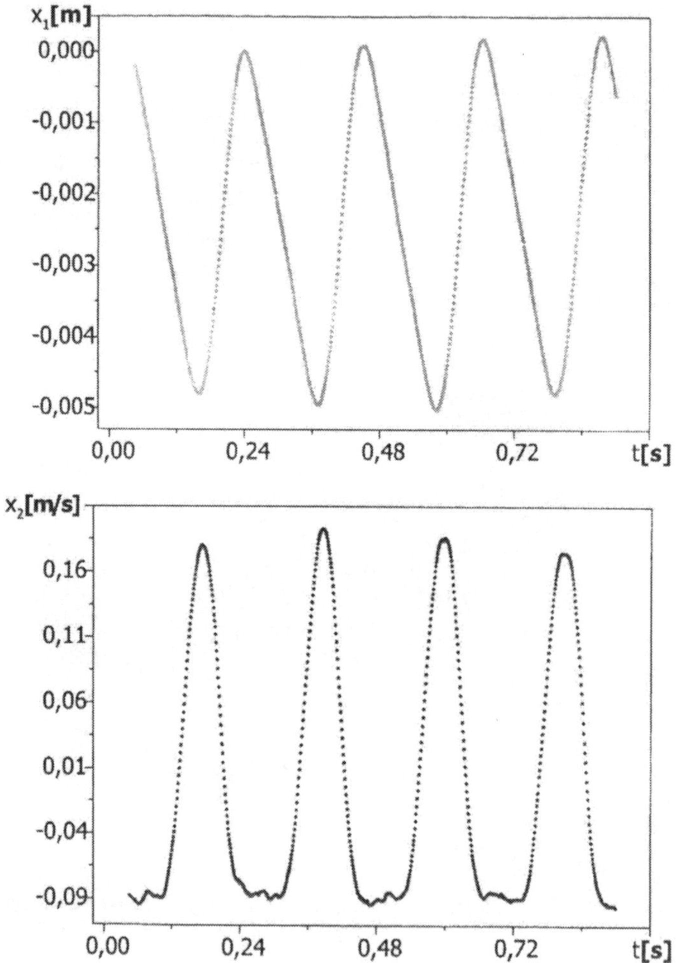

Figure 11.15. Real time histories of (a) displacement x_1, (b) velocity x_2, and (c) acceleration $\dot{x}_2$ of the block for velocity of the belt $v_b = -0.13\,m/s$.

A characteristic positive slope (slip phase of the block) of the time history and a negative slope (stick phase of the block) can be seen in Fig. 11.15a. A time dependency of velocity of the block is presented in Fig. 11.15b. There are some time intervals between nodes where velocity is almost constant and equal to the belt velocity. This happens if the considered block is in the stick

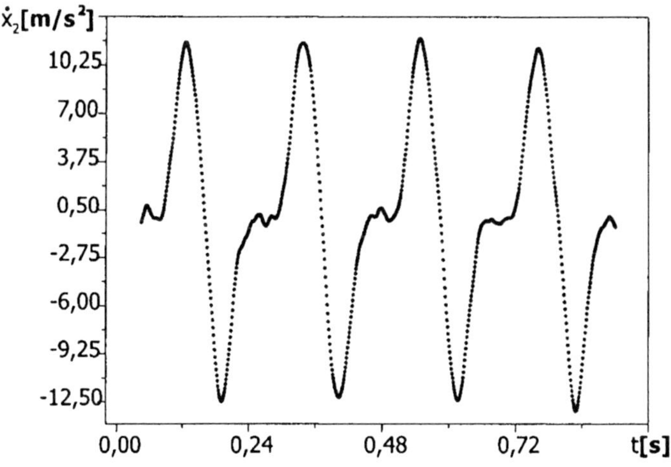

Figure 11.15. cont.

phase. Otherwise, the stick phase can be observed on a time history graph of acceleration, where some of the intervals (at the zero value of velocity) are parallel to t-axle.

The phase planes give an opportunity to explore the dynamics of the investigated system more comprehensively. The well-known shapes of phase curves usually visible in the stick-slip motion are presented in Fig. 11.16a-b. The stick (almost straight lines) and slip (arcs connecting the ends of the straight lines) phases can be easily observed.

4.2 Friction Force Model

Appropriately transformed equations of motion (see Eq. (11.1)) can be used to calculate the friction force after the state variables of the investigated system have been measured in real time. Characteristics of the friction force versus the relative velocity between the belt and the block for positive and negative velocities of the belt are shown in Fig. 11.17.

It can be observed that the zones occupied by the closed functions of the friction model differ significantly. It is a regularity since the angle body causes reinforcement of the friction force for the positive velocity of the moving base (see Fig. 11.2, Fig. 11.14b). This explains why the T_+ and T_- friction force characteristics are described by linearly and exponentially (of second order) decaying functions, respectively. The friction force model represented by the solid red curve and indicated in Fig. 11.17 is separated from the real friction force dependencies for all relative velocities v_{rel} as illustrated in Figs. 11.18–11.19.

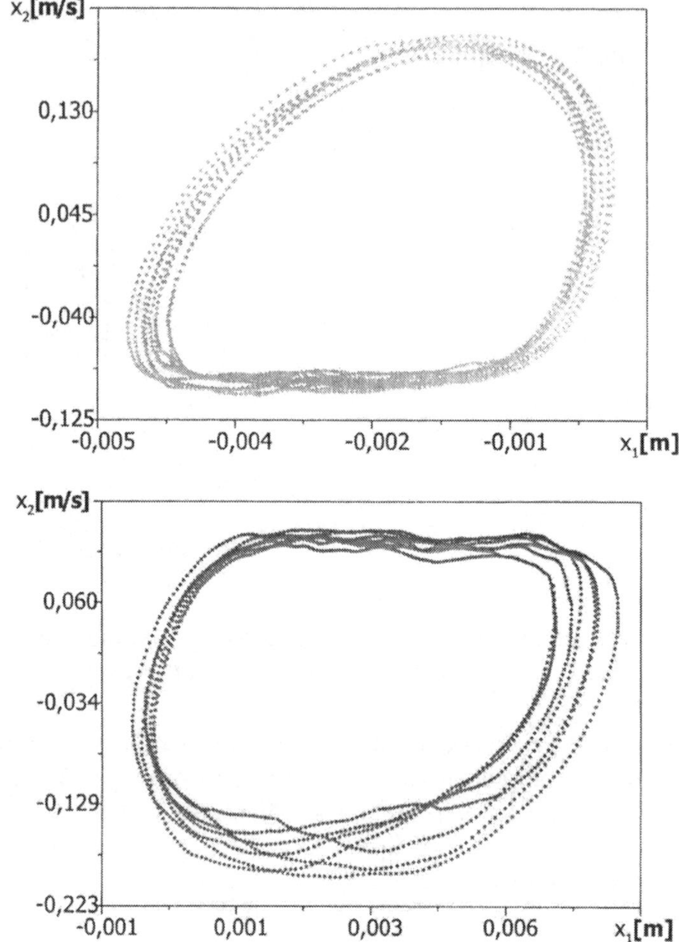

Figure 11.16. Phase planes of the block for (a) $v_b < 0$, (b) $v_b > 0$ and the phase plane of (c) the angle body for $v_b > 0$.

In the case of the T_+ branch, the equation of the friction force dependence describing the friction force model for the positive relative velocity have the following form:

$$T_+ = T_s - |v_{rel}| \frac{T_s - T_{\min}}{v_{rel,\max}}, \tag{11.6}$$

where: T_S is a static friction force, $v_{rel,max}$ is a maximum positive relative velocity. The T_- branch can be described by an exponentially decaying function of second order describing the friction force model for the negative relative

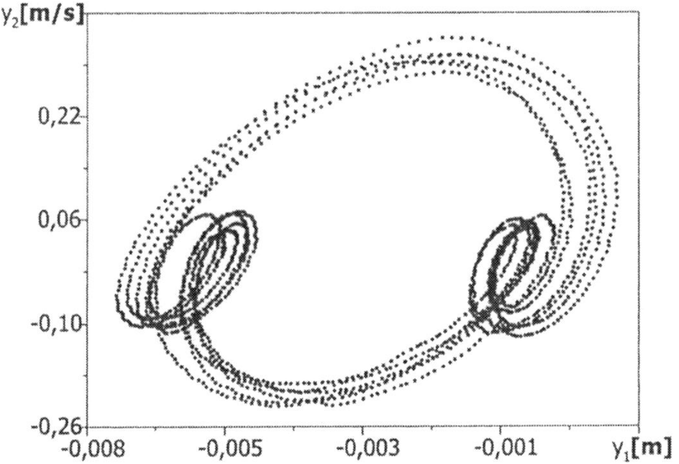

Figure 11.16. cont.

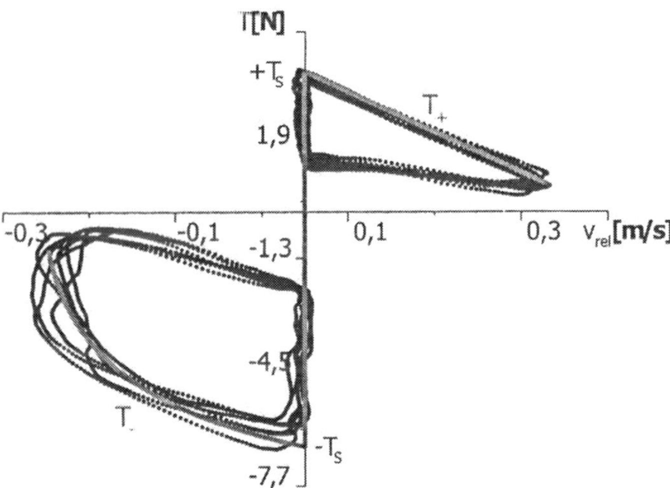

Figure 11.17. Friction force characteristics for positive (T_+) and negative (T_-) relative velocities.

velocity of the following form:

$$T_- = T_s + A_1 \exp\left(-\frac{|v_{rel}| - v_{rel,\min}}{t_1}\right) + A_2\left(-\frac{|v_{rel}| - v_{rel,\min}}{t_2}\right), \quad (11.7)$$

where: $v_{rel,min}$ is a maximum negative velocity, A_1, A_2, t_1, t_2 are constant values. The main multivalued function describing friction force changes (grey

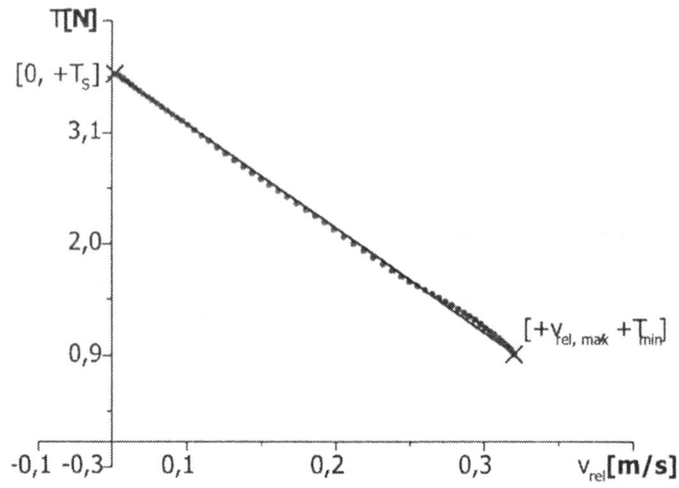

Figure 11.18. Linear fitting (solid black curve) of the branch of the real friction force characteristics (solid grey points) for the positive relative velocity.

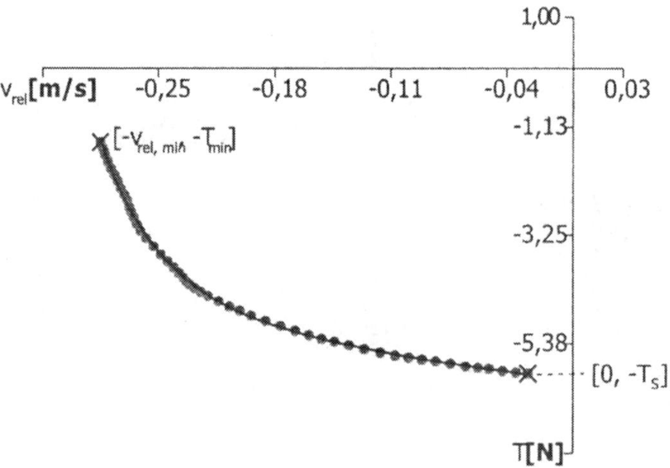

Figure 11.19. Exponential fitting (solid black curve) of the branch of the real friction force characteristics (solid grey points) for the negative relative velocity.

curve in Fig. 11.17) occurring in our investigated 2-DOF system with a variable normal force is determined from the following equation:

$$T = \begin{cases} \operatorname{sgn}(v_{rel})\, T_+ & \text{if} \quad v_{rel} > 0, \\ \operatorname{sgn}(v_{rel})\, T_- & \text{if} \quad v_{rel} < 0, \\ |T_S| & \text{if} \quad v_{rel} = 0. \end{cases} \tag{11.8}$$

4.3 Comparisons

The friction force model given by Eq. (11.8) is transformed to a non-dimensional one, and then a numerical analysis based on the T_+ and T_- friction force characteristics is carried out. The parameters of the two models are obtained by both measurement and identification: $T_s = 3.63$, $T_{min} = 0.86$, $v_{rel,max} = 0.27$ (T_+ branch); $T_s = -5.94$, $T_{min} = -1.42$, $-v_{rel,max} = 0.28$, $A_1 = 3.23453$, $A_2 = 2.87362$, $t_1 = 0.0342$, $t_2 = 0.30529$ (T_- branch). The numerical analysis with implementation of the introduced friction force dependency has yielded the results presented in Fig. 11.20.

The numerical trajectory (grey cruve) illustrated in Fig. 11.20a is satisfactorily close to its experimental counterpart recorded for the investigated dynamical system. The sticking velocity is almost the same, but in the sliding phase some distinguishable differences are observed. T_- friction force model can be used after the friction effects occurring in the systems with the variable normal force acting between surfaces have been analyzed.

A significant difference between the trajectories under consideration (cf. Fig. 11.20b) is visible, but the sticking phase still coincides for the positive relative velocity. The comparison of the results with those in Fig. 11.20b shows non-symmetry of the system under analysis, which is information of considerable usefulness.

5. Conclusions

The 2-DOF self-excited system with friction is analysed using numerical methods. A special numerical scheme based on the Hénon approach and exhibiting good suitability for investigations of non-smooth dynamical systems is applied. Many interesting dynamical non-linear behaviors are reported and analysed, including stick-slip periodic (Fig. 11.5, 11.7, 11.8), quasi-periodic (Fig. 11.9) and chaotic (Fig. 11.10) motions. In the analysis, all standard techniques are applied, i.e. time histories, phase planes, Poincaré maps, bifurcation diagrams and the Lyapunov exponents. The calculation of Lyapunov's exponents from an interpolated time series offers sufficient accuracy and correct values of its spectrum. The expected estimation accuracy of the Lyapunov exponents for each type of the motion yields different R relations between the number of the trajectory points solved by the Hénon method and its equivalent Lagrange interpolation.

In addition, the numerical analysis is supported by the investigation of a real laboratory object modeling the feedback reinforcement of the friction force (model of T_- branch) and the friction force without the feedback (model of T_+ branch). The numerical solution (grey curve) obtained using the T_- branch model has not proved a transition, which can be observed in our experimental

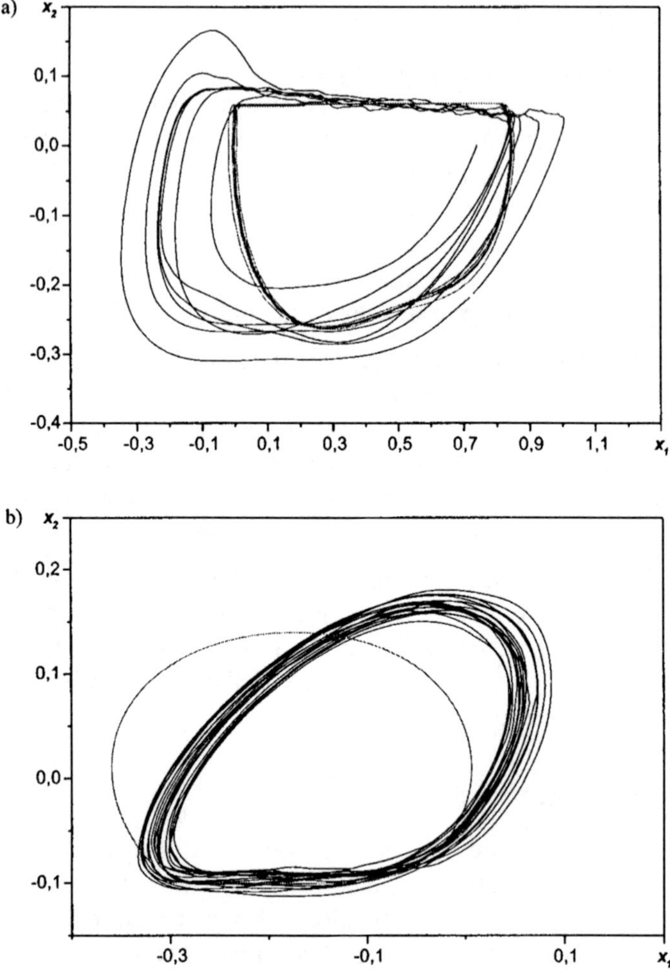

Figure 11.20. Verification of the results obtained via computer simulation (grey curve) with experimental measurement of state variables (black curve) of the analyzed system: friction force model: (a) T_-, (b) T_+; parameters: $\alpha_1 = 1.87$, $\alpha_2 = 0.72$, $\beta_1 = 2.87$, $\beta_2 = 1.62$, $\beta_3 = 2.47$, $\eta_1 = \eta_2 = \eta_{12} = 0$, $v_b = 0.06$, $\mu_0 = 1.1$; initial conditions: $\tau_0 = 400$, $\tau_k = 5000$, $x_1(0) = 0$, $x_2(0) = 0$, $y_1(0) = y_2(0) = 0$.

measurement (black curve). The sticking velocity is almost the same, but in the sliding phase some distinguishable differences are observed.

It is suggested that the T_- friction force model should be used after an analysis of the friction effects occurring in the systems where the normal force acting between cooperated surfaces fluctuates. Application of the branch friction force model leads to rapid entries on the stick phase and rather smooth backslides from it (see grey curve in Fig. 11.20b).

To conclude, a new idea for the friction pair modeling using both laboratory equipment and numerical simulations is proposed allowing for observation and control of the friction force. The experimental data are compared with those obtained via numerical simulations showing good agreement.

References

[1] Abe M. A theoretical prediction of subjective vehicle handling evaluation. VDI Berichte 1980; 368:261-266

[2] Abe M. et al. A Direct Yaw Moment Control for Improving Limit Performance of Vehicle Handling – Comparison and Cooperation with 4WS. Vehicle System Dynamics Supplement 1996; 25

[3] Abe M., Handling characteristics of four wheel active steering vehicles over full maneuvering range of lateral and longitudinal accelerations. Proceedings of 11th IAVSD Symposium. Vehicle System Dynamics 1989; 18:1-14

[4] Aiche N., Andrzejewski R. Braking of a roadtrain with ABS on nonhomogeneous pavement. Konmot; 1996; Krakow-Szczawnica.

[5] Alleyne A., Hedrick J.K. Nonlinear Adaptive Control of Active Suspensions. IEEE Transactions on Control Systems Technology 1995; 3(1): 94-101

[6] Altpeter F., Ghorbel F., Longchamp R., Relationship between two friction models: a singular perturbation approach. Proceedings of the 37th Conference on Decision and Control; 1998 December 16-18; Tampa. Florida

[7] Andreaus U., Casini P. Dynamics of friction oscillators excited by a moving base and/or driving force. Journal Sound and Vibration 2001; 245(4):685-699

[8] Andrzejewski R. *Analysis and synthesis of braking process of a vehicle.* Lodz: Scientific Bulletin of Lodz Technical University No 528, 1988

[9] Andrzejewski R. Stability of technical systems (TS). Stability of vehicle motion. The Archive of Mechanical Engineering 1989; 2-3:125-138

[10] Andrzejewski R., Technical stability of a vehicle, moving along a circle. The Archive of Mechanical Engineering 1990; 4:287-296

[11] Andrzejewski R. Quality criterion of wheel motion regulation. The Archive of Mechanical Engineering 1991; 4:255-276

[12] Andrzejewski R. *Stability of wheeled vehicles. Criterion. The methods of investigations.* Lodz: Scientific Bulletin of Lodz Technical University, 1993

[13] Andrzejewski R. Lateral kinematic of multibody vehicle. The Archive of Mechanical Engineering; 1996; 1:31-44

[14] Andrzejewski R. *Stability of Vheeled Vehicles Motion.* Warszawa: WNT, 1997, in Polish

[15] Andrzejewski R., Ormezowski J. Active moving control of movement articulated vehicle. Conference "Brake of Road Vehicles"; 2001 April 18-21; Lodz

[16] Andrzejewski R., Werner J. Vibrations of the non-linear three-body wheeled vehicle model. Bifurcations and chaos. The Archive of Mechanical Engineering; 2002; XLIX(1):81-85

[17] Antunes J., Axisa F., Beaufils B., Guilbaud B. Coulomb friction modeling in numerical simulations of vibrations and wear work rate of of multispan tube bundles. Journal of Fluids and Structures 1990; 4:287-304

[18] Armstrong-Hélouvry B. A perturbation analysis of stick-slip. Transactions of the ASME Friction-Induced Vibrations, Chatter, Squeal, and Chaos; 1992; 49:41-48

[19] Awrejcewicz J. *Deterministic oscillations of lumped systems*, Warsaw: WNT, 1996, in Polish.

[20] Awrejcewicz J. *Bifurcation and Chaos in Coupled Oscillators*. Singapore: World Scientific, 1991

[21] Awrejcewicz J., Delfs J. Dynamics of a self-excited stick-slip oscillator with two degrees of freedom, Part I, Investigation of equilibria. European Journal of Mechanics. A Solids; 1990; 9(4):269-282

[22] Awrejcewicz J., Delfs J. Dynamics of a self-excited stick-slip oscillator with two degrees of freedom, Part II, Slip-stick, slip-slip, stick-slip transitions, periodic and chaotic orbits. European Journal of Mechanics. A Solids; 1990; 9(5):397-418

[23] Awrejcewicz J., Holicke M.M. Melnikov's method and stick-slip chaotic oscillations in very weakly forced mechanical systems. International Journal of Bifurcation and Chaos; 1999; 9(3):505-518

[24] Awrejcewicz J., Kudra G., Lamarque C.-H. Analysis of bifurcations and chaos in three coupled physical pendulums with impacts. Proceedings of Design Engineering Technical Conferences; Pittsburgh, 2001, CD-ROM, 8 pages.

[25] Awrejcewicz J., Kudra G., Lamarque C.-H. Dynamics investigation of three coupled rods with a horizontal barrier. Special Issue of Meccanica; 2003; 38(6): 687-698

[26] Awrejcewicz J., Kudra G., Lamarque C.-H. Nonlinear dynamics of triple pendulum with impacts, Journal of Technical Physics; 2002; 43:97-112

[27] Awrejcewicz J., Lamarque C.-H. *Bifurcation and Chaos of Nonsmooth Dynamical Systems*. Singapore: World Scientific, 2003.

[28] Awrejcewicz J., Olejnik P. Calculating Lyapunov exponents from an interpolated time series. XX Symposium Vibrations in Physical Systems, Poznan, Blazejewko; 2002; 94-95

[29] Awrejcewicz J., Olejnik P. Stick-slip dynamics of a two-degree-of-freedom system. International Journal of Bifurcation and Chaos; 2003; 13(4): 843-861

[30] Ballard P. The dynamics of discrete mechanical systems with perfect unilateral constraints. Archives of Rational Mechanical Analysis; 2002; 154:199-274

[31] Bakker E., Pacejka H. B. Lidner L. A new tire model with an application in vehicle dynamics studies. SAE 890087 (1994); 83-95

[32] Bielicki M.,*Forgotten Summers' World.* PWN, 1973, in Polish.

[33] Bishop S.R., Clifford M.J. Zones of chaotic behaviour in the parametrically excited pendulum. Journal of Sound and Vibration; 1996; 189(1):142-147

[34] Bliman P-A, Sorine M. Easy-to-use realistic dry friction models for automatic control. Proceedings of the 3rd European Control Conference; 1995; Rome. Italy

[35] Bogacz R., Ryczek B. Dry friction self-excited vibrations; analysis and experiment. Engineerig Transactions 1997; 3-4:487-504

[36] Bogusz W. *Stability of Nonlinear Systems.* IPPT PAN, 1966, in Polish.

[37] Bogusz W. *Technical Stability.* Warszawa: PWN, 1972, in Polish.

[38] Brandl M, Pfeiffer F. Tribometer for dry friction measurement. ASME Design Engineering Technical Conferences; 1999 September 12-15; Las Vegas. Nevada

[39] Brogliato, Bernard, *Nonsmooth Mechanics: Models, Dynamics and Control.* London: Springer Verlag, 1999

[40] Brogliato, Bernard, *Nonsmooth Impact Mechanics: Models, Dynamics and Control.* London: Springer Verlag, 1996

[41] Canudas de Vit C., Olsson H., Aström K.J., Lischinsky P. A new model for control of systems with friction. Proceedings of the 34th IEEE Transactions Automatics Control

[42] Canudas de Wit C., Horowitz R., Tsiotras R.P. *Model-Based Observers for Tire/Road Contact Friction Prediction.* In New Directions in Nonlinear Observer Design, Nijmeijer, H. and T.I Fossen (Eds), Springer Verlag, Lectures Notes in Control and Information Science, May !999.

[43] Bo-Chiuan Chen, Huei Peng. A real-time rollover threat index for sports utility vehicles. Proceedings of the American Control Conference San Diego; 1999 1 June; California

[44] Choi SB, Choi YT, Park DW. A sliding mode control of a full-car electrorheological suspension system via hardware-in-the-loop simulation. Transactions ASME, Journal of Dynamic Systems, Measurements, and Control 2000; 122(2):114-121

[45] Clover C.L., Bernard J. Longitudinal tire dynamics. Vehicle System Dynamics 1998; 19(4):231-259

[46] Dahl P.R. Solid friction damping of mechanical vibrations. AIAA Journal 1976; 14(12):1675-1682

[47] Demidowicz B.P. *Mathematical Theory of Stability.* Warszawa: WNT, 1972, in Polish

[48] Donahue M.D. *Implementation of an Active Suspension.* Preview Controller for Improved Ride Comfort. The University of California at Berkeley, 2001

[49] Dugoff H., Fancher P.S., Segel L. *Tire Performance Characteristics Affecting Vehicle Response to Steering and Braking Control Inputs.* Ann Arbour: HSRI University of Michigan, 1996

[50] Friedland B., Park Y. On adaptive friction compensation. IEEE Transactions on Automatic Control 1992; 37(10): 1609-1612

[51] Galvanetto U., Bishop S. R., Briseghella L. Mechanical stick-slip vibrations. International Journal of Bifurcation and Chaos 1995; 5(3): 637-651

[52] Garrott R., Howe J.G., Forkenbrock G.. *An experimental examination of selected maneuvers that may induce on-road untripped, light vehicle rollover - phase II of NHTSA's 1997-1998 vehicle rollover research program (VRTC-86-0421).* Washington, DC. National Highway Traffic Safety Administration. 1999

[53] Genta, G. *Motor Vehicle Dynamics: modeling and simulation.* Singapore. World Scientific Publishing, 1997

[54] Gillespie T.D., *Fundamentals of Vehicle Dynamics.* Society of Automotive Engineers, 1992

[55] Hanley R.J., Crolla D. Tyre modelling for misuse situations. Exhibition of Automotive innovation for the New Millennium. SEOUL 2000 FISITA CONGRESS; 2000 June 12-15; Seoul. Kore

[56] Hanley R.J. *Impact Model.* Tire Technology International, June 2000.

[57] Hénon M. On the numerical computation of Poincaré maps. Physica D 1982; 5:412-413

[58] Higuchi A. *Transient Response of Tyres at Large Wheel Slip and Camber.* Doctoral thesis, Delft University of Tchnology, Delft, The Netherlands, 1997

[59] Horiuchi S. et al. Two degree of freedom /-/, controller synthesis for active four wheel steering vehicles. Vehicle System Dynamics Supplement 1996; 25

[60] Ibrahim R.A., Mechanics of friction. Trans. ASME Friction-Induced Vibration, Chatter, Squeal, and Chaos; Transactions of the ASME 1992; 49: 107-122

[61] Inoue H., Itimaru H., Kawakami H., Sato H., Tabata M., Development of integrated system between active control suspension, Active 4WS, TRC and ABS. SAE Paper No. 920271 (1992)

[62] Karnopp D. Computer simulation of slip-stick friction in mechanical dynamic systems. Transactions of the ASME, Journal of Dynamic Systems, Measurement, and Control 1985; 107(1):100-103

[63] Keum-Shik Hong, Hyun-Chul Sohn, Hedrick J. K. Modified skyhook control of semi-active suspensions: a new model, gain scheduling, and hardware-in-the-loop tuning. Transactions od the ASME, Journal of Dynamic Systems, Measurement, and Control March 2002; 124

[64] Kitajima K. *Control for Integrated Side-Slip, Roll and Yaw Controls for Ground Vehicles.* Master Thesis, University of Michigan, 2000.

[65] Krasowski N.M. *Some Problems of the Stability Theory of Motion.* Moscow: Fizmatgiz, 1959, in Russian.

[66] Kudra G. *Analysis of Bifurcation and Chaos in the Triple Physical Pendulum with Impacts.* PhD thesis, Technical University of Lodz, 2002, in Polish.

[67] Kunze M. *Non-Smooth Dynamical System.* Lecture Notes in Mathematics 1744, Berlin: Springer Verlag, 2000.

[68] Lanzendoerfer J., Szczepaniak C. *Theory of a Car Motion.* Warszawa: WKiL, 1980, in Polish.

[69] Wen-Hou Ma, Huei Peng. Worst-case vehicle evaluation methodology - examples on truck rollover/jackknifing and active yaw control systems. Vehicle System Dynamics 1999; 32(4-5):389-408

[70] Makris N., Constantinou M.C. Analisis of motion resisted by friction. Part I: constant Columb and linear/Columb friction. Mechanics of Structures and Machines 1991; 9(4):487-510

[71] Michelberger P., Bokor J., Palkovics L. Robust design of active suspension system. International Journal of Vehicle Design 1993; 14(2-3): 145-165

[72] Miege A.J.P, Cebon D. Design and implementation of an active roll control system for heavy vehicles. Proceedings of the 6th International Symposium on Advanced Vehicle Control (AVEC'02); 2002 September; Hiroshima. Japan: Society of Automotive Engineers of Japan

[73] Monteiro Marques M.D.P. An existence, uniqueness and regularity study of the dynamics of systems with one-dimensional friction. European Journal of Mechanics. A Solids 1994; 13(2):277-306

[74] Muszynska A., Radziszewski B. Exponential stability as a criterion of parametric modification in vibration control. Non-linear Vibration Problems 1986; 20:175-191

[75] Nakashima T. Promotion of the program of advanced safety vehicles for 21st century. Society of Automotive Engineers of Japan 1995; 16

[76] Nalecz A., Lu Z., d'Entremont K. Development of vehicle-terrain impact model for vehicle dynamics simulation. SAE Paper No. 930833; 1993; 201-214

[77] Oden J.T., Martins J.A.C. Models and computation methods for dynamic friction phenomena. Computer Methods Application Mechanical Engineering 1985; 52:527-634

[78] Oosten van J.J.M., Pacejka H.B. SWIFT-Tyre: An accurate tyre model for ride and handling studies also at higher frequencies and short road wavelengths. International ADAMS User Conference 2000, June 19-21, Orlando.

[79] Osorio C., Gopalasamy S., Hedrick J. Force tracking control for electro hydraulic active suspensions using output redefinition. Proceedings of the ASME Winter Annual Meeting, 1999, Nashville.

[80] Pacejka H.B. *Modelling of the Pneumatic Tyre and its Impact on Vehicle Dynamic Behaviour*. Research report no. i72B, 1988, TU Delft.

[81] Pacejka H.B. *Tyre and Vehicle Dynamics*. Butterworth-Heinemann, 2002.

[82] Pacejka H.B. *The Wheel Shimmy Phenomenon*. Doctoral Thesis, Delft University of Technology, Delft, The Netherlands, 1966.

[83] Pacejka H. B., Besselink I. J. M. Magic formula tyre model with transient properties. Vehicle System Dynamics. Supplement 1997; 27:234-249

[84] Pacejka H.B., Sharp R.S. Shear force generation by pneumatic tyres in steady state conditions: a review of modeling aspects. Vehicle System Dynamics 1991; 20:121-176

[85] Peterfreund N., Baram Y. Convergence alanysis of nonlinear dynamical system by nested Lyapunov function. IEEE Transactions on Automatic and Control 1998; 43(8):1179-1184

[86] Popp K., Hinrichs N., Oestreich M. *Analysis of a Self-Excited Friction Oscillator with External Excitation*. Dynamics with Friction, Guran A., Pfeiffer F., Popp K. (eds); 1996; 1-35

[87] Pfeiffer F. Unilateral problems of dynamics. Archive of Applied Mechanics 1999; 69:503-527

[88] Radziszewski B. *On the best Liapunov function and its application for investigation of stability of motion*. IFTR Reports, Warsaw, 1977, in Polish.

[89] Radziszewski B. "Study of stability of motion by the "Best" Liapunov function." In *Stability and Sensivity in Mechanical Systems*; Wroclaw: Ossolineum, 1978, in Polish.

[90] Radziszewski B., Slawinski A. Comparative analysis of some criteria of stability of motion. Non-linear Vibration Problems 1989; 23:123-135

[91] Ranganthan R., Ying Y., Miles B.J. Development of a mechanical analogy model to predict the dynamic behaviour of liquids in partially filled tank vehicles. SAE Paper 942307; 194:79-85

[92] Ryu C., Gerdes J.C. Vehicle sideslip and roll parameter estimation using GPS. AVEC 2002: Hiroshima. Japan

[93] Sayers M.W., Han D.S. A generic multibody vehicle model for simulating handling and braking. Vehicle System Dynamics 1996; 25:599-613

[94] Sampson D.J.M., Cebon D. Achievable roll stability of heavy vehicles. ImechE. Journal of Automobile Engineering 2003; 217:269-287

[95] Segel L. Research in the fundamentals of automobile control and stability. SAE Transactions 1957; 65

[96] Singer I.L., Pollock H.M. *Fundamentals of Friction: Macroscopic and Microscopic Processes*. Dordrecht: Kluwer Academic Publishers, 1992.

[97] Slawinski A. On the problem of perturbation of non-linear system of differential equations. Revista de Matematicas Aplicadas. Departamento de Matematicas Aplicadas, Facultad de Ciencias Fisicas y Matematicas. Universidad de Chile. Santiago; 1990; 12:45-62

[98] Slawinski A. On the Comparative Analysis of Some Criteria of Stability of Motion. International Journal of Nonlinear Mechanics; 1998; 33(5):783-799

[99] Slawinski A. On control of an autonomous system with application of Lyapunov function. Non-linear Vibration Problems 1991; 24:283-295

[100] Su-Hsin Yu, Moskwa John J. A global approach to vehicle control: co-ordination of four wheel steering and wheel torque's. Journal of Dynamic Systems, Measurement, and Control; 1994; 116:659-667

[101] Stone E., Cebon D. A preliminary investigation of semi-active roll control. Proceedings of the 6th International Symposium on Advanced Vehicle Control, AVEC2002; 2002 September; Hiroshima, Japan

[102] Strzalko J., Grabski J. *Introduction to Analytical Mechanics.* Technical University of Lodz, 1997, in Polish

[103] Swaroop D., Yoon S. Untegrated lateral and longitudinal vehicle control for an emergency lane change manoeuvre design. International Journal of Vehicle Design 1999; 21(2):161-174

[104] Szczepaniak C. et al. *Anti-blocking System. Study of Theory and Construction.* Scientific Bulletin of ITS 1992; 76, in Polish

[105] Szemplinska-Stupnicka W., Tyrkiel E. The oscillation-rotation attractors in a forced pendulum and their peculiar properties. International Journal of Bifurcation and Chaos 2002; 12(1):159-168

[106] Szpunar K. Modified concept of technical stability. Nonlinear Vibration Problems; 1983; 21:141-145

[107] Sygniewicz J. Modeling of cooperation of the piston with piston rings and barrel. Scientific Bulletin of Lodz Technical University 1991; 615(149); in Polish

[108] Tesi A., Villoresi F., Genesio R. On the stability domain estiamtion via a quadratic Lyapunov function: convexity and optimality properties for polynomial systems. IEEE Trans. on Automat Control 1996; 41:1650-1657

[109] Wagner J. Optimization of a tire traction model for antilock brake system simulations. Journal of Dynamic Systems, Measurement and Control 1995; 117(2): 199-204

[110] Winkler C. et al. *Cooperative Agreement to Foster the Deployment of a Heavy Vehicle Intelligent Dynamic Stability Enhancement System.* University of Michigan Transportation Research Institute, Interim report, 1998 January.

[111] Wolf A., Swift J.B., Swinney H.L., Vastano J.A. Determining Lyapunov exponents from a time series. Physica D 1985; 16:285-317

[112] Wösle M., Pfeiffer F. Dynamics of multibody systems containing unilateral constraints with friction. Journal of Vibration and Control 1996; 2:161-192

[113] Van Wyk M. A., Steeb W.-H. *Chaos In Electronics: Mathematical Modeling, Theory And Applications.* Kluwer Academic Publishers, 1997

Index

Lightning Source UK Ltd.
Milton Keynes UK
14 July 2010

156962UK00007B/180/P

9 780387 243580